D1119759

By Alastair Fuad-Luke

ECODESIGN
THE SOURCEBOOK

THIRD EDITION FULLY REVISED

CHRONICLE BOOKS
SAN FRANCISCO

Acknowledgments
My heartfelt thanks to all the designers, designer-makers, manufacturers and other organizations who supplied information and photographic or illustrative materials, without whose kindness this third edition would not have been possible. Thanks are also extended to all those who responded to my continuous stream of queries and requests for information. I would also like to acknowledge the continuing support of Lucas Dietrich, and the assistance of Diana Bullitt Perry and Philip Collyer at Thames & Hudson, and graphic designer Peter Dawson at Grade. Finally, a big thanks and hug to my wife Dina for her encouragement and support throughout the project.

I dedicate this book to all those who have contributed by their designs and products herein to inspire us towards a more sustainable future.

This revised edition first published in the United States in 2010 by Chronicle Books LLC.

First edition first published in the United Kingdom in 2002 by Thames & Hudson Ltd, 181A High Holborn, London WC1V 7QX
thamesandhudson.com

ISBN-13: 978-0-8118-7129-7

The Library of Congress has catalogued the previous edition under the following ISBN: 978-0-8118-3548-0

Manufactured in Singapore.

Cover design by Sarah O'Rourke

Distributed in Canada by Raincoast Books
9050 Shaughnessy Street
Vancouver, British Columbia V6P 6E5

10 9 8 7 6 5 4 3 2 1

Chronicle Books LLC
680 Second Street
San Francisco, California 94107

www.chroniclebooks.com

This book was printed on 115 gsm Munken Premium Cream paper, produced by the Swedish mill Munkedals, which has a long tradition of environmental responsibility. Munkedals' papers are created to minimize the impact on the environment. For example, they have not been subject to chlorine bleaching; pulp is taken from sustainable forests; and production processes have been developed to reduce harmful effects and effluents.

Preface

This third edition of *EcoDesign*, seven years after the first edition in 2002, continues to take an eco-pluralistic approach to design, celebrating diverse thinking by designers and manufacturers around the globe who are challenging the direction of contemporary design in order to encourage transition towards more sustainable production and consumption. Eco-pluralist designs range from those that embrace minor modifications of existing products (such as the use of recycled rather than virgin materials) to radical new concepts and/or the 'dematerialization' of existing products into services.

Over the last decade the eco (r)evolution has undoubtedly progressed and is pushing at the mainstream consumer markets. Certain 'eco' products are now accepted as the cultural norm, such as the Smart Car, Ecover low-impact household cleaning products, Interflor carpet tile flooring systems and more. A growing number of young designers/designer-makers continue to show imaginative approaches to the eco challenge. Renewable energy technologies are appearing in everything from electronic goods to cars and buildings. Leading brands that have pioneered ethical, sustainable shopping, such as The Body Shop, Ben & Jerry's and Innocent, have been acquired by huge global brands indicating that the latter see these companies as a safe bet for the (sustainable) future.

A common thread runs through the selection of products, buildings and materials in this book. Each is an attempt to improve on the status quo, in small or large increments, and to increase our well-being while reducing the inherent environmental impact of these 'designed' things. Yet, the challenge to change our consumption habits is huge – climate change, the global financial crisis and looming resource depletion present a triad of interlinked challenges. So, before you rush to purchase that eco-efficient car illustrated in this book, ask yourself, 'Do I really need it? Or shall I use the car I already have for only half the time?' More sustainable production and consumption requires a concomitant change in consumer behaviour and ways of living, especially from the 20% of the global population that use 80% of the world's resources (that's us!).

How to use the tables in this book

Each product or material is accompanied by a description and a table. The page numbers in the tables permit rapid cross-referencing with the resources at the end of the book, enabling the reader to find designers/designer-makers, manufacturers and eco-design strategies, represented by the following icons:

The name and nationality of the designer/designer-maker (see pages 304–311), or manufacturer if the product/material is designed in-house (see pages 312–323).

The name and country of the manufacturer (see pages 312–323) if relevant.

The main materials and/or components. (See Glossary 338–341 and Eco-Design Strategies 324–330 for more about materials with reduced environmental impacts.)

The main eco-design strategies applied to the design of the product (see pages 324–330).

Additionally, if the product has won an important design award recognizing eco-design (see page 332) that will be included.

Contents

2.0 Objects for Working 190

Work: An Evolving Concept

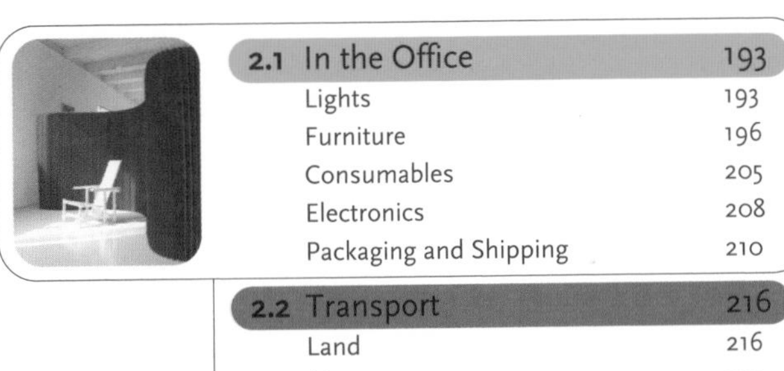

3.0 Materials 276

It's a Material World

4.0 Resources 302

Design for Sustainable Futures

Nearly five decades ago Rachel Carson documented the devastating effects of pesticide use on mammals and birds in the USA in her seminal book *Silent Spring*. Today traces of organophosphorus pesticides are found in organisms – including humans – around the world. At the 1968 UNESCO Intergovernmental Conference for Rational Use and Conservation of the Biosphere, the concept of ecologically sustainable development was first mooted. Paul Ehrlich's 1968 book *The Population Bomb* linked human population growth, resource degradation and the environment, and pondered the carrying capacity of the planet. By 1973 the Club of Rome, in its controversial report *Limits to Growth,* was predicting dire consequences for the world if economic growth was not slowed down. This report accurately predicted that the world population would reach 6 billion by the year 2000. More frightening predictions of the exhaustion of resources such as fossil fuels were less accurate, but recent evidence about the phenomenon of 'peak oil' – the observation that in the last 150 years we've consumed half the world's oil reserves (and therefore only have one half remaining) – would indicate that precautionary tales from the 1970s are looking more and more realistic. This reality is further compounded by the burning of these fossil fuels, which has substantially increased emissions of carbon dioxide (CO_2), a key greenhouse gas contributing to global warming and climate change. Using evidence from scientists and observers working with the Intergovernmental Panel on Climate Change, Mark Lynas has outlined devastating scenarios of life on a hotter planet in his book *Six Degrees*. Significant minorities around the globe face the real risk that their land will be inundated by rising sea levels.

The statistics that chart this unprecedented challenge are startling. In 1950 the world car fleet numbered 50 million vehicles and global fossil fuel use was the equivalent of 1,715 million tonnes of oil. Today there are more than 800 million cars, 60–90 million motorcycles and scooters, and 100–125 million other vehicles. Consumption of oil is about 9 billion barrels per year. As developing nations such as India and China begin to produce small but relatively efficient cars for an emerging middle class, the car fleet and oil demand is set to expand and continue to adversely impact the environment, specifically in the balance of gases, particulate matter and carcinogens in the atmosphere.

For every one of the millions of products we use to 'improve' the quality of our lives there are associated environmental impacts. While some products have a small impact, others consume finite resources in vast quantities. The ultimate design challenge of the 21st century is to avoid or minimize the adverse impacts of all products on the environment. Like all challenges, this constitutes both a demand and an opportunity – to steer the debate on more sustainable patterns of production and consumption. Designers, producers and consumers must be an integral part of the debate rather than being subject to the whim of the political and commercial forces of the day.

A brief history of green design

Green design has a long pedigree and before the Industrial Revolution it was the norm for many cultures. Goods like furniture and utility items tended to be made locally by craftsmen such as blacksmiths, wheelwrights and woodland workers, from readily available local resources. Innovation in farming machinery in Europe, particularly Britain, destabilized the natural employment structure of rural areas, and in the first half of the 19th century almost half of the rural British population migrated to towns to work in factories. Throughout the 20th century this pattern was repeated around the world as countries industrialized and created new urban centres.

The founders of the British Arts and Crafts movement (1850–1914) were quick to note the environmental degradation associated with the new industries. Their concerns about the poor quality of many mass-manufactured goods and the associated environmental damage prompted them to examine new methods combining inherently lower impact with increased production. For various social and technical reasons, only a small section of society reaped the benefits of the Arts and Crafts movement but the seeds were sown for development of the early modernist movements in Europe, notably in Germany (the Deutscher Werkbund and the Bauhaus), Austria (the Secession and the Wiener Werkstätte) and the Netherlands (De Stijl). The modernists insisted that the form of an object had to suit its function and that standardized simple forms facilitated the mass production of good-quality, durable goods at an affordable price, thus contributing to social reform.

Economy of material and energy went hand in hand with functionalism and modernism. Marcel Breuer, an eminent student at the Bauhaus between 1920 and 1924, applied new lightweight steel tubing to furniture design, arriving at his celebrated Wassily armchair and Model B32 cantilever chair. Breuer's 1927 essay 'Metal Furniture' conveys his enthusiasm for the materials and reveals his green credentials. He saw the opportunity to rationalize and standardize components, allowing the production of

'flat pack' chairs that could be reassembled to save transport energy, and were durable and inexpensive.

The early proponents of organic design promoted a holistic approach, borrowing from nature's own model of components within systems. In the USA the architect Frank Lloyd Wright was the first to blend the functionality of buildings, interiors and furniture into one concept. In the 1930s the Finnish architect and designer Alvar Aalto also achieved a synergy between the built environment and his curvilinear bent plywood furniture that evoked natural rhythms. At a landmark competition and exhibition organized by the Museum of Modern Art, New York, in 1942 called Organic Design for Home Furnishings, the winners, Charles Eames and Eero Saarinen, firmly established their biomorphic plywood furniture as a means of satisfying the ergonomic and emotional needs of the user. These designs often incorporated laminated wood or plywood to obtain more structural strength with greater economy. More ambitious expressions of biomorphism were achieved with the rapid evolution of new materials such as plastics in the 1960s and 1970s.

One of the early advocates of a more sustainable design philosophy, Richard Buckminster Fuller, came from the USA. One of Buckminster Fuller's early ventures, the Stockade Building System, established a method of wall construction using cement with waste wood shavings. Building inspectors of the day did not approve of this innovation and the venture faded. Not easily deterred, he soon set up a new design company, 4D, whose name makes reference to the consequence (to humanity) of 3D objects over time. 'Dymaxion' was the term he coined for products that gave maximum human benefit from minimal use of materials and energy. His 1929 Dymaxion House – later developed as a commercial product in the metal prefabricated Wichita House (1945) – and 1933 teardrop-shaped Dymaxion Car were both radical designs. The car had room for 11 adults, fuel consumption of 10.7 km/litre (30 mpg) and the ability to turn within its own length thanks to the arrangement of the three wheels. Remarkable as it was, the car was plagued with serious design faults and never became a commercial reality. The Wichita House could have been a runaway commercial success as nearly 40,000 orders poured in, but delays in refining the design led to the collapse of the company. Fuller persevered and in 1949 developed a new method of construction based on lightweight polygons. The geodesic dome was suitable for domestic dwellings or multipurpose use and its components were readily transported, easily erected and reusable. Fuller's legacy has inspired new endeavours such as the Eden Project in Cornwall, UK (opened in 2001), in which the world's largest biomes house 80,000 plant species from tropical to temperate climates.

From 1945 to the mid-1950s most of Europe suffered from shortages of materials and energy supplies. This austerity encouraged a rationalization of design summed up in the axiom 'less is more'. The 1951 Festival of Britain breathed optimism into a depressed society and produced some celebrated designs including Ernest Race's Antelope chair, which used the minimum amount of bent steel in a lightweight curvilinear frame.

During the 1950s–60s European manufacturers such as Fiat, Citroën and British Leyland extolled the virtues of small cars. Economical to build, fuel-efficient (by standards of the day) and accessible to huge mass markets, these cars transformed the lives of almost 9 million owners. In the US gas-guzzling, heavyweight, shortlived Buicks, Cadillacs and Chevrolets may have celebrated American optimism but were the very antithesis of green design.

The hippie movement of the 1960s questioned consumerism and drew on various back-to-nature themes, taking inspiration from the dwellings and lives of nomadic peoples. Do-it-yourself design books sat alongside publications like the California-based *Whole Earth Catalog*, a source book of self-sufficiency advice and tools. Out of this era emerged the 'alternative technologists' who encouraged the application of appropriate levels of technology to the provision of basic needs such as fresh water, sanitation, energy and food for populations in developing countries. Young designers experimented with new forms using recycled materials and examined alternative systems of design, production and sales.

In 1971 the rumblings of the first energy crisis were felt and by 1974, when the price of a barrel of oil hit an all-time high, the technologists began designing products that consumed less energy and so decreased reliance on fossil fuels. This crisis had a silver lining in the form of the first rational attempts to examine the life of a product and its consequent energy requirements. Lifecycle analysis (LCA), as it became known, has since been developed further into a means of examining the 'cradle to grave' life of products to determine not only energy and material inputs but also associated environmental impacts.

In his 1971 book *Design for the Real World*, Victor Papanek confronted the design profession head on, demanding that designers assume social responsibility instead of selling out to commercial interests. Although he was pilloried by most design establishments of the day, his book was translated into 21 languages and remains one of the most widely read books on design. Papanek believed that designers could provide everything from simple, 'appropriate technology' solutions to objects and systems for community or society use.

By the 1980s three factors – improved environmental legislation, greater public awareness of environmental issues and private-sector competition – ensured that green consumers became a visible force. *The Green Consumer Guide* (1988) by John Elkington and Julia Hailes was purchased by millions keen to understand the issues and exercise their 'consumer power'. Designers and manufacturers worked to make their products 'environmentally friendly', yet not always with genuine zeal or success. Unsubstantiated claims on product labels soon disillusioned an already sceptical public, and green design got buried in an avalanche of market-driven, environmentally unfriendly products from the emerging capitalist-driven 'global economy'. Then the pendulum swung back, resulting in more stringent environmental legislation, greater regulation and more uptake of eco-labelling, energy labels and environmental management standards.

Against the grain of the high-tech, matt-black 1980s, a few notable designer-makers blended post-modernism with low environmental-impact materials and recycled or salvaged components. In London Ron Arad produced eclectic works ranging from armchairs made of old car seats to stereo casings of reinforced cast concrete, and Tom Dixon created organic chair forms using welded steel rods covered with natural rush seating, a design still manufactured by Cappellini SpA, Italy.

The green design debate gathered momentum following publication in 1987 of the Brundtland Report, *Our Common Future*, prepared by the World Commission on Environment and Development, which first defined 'sustainable development', and also as a result of collaboration between governments, industry and academia.

Dorothy MacKenzie's 1991 book *Green Design* reported initiatives by individual designers and the corporate world to tackle the real impact of products on the environment.

In the early 1990s in the Netherlands, Philips Electronics, the Dutch government and Delft University of Technology collaborated to develop lifecycle analysis that could be widely used by all designers, especially those in the industrial sector. Their *IDEMAT* LCA software provided single eco-indicators to 'measure' the overall impact of a product. *IDEMAT* was rapidly followed by three commercial options, *EcoScan*, *EcoIT* and *SimaPro*. Today there are many different LCA and lifecycle inventory (LCI) packages that can help designers to minimize the impact of their designs from 'Cradle 2 Cradle', encouraging recycling of materials at the end of a product's useful life.

Over the past fifteen years academic communities around the world have evolved new terminology to describe particular types of 'green' design, such as Design for environment (DfE), DfX – where X can be assembly, disassembly, reuse and so on – eco-efficiency, eco-design and EcoReDesign. (See the Glossary for full definitions.)

The concept of sustainable product design (SPD) has emerged from the vigorous debate around sustainable development. Most definitions of SPD embrace the need for designers to recognize not only the environmental impact of their designs over time, but also the social and ethical impact. Buckminster Fuller and Papanek would perhaps wonder why it took so long for the design, manufacturing and consumer communities at large to take up these issues.

Our imperilled planet

Twenty-five per cent of the world's population of 6.75 billion people account for 80% of global energy use, 90% of car use and 85% of chemical use. By 2050 there may be 20 billion people on the planet, ten times more than at the beginning of the 20th century. Scientists estimate that human activities over the last 100 years have been responsible for an average global temperature rise of 0.8° C (1.4° F) and the IPCC predicts a temperature rise of between 1.4° C and 5.8 °C (2.5° F and 10.4° F) by 2100 depending on future emissions of greenhouse gases. Global warming on an unprecedented scale is melting the ice caps and permafrost, with consequent rises in sea level by 2100 up to 90–880 mm (3.5–35 in), with global ocean expansion, due to the heating, creating a possible rise of 110–430 mm (4.3–17 in). Climate change refugees are *already* on the move.

It is not an equable world. A typical consumer from the developed 'North' consumes between ten and twenty times more resources than a typical consumer from the developing 'South'. Both types of consumer can sustain their lives but the quality of those lives is substantially different. Almost 1 billion people suffer from poverty, hunger or water shortages. At present rates of production and consumption the earth can sustain 2 billion people at Northern standards of living. Could it support 20 billion people at Southern standards of living? Or is there an urgent need to address the way Northern populations consume and examine the true impact of each product's life?

The impact of global production and consumption

Between 1950 and 1997 the production of world grain tripled, world fertilizer use increased nearly tenfold, the annual global catch of fish increased by a factor of five, global water use nearly tripled, fossil-fuel usage quadrupled and the world car fleet increased by a factor of ten. During the same period destruction of the environment progressed on a massive scale, with reduction in biodiversity: the world elephant population decreased from 6 million to just 600,000 and total tropical rainforest cover decreased by 25%. Average global temperature rose largely due to an increase in CO2 emissions from 1.6 billion tonnes per annum in 1950 to 7 billion tonnes in 1997. Chlorofluorocarbon (CFC) concentrations rose from zero to three parts per billion, causing holes in the protective ozone layer at the North and South poles.

In the North ownership of products such as refrigerators and televisions has reached almost all households. More than two in three households own a washing machine and a car. The North is indeed a material world, and generates huge quantities of waste. According to *The Green Consumer Guide*, back in 1988 an average British person generated two dustbins of waste per week, used two trees a year in the form of paper and board, and disposed of 90 drink cans, 105 food/petfood cans, 107 bottles/jars, and 45 kg (99 lb) of plastics. By 2007 the average British citizen was generating 495 kg (1,091 lb) of household waste per annum. Local authorities in Britain recycled just 34.5% of domestic waste on average, with a further 10.6% used for generating energy or fuel manufacture and the remaining 54.5% of waste going to landfill.

The big environmental issues

In 1995 the European Environment Agency defined the key environmental issues of the day as climate change, ozone depletion, acidification of soils and surface water, air pollution and quality, waste management, urban issues, inland water resources, coastal zones and marine waters, risk management (of man-made and natural disasters), soil quality and biodiversity. Today those issues have become more persistent and, as has been noted by the World Resources Institute from 2000 onwards, the health of all the world's major ecosystems is declining.

Recognition that the planet was fast reaching a perilous state galvanized more than 100 heads of state to gather in Rio de Janeiro, Brazil, in 1992 for the United Nations Conference on Environment and Development. The achievements of the 'Earth Summit' were considerable: the Rio Declaration on Environment and Development set forth a series of principles defining the rights and responsibilities of states; a comprehensive blueprint for global action called Agenda 21 was published; guidelines for the management of sustainable forests (Forest Principles) were set; and the UN Convention on Biological Diversity and the UN Framework Convention on Climate Change were both ratified. The conference set the foundations for establishing the UN Commission on Sustainable Development, which produces annual progress reports, and adopted the Precautionary Principle that states: 'lack of full scientific certainty shall not be used as a reason for postponing cost-effective measures to prevent environmental degradation'.

Europe's cutting-edge environmental legislation

In 1972 the then members of the European Economic Community (now the European Union), recognizing that environmental damage transgresses national boundaries, agreed that a common transnational policy was required in Europe. Since then the European output of legislation and regulatory measures to combat environmental degradation has been prolific.

Regulations passed by the European Council become effective law for all member states immediately, whereas directives, which are also legally binding, do not come into force in the member states until carried into national law by individual governments. Important legislative advances include the Directive on the Assessment of the Effects of Certain Public and Private Projects on the Environment 1985, the Directive on the Conservation of Natural Habitats and of Wild Flora and Fauna 1992, and the Directive on Integrated Pollution Prevention and Control 1996. A range of other directives is of great relevance to manufacturers and designers, including on vehicles, electronic equipment, toxic and dangerous waste, and packaging and packaging waste. More recently the Directive on Energy using Products sets a legal requirement to embed eco-design considerations into the design of many types of electrical products. The effect of these regulations is felt well beyond Europe, as transglobal companies manufacturing cars, electronic goods, packaging and chemical products have to meet these stringent standards.

Europe's collaborative efforts to introduce environmental regulation provide a model for other regions of the world for international cooperation, for example, North America and the 'Tiger' economies of Southeast Asia.

Fiscal stimulation packages for a green economy

The 'credit crunch', and ensuing financial crises to reverberate globally in 2008/2009, has led governments worldwide to review how they re-stimulate the economy and help job creation. While there is much rhetoric about encouraging growth of the green technology and manufacturing sectors, the facts are more sobering. Only South Korea and Japan have endorsed fiscal packages including substantial budgets for green technologies and jobs, the measures being worth hundreds of billions of dollars. In contrast, the USA and UK have only earmarked 10% or less of their fiscal stimulus packages specifically to the green economy, with a message that it is predominantly business-as-usual. In 2009 BP Solar, a unit of the major oil company that had heavily invested in solar technology manufacturing in the last decade, actually sold some of its capacity. The signals for a full-steam-ahead green economy are mixed. Yet, as the model of the global economy and market mechanisms that underpinned it are re-evaluated, eco-design and sustainable product design seem needed more than ever as a means to create new green and social enterprises.

The real lives of products

Freedom and death

The car is the ultimate symbol of personal freedom for the 20th century. It liberates users but condemns many to death, directly as accident victims and indirectly as the recipients of pollutants causing asthma (from particulate matter), brain damage (from lead) and cancer (from carcinogens). Cars also contributes towards climate change via emissions of CO_2, marine pollution through oil tanker spillage or accidents, and noise pollution. Most societies feel that the individual freedom outweighs the collective drawbacks, but many cities worldwide are examining congestion charges, car-free days and other measures to combat the real human and environmental costs of urban overcrowding.

One-way trip

Some products lead short, miserable lives, destined for a one-way trip between the retail shelf and burial in a landfill. Packaging products are a prime example of one-way trip products but there are many others – kitchen appliances, furniture, garden accessories. The quiet but uplifting example of Modbury, a small village in south Devon, UK, that successfully banned plastic bags for shopping in 2007 shows that positive actions, collectively made, can make a real difference.

Everyday products quietly killing

Quietly humming away in the corner of millions of kitchens worldwide is the humble refrigerator. It keep food fresh, but it is a killer too. Coolants using CFCs or HCFCs (hydrochlorofluorocarbons) cause rapid degradation of the protective ozone layer and have been largely excluded from many global markets, but decommissioning of old refrigerators is often unregulated. Even as some recovery of the ozone 'holes' at the North and South poles has been reported, the layer continues to thin in other parts of the world and is not expected to 'heal' until 2050. Thus inhabitants receive higher doses of radiation with an increased risk of contracting skin afflictions and cancer.

Everyday inefficient products

The efficiency of products that have become a way of life needs to be challenged continually. The European Energy-label, specifying 'A' efficient to 'G' least efficient for washing machines, has received substantial industry support from manufacturers. The A rating specifies an energy efficiency index of less than 0.19 kWh, but to achieve the eco-efficiency targets needed for the future the manufacturers will have to strive to halve this in the next 20 years. Failure to apply the best technology available means unnecessary daily consumption of massive quantities of electricity and water.

Occasional use

The developed world's preoccupation with DIY home improvements means that many households own specialist tools, such as electric drills and screwdrivers, which are rarely used.

Novelties and gimmicks

Many of the products available through mail order catalogues are in fact gimmicks that will do no more than provide temporary amusement.

Small but dangerous

Many small electronic devices, such as MP3 players and mobile phones, have a voracious appetite for batteries. While more devices are offered with rechargeable batteries, the older models still consign millions of batteries to landfill sites, where cadmium, mercury and other toxic substances accumulate. In the European Union the disposal of certain battery types is illegal but in many parts of the world it continues unabated. Even compact fluorescent lights, with their significantly improved energy efficiences over tungsten lights, pose a problem as they contain small traces of mercury. Only substantial 'product take-back' schemes will prevent this mercury from ending up in landfills.

Industry visions and reality

Although the wastage of resources associated with the planned obsolescence in the US car industry in the 1950s is no longer tolerated, the lifetime of the average family vehicle remains less than ten years. Furthermore, the global car industry is geared up to keep adding to the existing 800 million cars worldwide at the same level of production. More fuel-efficient cars are now becoming a necessity for the big manufacturers to compete globally. The global financial crisis and the threatened demise of American car manufacturers is forcing the industry to rapidly re-think its future. Fiat has maintained its position as the European manufacturer with the lowest CO_2 output across their car range for the last two years. While Indian manufacturers like Tata recently introduced the small car Nano, it has relatively low fuel consumption when compared to the average family car in developed Northern nations. Despite obvious improvements to fuel-efficient cars, the gains in efficiency must be measured against sales of new units to ascertain whether total global car emissions can be decreased in the next decade.

Both hardware and software companies are obsessed with doubling the speed of personal computers every eighteen months as chip technology continues its meteoric development. Users are seduced into buying faster machines even though they likely use a small fraction of the computing power. Basic functionality, such as being able to adjust the height of a monitor or arrange a keyboard to suit individual needs, remains inadequate. Yet the computer industry conjures up a vision of a future in which we can programme our house to cook dinner before we arrive back from work, of a wired-up 'information age' in which everyone has access to the Internet. The reality is that only 1.6 billion people worldwide have Internet access, so 76% of the world's population is still not online.

The brand thing

Companies with internationally recognized brands aspire to increase their market share in individual nations in order to claim world dominance. Expectation, in the form of the brand promise, often delivers a transient moment of satisfaction for the purchaser. Whatever happened to products that were guaranteed to 'last a lifetime'? Where is the long view in the companies that sell these brands? The big brands have the potential to reduce the environmental impact of their activities, but not if they persist in encouraging their customers to consume more, not less.

Introduction

Moving commerce towards sustainability

Evolving environment management systems (EMS)

The flagship international standard that encourages organizations to examine their overall environmental impact arising from production (but not the impact of their products during usage) is ISO 14001, compiled by the International Organization for Standardization in Switzerland. Companies that achieve this independently certified EMS have integrated management systems into their business to reduce environmental impacts directly, and have agreed to publication of an annual environmental report from an audited baseline to measure reductions in impact. Other independently certified standards exist, such as the Eco-Management and Audit Scheme (EMAS) for companies in EU member states.

Sustainable production and consumption

In 1995 the World Business Council for Sustainable Development (WBCSD), a coalition of 120 international companies committed to the principles of economic growth and sustainable development, published a report entitled *Sustainable Production and Consumption: A Business Perspective*. It defined sustainable production and consumption as 'involving business, government, communities and households contributing to environmental quality through the efficient production and use of natural resources, the minimization of wastes and the optimization of products and services'. The WBCSD continues to challenge the global business community to look at the real risks involved in not valuing 'free' ecosystem services and issues around resource depletion. The UN Commission on Sustainable Development sees the roles of business and industry as crucial. It requires important changes to the way businesses operate, through the integration of environmental criteria into purchasing policies (green procurement), the design of more efficient products and services – including a longer lifespan for durable goods, better after-sales service, increased reuse and recycling, and promotion of more sustainable consumption through improved product information – and the positive use of advertising and marketing.

Model solutions

WBCSD members are encouraged to adopt measures to improve their eco-efficiency – greater resource productivity – by maximizing the (financial) value added per unit of resource input. This means providing more consumer performance and value from fewer resources with less waste. Amory Lovins *et al* of the Rocky Mountain Institute in the USA proposed the concept of 'Factor 4' – a doubling of production using half the existing resources, with a consequent doubling of the quality of life. Researchers at the Wuppertal Institute in Germany found this inadequate to deal with the expected doubling or trebling of world population by 2050 and proposed Factor 10 as a more appropriate model for the developed North to achieve equable resource use for the North and developing South.

Another model that is finding favour with businesses is called The Natural Step (TNS). It sets out four basic 'system conditions' for businesses to adopt. First, substances from the earth's crust, the lithosphere, must not be extracted at a greater rate than they can re-accumulate – thus there must be less reliance on 'virgin' raw materials. Second, man-made substances must not systematically increase but should be biodegradable and recyclable. Third, the physical basis for the productivity and diversity of nature must not be systematically diminished – renewable resources must be maintained and ecosystems kept healthy. Fourth, we must be fair and efficient in meeting basic human needs – resources should be shared in a more equable manner. Companies as diverse as carpet manufacturers, water suppliers and house builders have taken up TNS.

Early adopters and new business models

International companies from Europe, the USA and Japan are exploring new business models that take a long view enmeshed with the concept of sustainable development. For example, Mitsubishi considered the ecology of the tropical rainforest system, which is highly productive in terms of biomass on a fixed amount of nutritional resources. Waste becomes other organisms' food in the rainforest. Mitsubishi mimics this ecology by ensuring its industrial system meets eco-efficient parameters. The concept of the 'low carbon economy' is also challenging business. Low carbon implies reduced energy use, increased energy quotient from renewable sources and decreased dependence on oil and other fossil fuels. The Carbon Trust in the UK is experimenting with labelling everyday products with their carbon footprint. Cradle 2 Cradle, a protocol developed by the US company MBDC, is an independent certification scheme that ensures materials from one generation of products are reused for the next, encourages closed loop production systems to minimize environmental impacts and re-sets the business model. In the next decade expect businesses to emerge that address issues as diverse as 'virtual water', 'transport miles' and supply chain transparency.

Triple bottom line

Companies engaging in sustainability have a healthy 'triple bottom line', as improved environmental and social performance mirrors itself in increased profits. It is not surprising that there is an increase in the number of companies seeking membership in the Dow Jones Sustainability Indexes, where financial performance often outstrips the more conventional market indexes. Sustainability is, at last, being seen as an opportunity to do business differently. A key catalyst is design, the faithful translator of human, financial, natural and social capital into goods since the start of the Industrial Revolution. Design is critical to consumer acceptance and product success in the marketplace.

Co-designers save the earth

The power of designers is catalytic. Once a new, more environmentally benign and socially positive design penetrates markets its beneficial effects multiply: businesses spend less on raw materials and production and realize better profits; users enjoy more efficient, better-value products; and governments reduce spending on regulatory enforcement. The net gain is an improved environment and quality of life. The vivid examples in this book demonstrate the capability of design and designers to shape the future and save the earth. However, the complexity of the sustainability challenge means that designers have to co-design and co-create with a wider variety of stakeholders to achieve the huge transformations we need to move towards genuine sustainable consumption and production.

A robust tool kit

Today's designer has a powerful array of tools to assist him or her to meet the challenge of reducing environmental impacts at the design stage, such as simple checklists, impact matrices, lifecycle matrices, eco-wheels, Lifecycle Inventory (LCI) and Lifecycle Analysis (LCA) software. A checklist of Eco-Design Strategies can be found on p. 324 and a full list of organizations and agencies offering information and software to assist designers is given in the Green Organizations section (p. 331).

An evolving manifesto for eco-pluralistic design ... designs that tread lightly on the planet

The thoughtful designer of the 21st century will design with integrity, sensitivity and compassion. He or she will design products/materials/product service systems and services that are sustainable, i.e. they serve human needs without depleting natural and man-made resources, without damage to the carrying capacity of ecosystems and without restricting the options available for present and future generations. An eco-pluralistic designer will:

1. Design to *satisfy real needs* rather than transient, fashionable or market-driven needs.
2. Design to *minimize the ecological footprint* of the product/material/service product, i.e. reduce resource consumption, including energy and water.
3. Design to *harness solar income* (sun, wind, water or sea power) rather than use non-renewable natural capital such as fossil fuels.
4. Design to *reduce global warming potential* by minimizing CO_2 and other greenhouse gases associated with the production, use and disposal/reuse of a product.
5. Design to *encourage Cradle 2 Cradle and lifecycle thinking* to improve resource efficiency and reduce ecosystem damage.
6. Design to *enable separation of components* of the product/material/service product at the end of its life in order to encourage recycling or reuse of materials and/or components.
7. Design to *exclude the use of substances toxic or hazardous* to human and other forms of life at all stages of the product/material/service product's lifecycle.
8. Design to *engender maximum benefits to the intended audience* and to educate the client and the user and thereby create a more equable future.
9. Design to *use locally available materials and resources* wherever possible (thinking globally but acting locally).
10. Design to *exclude innovation lethargy* by re-examining original assumptions behind existing concepts and products/materials/service products.
11. Design to *dematerialize products into services* wherever feasible, providing overall negative environmental and social impacts are less than individually used products.
12. Design to *help regenerate social capital* by maximizing product/material/service benefits to communities.
13. Design to *encourage modularity and adaptability* in design to permit sequential purchases, as needs require and funds permit, to facilitate repair/reuse and to improve functionality.
14. Design to *foster debate and challenge the status quo* surrounding existing products/materials/service products.
15. Design *for the social commons* by publishing eco-pluralistic designs in the public domain for everyone's benefit, especially those designs that commerce will not manufacture.
16. Design to *create more affordable sustainable* products/materials/services for mass markets to encourage transition towards a more sustainable future.

1.0 Objects for Living

Living or Lifestyles?

In a media-driven world, where brands promise a lifestyle guaranteed to satisfy your desires, it is difficult to step back and honestly appraise your real needs for living. The word 'lifestyle' implies not just a way of life but also choice. For many people around the globe lifestyle choices are simply not available, as the basic needs of life – clean water, clean air, sufficient food, shelter and medical care – are not always met. In today's global economy international brands, such as Coca-Cola soft drinks and Nike trainers, exist cheek by jowl with locally or nationally made products. We are all invoked to balance our local, national and international consumption, to understand how this affects the bigger picture of climate change, social equity and inter-generational futures. Consumers need to pause and reflect on each purchase, manufacturers need to become leaner, greener and more ethically aware and designers need to reappraise their role in the production and consumption cycle.

Design enabling new freedoms

The car has become the ultimate symbol of our freedom to move around, yet this 'impact-use' product, which only 20% of the world's population own, impinges on the collective freedom of all people to enjoy clean air and unpolluted water. Over the last 25 years the fuel efficiency of the average car has improved only 18%, although recent trends indicate efficiency improvements of up to 30% are emerging from the new generation of hybrids and super-efficient engines. The eco-design innovation lethargy that is embedded in the car industry is being challenged by more powerful legislation and regulation, improved consumer awareness and the emergence of a new breed of car manufacturers – engineering companies such as Venturi in France, and open source design groups like c,mm,n, conceived by The Netherlands Society for Nature and Environment. Perhaps the financial and peak oil crises are priming us for a paradigm shift in the design of cars aimed at significantly removing the key environmental burdens of this product. We are also seeing improvements in alternative modes of mobility – the push scooter, bicycle and motorbike – but these must also be accompanied by radical improvements in systems of public transport to provide 'flexible mobility paths' for individual and group use for commuting and leisure.

Our dependency on electronic and telephony networks for our work, social and leisure lives is now deeply embedded in our way of life. The real cost of these services in terms of energy and associated environmental impacts is hidden, but some studies have shown that Internet-savvy occupants of the virtual world Second Life have the same energy footprint as a real Brazilian person. In short, life in the Information Age has a significant impact on energy use, carbon footprint and, therefore, climate change. Projects like the One Laptop Per Child (see p. 126) indicate that low-energy personal computers can challenge the orthodoxy of the global brands and help wire in the 'under-consumers' of the South to benefit from Internet connectivity. What lessons do these innovations provide for the North?

Each individual requires different products to sustain life. Aside from the essential physical resources, humans need 'comfort' products to achieve a level of emotional, spiritual and social well-being. These products may permit or provide improved mobility, specialist recreational activities, communal meeting-places or spiritual contemplation. Since comfort products tend to be used over a longish time, rather than being ephemeral, the design parameters can embrace durability, and therefore judicious use of resources.

Living lightly – a sustainable day

As the products in this section illustrate, it is possible to tread more lightly on the planet, to consume and waste less, yet to maintain or even improve the quality of life. A double responsibility falls on the consumers, manufacturers and designers of developed countries of the North. The North must rapidly evolve more sustainable patterns of consumption and production, models of enterprise that regenerate rather than exploit, that encourage equity not division. Additionally, the North must offer the South the assistance and the means to avoid bad practice and reap the benefits of a more sustainable way of life, sooner rather than later. A sustainable day in 2025 might involve the following products, Product Service Systems (PSS) and services...

Lost & Found

Here's a fresh examination of a humble yet resilient material. This quirky leather stool is made with traditional heavyweight shoe-stitching machinery to ensure strong, thick, stitched seams that provide an interesting aesthetic detail. The tough filling will contribute to the stool's long life, and as countless people grace its seat, the leather will develop an interesting patina.

✎	*Demakersvan, the Netherlands*	305
⚙	*Montis, the Netherlands*	318
▬	*High-density PU, leather*	
♻	*• Durability*	325

Isabella

The Isabella stacking stool was inspired by traditional African carved seating, but uses strawboard (from waste agricultural straw) rather than wood to achieve the quirky animistic shape. A tactile and luxurious finish is produced by encasing the strawboard in brightly coloured felt. This exploration hints at possibilities for manufacturing more goods 'at source' on the farm, providing revitalization opportunities for rural communities.

✎	*Ryan Frank, UK*	306
⚙	*One-offs and small batch production*	
▬	*Formaldehyde-free strawboard, 100% wool felt, natural non-toxic adhesive*	
♻	*• Low-embodied-energy materials* *• Renewable materials* *• Non-toxic, low-energy production*	325 325 326

Favela

Each Favela is unique, handmade from hundreds of pieces of recycled wood. With typical exuberance the Campana brothers focus the viewer on the nature and origins of the object while suggesting a whole world of questions beyond it: it is a commentary, a work of art, a crafted one-off and a manufactured product. It challenges our perceptions of manufacturing, renews our faith in the handmade and poses a final question: should we make our own chairs?

✎	*Fernando and Humberto Campana, Brazil*	304
⚙	*edra, Italy*	314
▮	*Recycled wood, nails, glue*	
♺	• *Recycled materials* • *Handmade, low-energy production*	325 326

Slick Slick

Stackable, injection-moulded, polypropylene chairs are produced by numerous manufacturers for the contract furniture market. Unfortunately ugliness is often a common design denominator of this genre. Starck rescues the concept with this elegant design requiring minimal materials, creating a chair suitable for conference or office seating and domestic use.

✎	*Philippe Starck, France*	309
⚙	*XO, France*	323
▮	*Polypropylene*	
♺	• *Single recyclable material* • *Multifunctional*	325 328

Sushi sofa and chair

Design is the transformation of financial, natural, social and human capital into saleable products, producing economic gain, providing employment, and, occasionally, meeting our real, human, needs. Today, there is an expectation that design will, unfailingly, deliver the (consumer) goods, and so the excitement from design can be short-lived. Campana Brothers' designs deliver much more. Their use and understanding of a wide palette of materials is uplifting. The Sushi sofa and Sushi chair celebrate everyday materials by repositioning them centre stage. Offcuts of carpet, felt, rubber, cotton and plastic find a new life. There's a sterility with certain types of mass production that leaves a gap in our souls; Sushi and other unique low-volume manufacturing fill that void.

✎	*Fernando and Humberto Campana, Brazil*	304
⚙	*edra, Italy (Sushi chair), Studio Campana (Sushi sofa)*	314 304
	Plastics, various textiles, rubber, felt, EVA, carpet, reused offcuts	
	• *Recycled materials* • *Low-energy production*	325 326

Eco-recliner

Simon White's Eco-recliner is made by joining identical profiles of locally sourced, steam-bent Cornish green ash, without glues or solvents, and then finished in Danish oil. The construction method ensures it can be disassembled for repair or reuse. This project demonstrates the feasibility of small- or large-scale manufacturing in rural areas – a model that will undoubtedly be adopted as the concept of localization gains ground.

Simon White Design, UK	309
Simon White Design, UK	309
Ash, Danish oil	
• *Local renewable materials* • *Easy to disassemble or repair*	325 329

Tetris

As the price of recycled cardboard hits all-time lows, now is perhaps the time to experiment with how eco-friendly consumers can make use of this resource themselves. WEmake has created a cutting pattern used to turn cardboard waste into a sturdy armchair, and offers the 'downloadable design' on their website. Tetris refers to a modular furniture system based upon grid patterns of 10 cm (4 in) that enables anyone to create an individual template.

Once your cardboard is cut, glued and stacked, it can then be customized with your own surface treatment (graffiti, paint, textile), allowing room for personalization and experimentation.

WEmake, UK	311
DIY	
Cardboard, PVA glue	
• *Recycled materials* • *DIY*	325 326

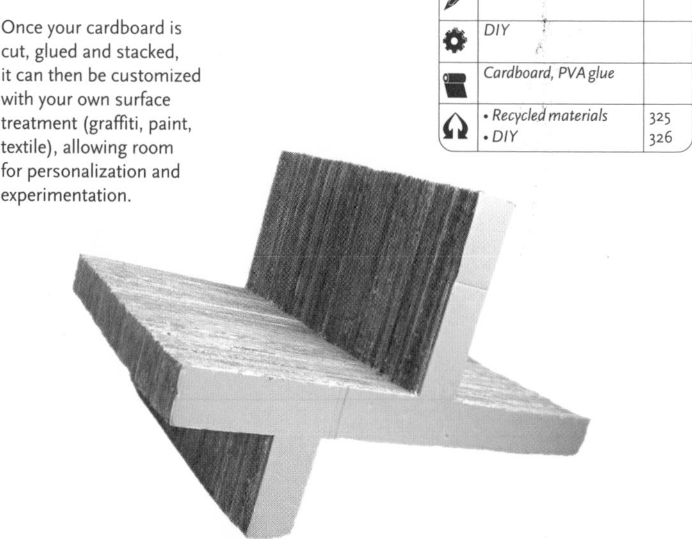

Airbag

Since the 1960s inflatable chairs have come and gone but Suppanen and Kolhonen have added an extra comfort dimension by placing balls of EPS inside the nylon outer cover, at the same time allowing the chair to be deflated when not in use. Nylon is tough and resists puncturing better than other polymers.

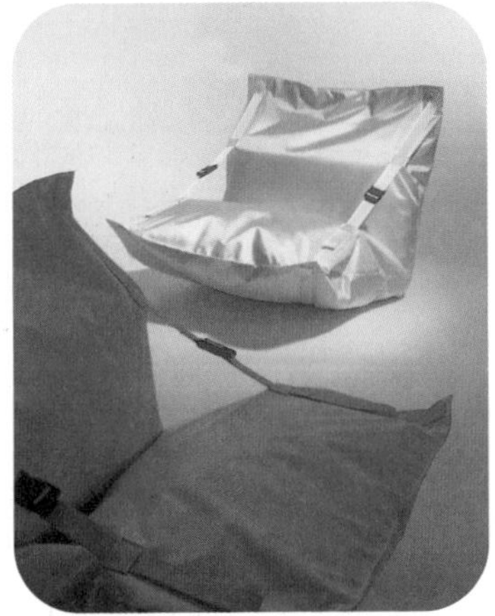

Designer	*Ilkka Suppanen and Pasi Kolhonen, Finland*	310
Manufacturer	*Studio Suppanen, Finland*	310
Materials	*Expanded polystyrene, nylon*	
Sustainability	*• Lightweight materials* *• Reduced energy use during transport*	326 327

Body Raft

Local wych elm is bent with steam to create a curved frame to which further curved lathes are attached. This organic shape is visually appealing. Hand-crafted furniture of this kind can contribute to sustaining local economies.

Designer	*David Trubridge, New Zealand*	310
Manufacturer	*Cappellini SpA, Italy*	313
Materials	*Wych elm*	
Sustainability	*• Renewable materials*	325

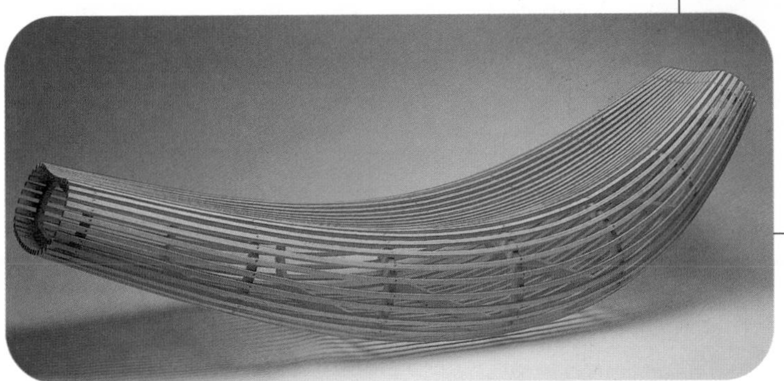

bolla 10, 11, 15, 16

Rattan harvested from forests in Southeast Asia is a wonderfully flexible and expressive material. More uniform than European willow, woven rattan produces smooth surfaces, making it a perfect material for sculptural and modernist gestures in furniture and lighting. Bolla 10 and 11 are combined side tables cum ottomans made entirely of rattan, while bolla 15 and 16 offer an additional function with a storage volume underneath the MDF white laminate lid. Companies like Gervasoni offer a valuable outlet for the consummate skills of the rattan weavers and in doing so participate in the on-going evolution of the craft.

Designer	*Michael Sodeau, UK*	309
Manufacturer	*Gervasoni SpA, Italy*	316
Materials	*Rattan, MDF*	
Sustainability	*• Renewable materials* *• Retention of craft skills*	325 325

Waka, Ruth 1 and 2, and Glide

David Trubridge continues his experimentation with new forms using native New Zealand or Australian timber from sustainably managed plantations. Waka is a bench seat made from plantation-grown Australian hoop pine plywood and is finished in a mixture of natural oil and earth pigment. The rocking chairs, Ruth 1 and 2, are made from the same plywood in combination with plantation-grown ash or oak. Glide is made from a mono-material, plantation-grown ash. As a designer Trubridge is concerned with the footprint he leaves and has embarked on a new project called 'Structures for Survival' that addresses the issue of design ethics and an artefact's lifecycle.

✎	*David Trubridge, New Zealand*	310
⚙	*One-offs and small batch production*	
▬	*Various timbers*	
♻	• *Renewable materials from managed sources* • *Natural finishes*	325

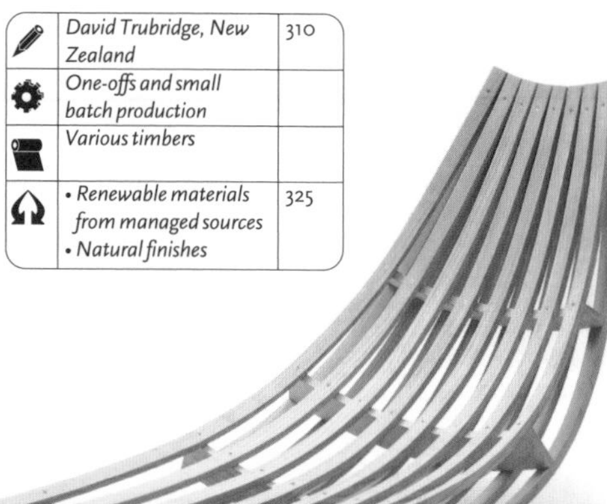

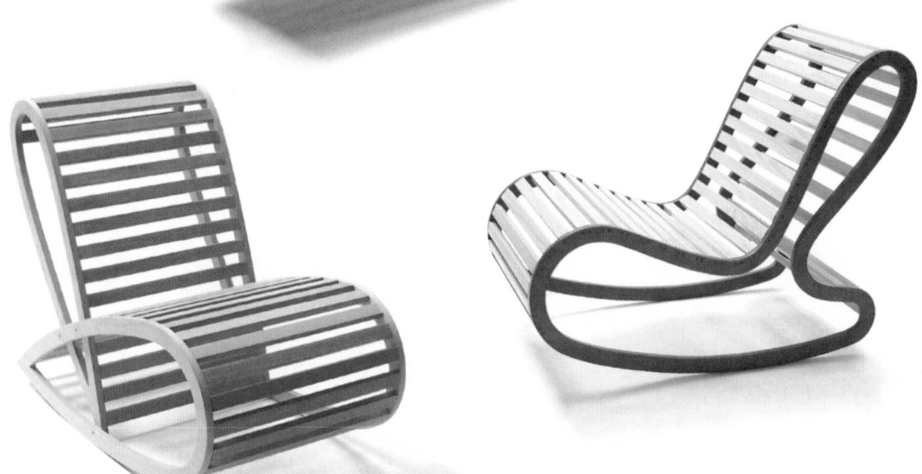

Bluebelle

Designing a chair remains the quintessential test of any furniture designer. The imaginative form of the Bluebelle chair carefully models itself like a prosthesis, supporting and caressing the seated body. Rigid seat and arms are made from one type of polypropylene to which is clipped a more flexible polypropylene forming the comfortable backrest.

Easily disassembled into component parts, the PP and metal frames can be recycled.

✎	*Ross Lovegrove, UK*	308
⚙	*Driade SpA, Italy*	314
▬	*Polypropylene, metal*	
↻	*• Improved ergonomics* *• Design for disassembly* *• Recyclable*	328 326 325

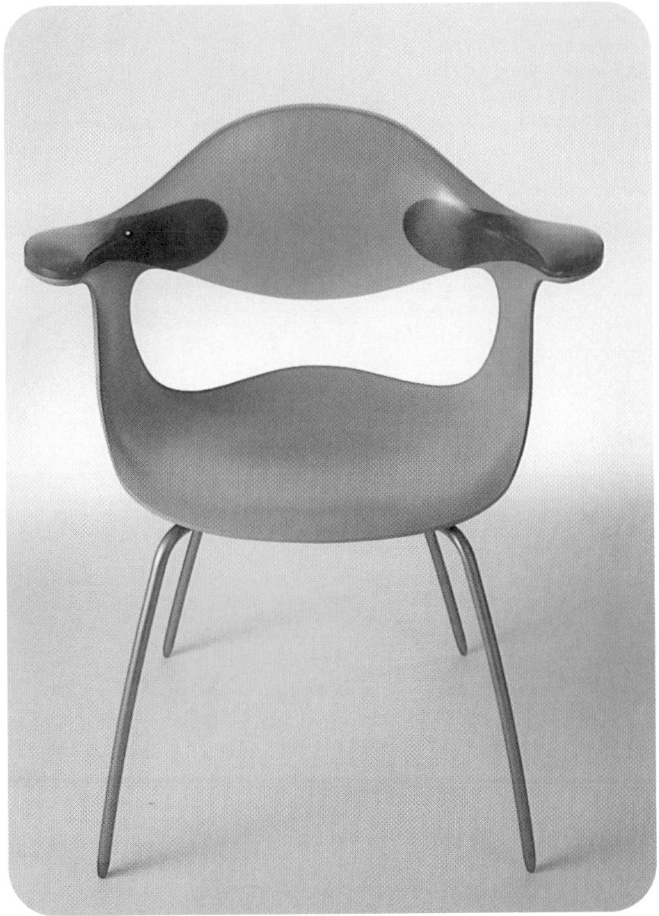

Agatha Dreams

Pillet's elegant chaise longue combines the eclecticism of craft with the technical skills of the workers at Ceccotti's factory, a labour force with a long history of 'craft technology'. Renewable materials are brought to a state of refinement that will encourage owners to cherish this design and confer a degree of longevity. High-quality manufacturing using nature's materials will always be a sustainable business model, as long as raw materials are procured from managed forests.

✎	*Christophe Pillet, France*	309
⚙	*Ceccotti Collezioni, Italy*	313
▬	*Layered timber, solid cherry wood*	
↻	*• Renewable materials*	325

Model 290F

For over 150 years the manufacturer Gebrüder Thonet has mass-produced elegant bentwood chairs, combining good design with economical use of local (European) materials to produce modular 'flat-pack' designs facilitating distribution with basic, yet customizable, options. In 1849 at Michael Thonet's factory in Vienna 'Chair No. 1', the Schwarzenberg chair, made of four prefabricated components that could be reassembled in different configurations, was the precursor of a design ideally suited to industrial production. Thonet chairs graced many a café and restaurant from Paris to Berlin and London in the late 19th and early 20th centuries and created the definitive archetype for the café chair. 'Chair No. 14', later known as the 'Vienna coffee-house chair', was one of the most successful products of the 19th century and probably remains the world's best-selling chair, with over 50 million sold in 1930 alone. The roll-call of iconic designers, such as Mies van der Rohe, Mart Stam, Marcel Breuer and Verner Panton, ensured that Thonet always explored designs driven by new movements and schools of thought. Yet Thonet remain aware of their traditions and currently produce modern variants using well-tested principles and materials, such as steamed and bent solid beech wood. Model 290F epitomizes the Thonet philosophy: the designers, Wulf Schneider and Partners, use three pieces to create a robust, durable and repairable chair. A single piece of solid bent beech forms the front legs and back stay, a cut-and-drilled, moulded, laminated beech section forms the back legs and back rest, and both pieces are fixed to the laminated seat with cast-aluminium angled brackets and screws. Nineteenth-century examples of Thonet chairs turn up in the prestigious sale rooms of Sotheby's, Christie's and Bonhams, attesting to their durability. It is quite likely that Model 290F will become a sought-after antique, validating it as a good and green design.

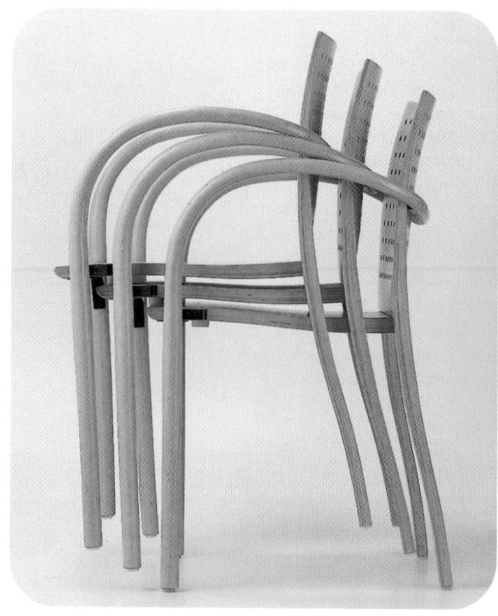

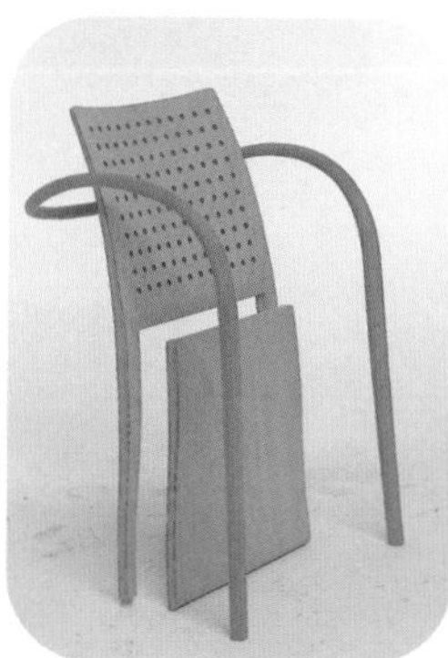

✎	*Prof. Wulf Schneider and Partners, Germany*	309
⚙	*Gebrüder Thonet GmbH, Germany*	322
▯	*Beech wood, aluminium*	
♻	*• Renewable materials* *• Reduced-energy transport and assembly*	325 327
✪	*iF Ecology Design Award, Germany, 1999*	332

Stoolen

Every time Bill Hilgendorf of the Uhuru design team, a small design and build furniture collective, has enough reclaimed or re-used wood scraps, another Stoolen arrives in the world. The hand-finished wooden pieces are snugly fitted into an old bicycle rim to create this sturdy and uniquely textured stool that also doubles as an end table.

✎	*Uhuru Design, USA*	310
⚙	*Uhuru Design, USA*	310
	Wood, steel	
	• Reclaimed materials *• Durable design*	325 325

Stoop™ Social Seating

Sitting on a door stoop (door step in the UK) in Philadelphia's row homes inspired Jaime Salm to examine the social role of these stoops in city life. Stoop™ takes the opportunity to 'stoop' to a multitude of indoor and outdoor environments. Stackable, portable and lightweight, this seating is suitable for small studios,

apartments, balconies or patio rooftop gardens, where it is easily stowed away when not needed. A rubber grommet lines the finger hole for easy transport while the brightly coloured powder-coated spun aluminium ensures durability and easy cleaning. Although designed to encourage better body posture, it could be argued that the main benefit is to foster an informal body language that encourages social discourse.

✎	*Jaime Salm, MIO, USA*	309
⚙	*MIO, USA*	308
	Spun aluminium, rubber	
	• Multifunctional *• Improved social well-being*	328 328

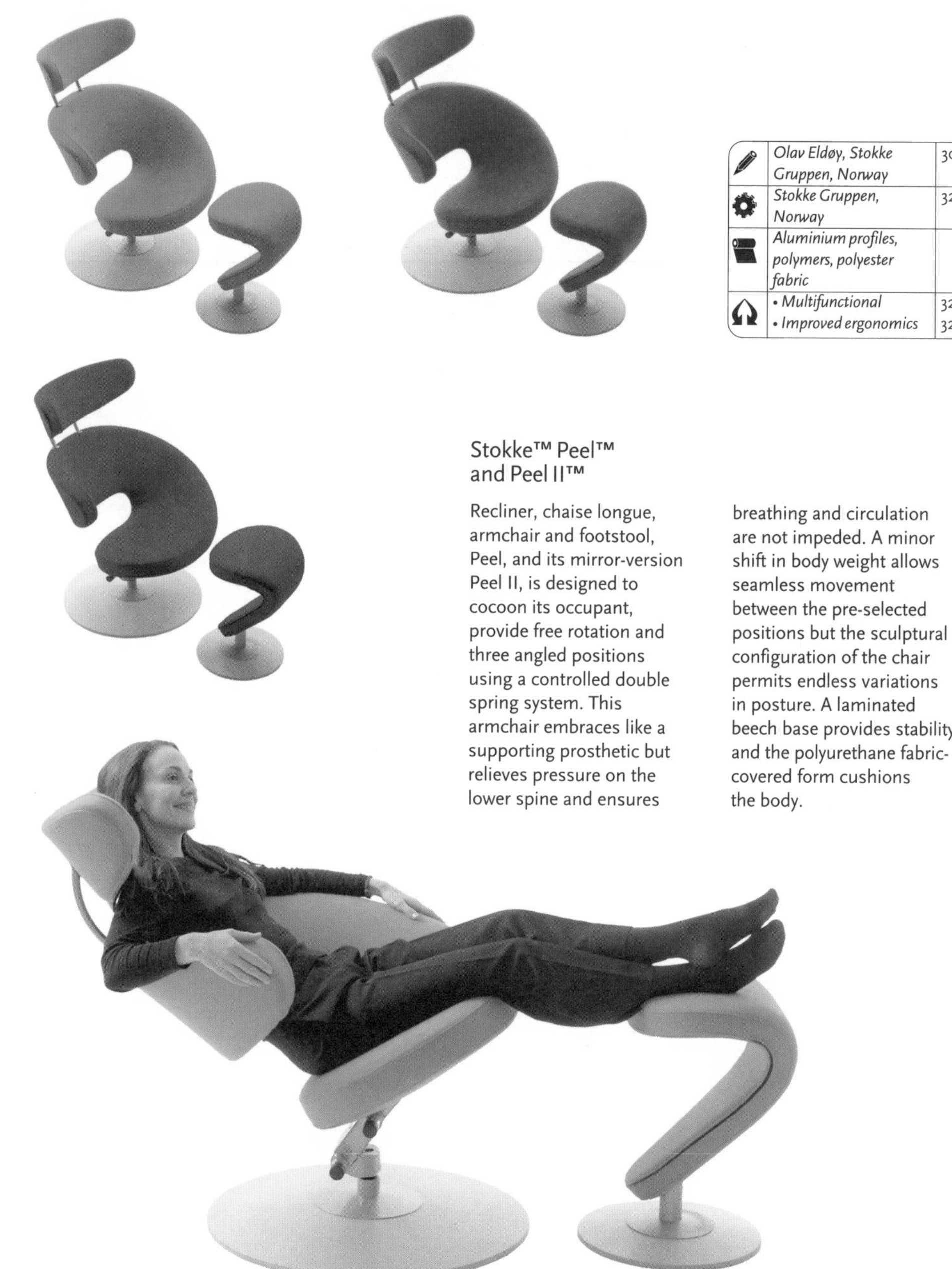

✎	*Olav Eldøy, Stokke Gruppen, Norway*	305
⚙	*Stokke Gruppen, Norway*	321
	Aluminium profiles, polymers, polyester fabric	
	• *Multifunctional* • *Improved ergonomics*	328 328

Stokke™ Peel™ and Peel II™

Recliner, chaise longue, armchair and footstool, Peel, and its mirror-version Peel II, is designed to cocoon its occupant, provide free rotation and three angled positions using a controlled double spring system. This armchair embraces like a supporting prosthetic but relieves pressure on the lower spine and ensures breathing and circulation are not impeded. A minor shift in body weight allows seamless movement between the pre-selected positions but the sculptural configuration of the chair permits endless variations in posture. A laminated beech base provides stability and the polyurethane fabric-covered form cushions the body.

Cocochair

Adding more natural latex to coconut fibre changes the density, allowing the fibre to be used for self-supporting structures. This armchair uses fibre with latex for the main structure, with a seat and back pad made of natural latex foam. The simple abstract shape is reminiscent of late 1960s polyurethane foam-block seating from the likes of Verner Panton, Roberto Sebastian Matta and Cini Boeri. The Cocochair, however, treads rather more lightly on the environment and is more easily sent around the materials loop again.

	Alessandro Zampieri Design, UK	304
	One-offs and small batch production	
	Coconut fibre, natural latex, natural latex foam, organic wool textile	
	• Renewable materials *• Compostable and/or reusable materials* *• Cold, low-energy, production processes*	325 325 326

Bilge Lounge

Old staves from reclaimed bourbon barrels are reincarnated in this sprung lounge chair that utilizes old vehicle springs for the base. The natural colouring of the staves – obtained by charring the inside of the white oak barrels before aging the bourbon – is retained as a memory of the wood's previous life. The designers have struck a fresh balance with this unorthodox and low-energy combination of recycled materials, creating a new vernacular aesthetic that plays off the naturally curved forms of both wooden staves and steel springs.

	Uhuru Design, USA	310
	Uhuru Design, USA	310
	Wood, steel	
	• Reclaimed material *• Durable design*	325 325

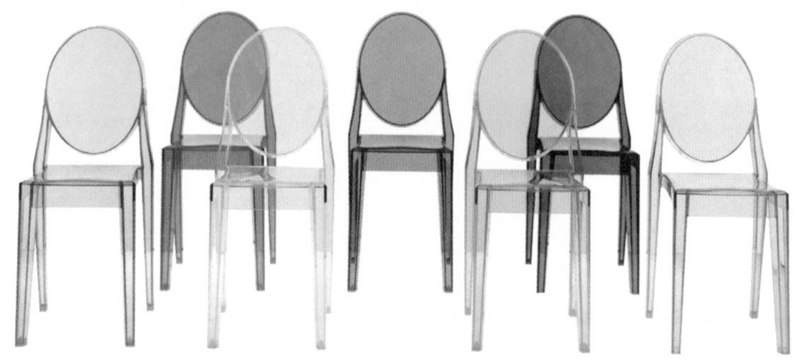

Victoria Ghost, Louis Ghost and La Marie

Philippe Starck has apparently turned over a new leaf in the last couple of years claiming that we need to imagine more 'ecologically democratic' artefacts. Just what he means by this isn't yet totally clear, but his legacy as a designer is to bring life to materials and forms so that we *care* about the objects. An upshot of this attachment is that it is easier to ignore the next trend and be satisfied with what you already have. The sophisticated production process for this range of imaginative transparent chairs reveals the amazing technical possibilities inherent in polycarbonate. All synthetic plastics involve industrial chemistry in their manufacture, so environmental impacts are implicit, but let us remember that we are all invoked – the entire global economy and its mobilization depends on oil, the raw material from which these polymers emerge. Fortunately, these chairs will be easier to recycle at the end of their long lives because of their mono-material construction of polycarbonate.

✎	*Philippe Starck, France*	309
⚙	*Kartell, Italy*	317
	Polycarbonate	
	• *Single material* • *Durable design*	325 325

Supernatural

Ross Lovegrove's skill for combining his aesthetic visions, inspired by organic and biomimetic forms and structures, with the latest in digital and production technologies has resulted in this animated lightweight indoor/outdoor stackable chair. Structural integrity is achieved by an advanced process of injection-moulding two layers of polypropylene with glass fibres. The result is strong and beautiful, making it likely this object will have physical and cultural durability.

✎	*Ross Lovegrove, UK*	308
⚙	*Moroso, Italy*	318
	Reinforced fibreglass with integrally coloured polypropylene	
	• *Lightweight, strong and durable materials* • *Certified to ISO 14001 since 1999*	325 330

Origami Zaisu

A single sheet of plywood is bent and cut to form a simple floor seat. In Japanese culture sitting on the floor is the norm but perhaps the practice should be adopted more widely, since the omission of legs that form a conventional chair saves materials and energy.

	Mitsumasa Sugasawa, Japan	310
	Tendo Co. Ltd, Japan	321
	Plywood	
	• *Reduction in materials used* • *Renewable materials*	325 325

Mirandolina

Reviving a technique first used by the designer Hans Coray for his pressed-aluminium 'Landi' chair designed in 1938, Pietro Arosio has produced an economical yet elegant stacking chair from a single sheet of aluminium. Cut and pressed into its final form, the Mirandolina shouts efficiency. The use of one material, aluminium, facilitates recycling waste offcuts and ensures it is easy to recycle or repair.

	Pietro Arosio, Italy	304
	Zanotta SpA, Italy	323
	Aluminium	
	• *Recyclable single material* • *Efficient use of raw materials*	325 326

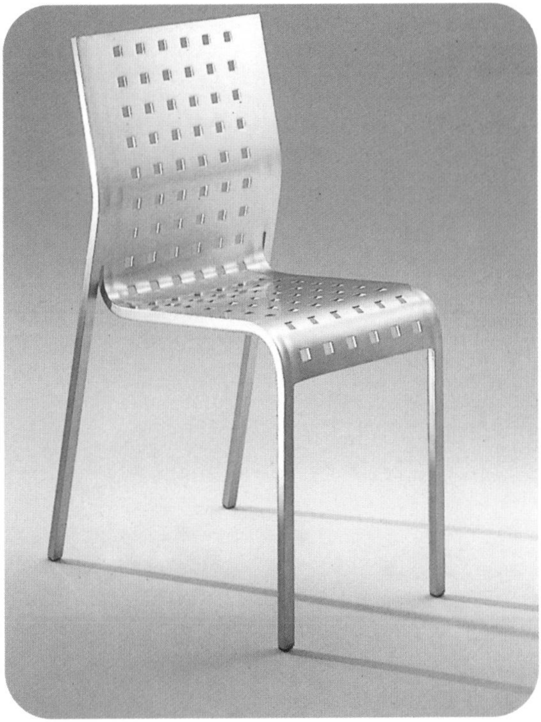

Superstructure

Attention to detail is combined with a love of nature's raw materials in this easy chair made with ash (white version) or oak (black version). The design evokes the aesthetic of Scandinavian modern furniture of the mid-20th century, recalling the ubiquitous 'capsule chairs', in this case using stronger, more durable materials subtly balanced in a contemporary form. This is a product designed for a long life that avoids the cycles of fashion and enables a new generation of trees to grow to ensure continuity of production.

✎	*Björn Dahlström, David design, Sweden*	305
⚙	*David design, Sweden*	314
▰	*Oak or ash wood, water-based lacquer, stain, leather or textile cushion*	
♻	*• Renewable materials* *• Durable design*	325 325

Teddy Bag

Is it a storage bag or is it a chair? The Teddy Bag can be both, and its tactile design allows you to store old (redundant) clothes or soft toys, and then fold the top to transform it into a seat. Sit for long enough and you might just think about the 'waste' you are sitting on and wonder where it all came from. The rectilinear design reduces waste on cutting and the label is laser cut, so no printing inks are required. The whole product is made of just one material – wool.

✎	*Andrew Millar, UK*	308
⚙	*One-offs and small batch production*	
▰	*100% wool felt, wool thread*	
♻	*• Renewable materials* *• Multifunctional*	325 328

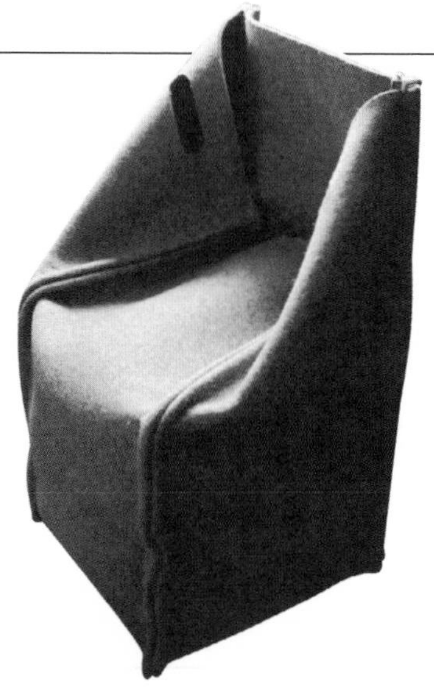

Honey-pop chair

First presented at the 2002 Milano Salone, the Honey-pop delights and surprises. Made entirely of paper, folded and joined, the flat-pack, planar chair unfolds like a magic lantern. The complex structure responds to support each individual. Yoshioka reveals a knowledge and respect for his chosen material and imbues the finished product with lightness, fun and a sculptural aesthetic. Driade invested in the prototype to create a unique manufactured, lightweight paper armchair delivering new experiences to the concept of seating.

✎	*Tokujin Yoshioka, Japan*	311
⚙	*Driade SpA, Italy*	314
▬	*Paper*	
♻	• *Recyclable* • *Lightweight materials* • *Low-energy materials and transport*	325 325 327

Tula

Made predominantly from ash thinnings harvested from small local woodlands in Somerset, Tula is part of the evolving story of 'stick and nail' furniture. Ash slats and boards are steam-bent, scorched black, then wire-brushed and waxed. Smaller diameter ash is worked for the legs and two-year-old cultivated willow withies are used for the hoop and struts. Economical, elegant and eclectic, this type of small-scale production is inherently sustainable by encouraging management of local resources and generating designs with regional identity.

✎	*Guy Martin Furniture, UK*	306
⚙	*One-offs and small batch production*	
▬	*Ash, willow withies, copper nails*	
♻	• *Renewable and locally-sourced materials*	325

Tokyo-Pop sofa and table-stools

As plastics continue to form an integral part of modern living it is important that we elevate the cultural value of plastic artefacts, so we cherish them rather than carelessly dispose of them in landfill sites. Over recent years the Italian manufacturers Driade have encouraged responsible, yet innovative, designs of one-piece furniture made from single synthetic polymers. The respect for material and obvious technical expertise in rotational- and heat-moulding shown by the designer and manufacturer somehow elevate the status of the polyethylene to a valued material once it emerges as a sculptural sofa and as multifunctional table-stools. The Tokyo-Pop range is versatile indoor or outdoor furniture. An indoor version of the sofa, Tokyo-Soft, is available with Trevira or wool fabric removable and washable covers. These objects are inherently durable; they represent a temporary materialization of a unique man-made resource that is destined to run out one day. Will furniture manufacturers operate a take-back policy when Tokyo-Pop and similar items end their useful life? It makes environmental sense to and, one day, it may be a legal requirement to 'close the loop' and reclaim plastic for future manufacturing.

✎	*Tokujin Yoshioka, Japan*	311
⚙	*Driade SpA, Italy*	314
▬	*Polyethylene, Trevira or wool fabric*	
♻	• *Recyclable* • *Single material* • *Multifunctional*	325 325 328

REEE® chair

Nine redundant video-game consoles, equivalent to 2.4 kg (5.3 lb) of polycarbonate or ABS plastic, are recycled to make one REEE® chair. The injection-moulded ribs and end caps enable easy assembly, disassembly and, most importantly, repair, making this an ideal chair for heavy domestic, academic and commercial use. Here is an excellent example of how legislation – in this case the European Waste Electrical and Electronic Equipment (WEEE) Directive – has encouraged the electronics industry to look for feasible outlets for using materials from 'take back' products. This process is known as 'closing the loop' to ensure that valuable man-made materials can find a new life.

✎	*Sprout Design, UK*	309
⚙	*Pli Design, UK*	319
▬	*Post-consumer games console plastics, powder-coated steel*	
♻	• *Recycled and recyclable* • *Ease of assembly, disassembly & repair*	325 326/ 329

Stokke™ Gravity II™

Half rocking chair, half armchair, the original Gravity by Stokke challenges our perceptions. Constructed of laminated wooden runners, this unique form enables the user to move seamlessly from an upright position, with the knees tucked behind the kneepads, through two intermediate positions to the fully inclined position where the blood circulation is encouraged in the legs and the back is completely supported. Ideal for those who spend long hours at a PC, the Gravity II offers instant opportunities for reducing stress and improving overall well-being. The ideal all-in-one solution to instant cat naps!

✎	*Olav Eldøy, Stokke Gruppen, Norway*	305
⚙	*Stokke Gruppen, Norway*	321
▮	*Aluminium profiles, polymers, polyester fabric*	
♺	*• Multifunctional* *• Improved ergonomics*	328 328

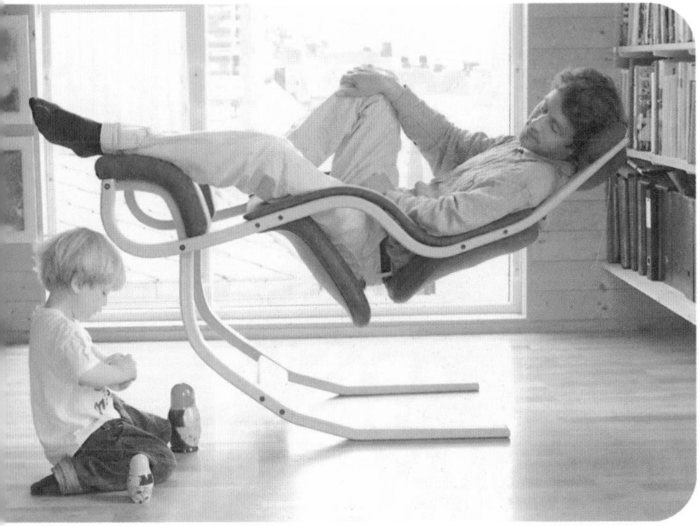

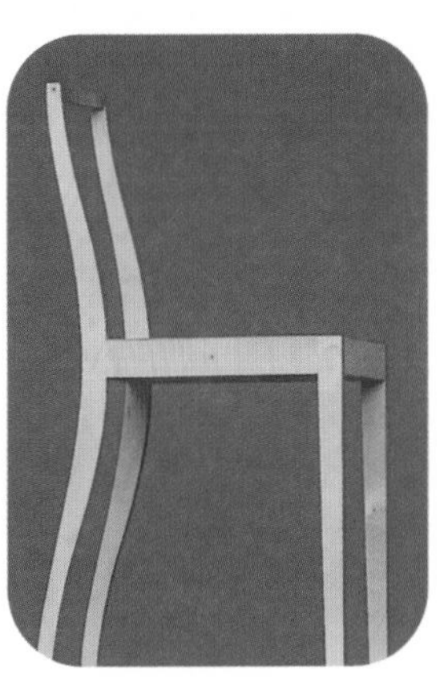

Ply Chair

Avoiding excessive usage or wastage of materials should be a guiding principle of any design in the 21st century. The Ply Chair is the latest answer, demonstrating restraint, grace, economy, strength and character.

✎	*Jasper Morrison, UK*	308
⚙	*Vitra, Switzerland*	322
▮	*Aeronautical-quality plywood*	
♺	*• Reduction in materials used*	325

Vuw

Part of the Iform 'Voxia', the 'voice of nature' collection (see Eco, p. 40), Vuw is a tour de force in 3D curved wood surfaces, made from one continuous piece carefully bent and formed. Beech wood originates from a local, sustainably managed forest and is rotary-cut to minimize waste from the harvested trees. Cut-away waste is also minimized by planning the veneer pattern in the laminated wood. The result is a thin, sensuously curved surface efficient in materials consumption and energy of manufacture.

✏	*Peter Karpf, Sweden*	307
⚙	*Iform, Sweden*	316
▬	*Beech veneer laminate*	
♻	*• Renewable materials from managed sources* *• Reduction in materials and energy used*	325 325

RD4S

Your old high-density polyethylene (HDPE) milk bottles, and other everyday HDPE waste, find new life in the RD4S chairs, an exuberant expression of material reincarnation. The process of production ensures that each chair is unique: the recyclate, available in flake form, is melted and extruded in a continuous ribbon that is woven over a mould. The result is a lightweight, rigid and strong chair that is distinctive and whose manufacture uses a fraction of the embodied energy associated with virgin plastics.

✏	*Richard Liddle, Cohda Design Ltd, UK*	308
⚙	*Cohda Design Ltd, UK, small batch production*	305
▬	*Recycled post-consumer HDPE*	
♻	*• Single material* *• Recycled and recyclable materials* *• Low-energy production*	325 325 326

100 Chairs in 100 Days

Martino Gamper's '100 Chairs in 100 Days' project involved collecting discarded or unused chairs from London's streets and friends' homes and then generating one hundred chair transformations over one hundred days. His 'mongrel morphology' is evident in the eclectic hybrid re-combinations that emerged. Gamper playfully questions relationships between people and objects, and explores the notions of value, function and status. With spontaneity and elements of randomness, Gamper reminds us of a diverse array of typeforms that can be re-blended into something meaningful to each of us in a personal way. He contests the notion of the 'best' chair design and suggests experimental pluralism as a viable way forward. Bonding with fewer objects in our lives is an effective strategy for reducing resource flow, but it is a growing challenge as the human population, and per capita consumption, continues to increase.

Designer	*Martino Gamper, UK/Spain*	306
Manufacture	*One-offs*	
Materials	*Various woods, metals, polymers, textiles*	
Strategy	*• Reclaimed material* *• Personalization for better emotional bonds*	325 328

Yolanda Collection

Creating a welcome relief from steam-bent and moulded temperate plywoods, Minakawa's design celebrates the aesthetic of bamboo in an unexpected form. Here laminated bamboo sheets create a striking sculptural form so different from traditional round, split and woven bamboo chairs. Finished surfaces are protected using non-biocide, plant-based oils and sealers.

Designer	*Gerard Minakawa, USA*	308
Manufacture	*Small batch production*	
Materials	*Bamboo laminate*	
Strategy	*• Renewable material*	325
Award	*IDRA award, 2002–2003*	

BUNSON chair and NYA armchair

BUNSON and NYA are by Lino Codato, whose motivation and ethos is to investigate new applications of natural materials for contemporary furniture. Water hyacinth and Yan lipao are abundant plants in tropical climates, the former being a native of South America that has become a weed and has invaded watercourses in the southern United States and other warm climes. Harvested, prepared and dried, it transforms into an excellent weaving material. Yan lipao, a fern vine, is harvested from the jungles of Thailand, its outer pith peeled off then polished and smoothed before weaving. As a traditional material for basketry it ensures that a skilled labour pool is available for creating other artefacts. This furniture represents that wonderful boundary between mass production and one-off craft, supporting use of local materials and skills.

	Lino Codato Collection, Italy	317
	Lino Codato Collection, Italy	317
	Water hyacinth, Yan lipao, oak	
	• *Renewable materials* • *Conservation work by weed removal*	325 330

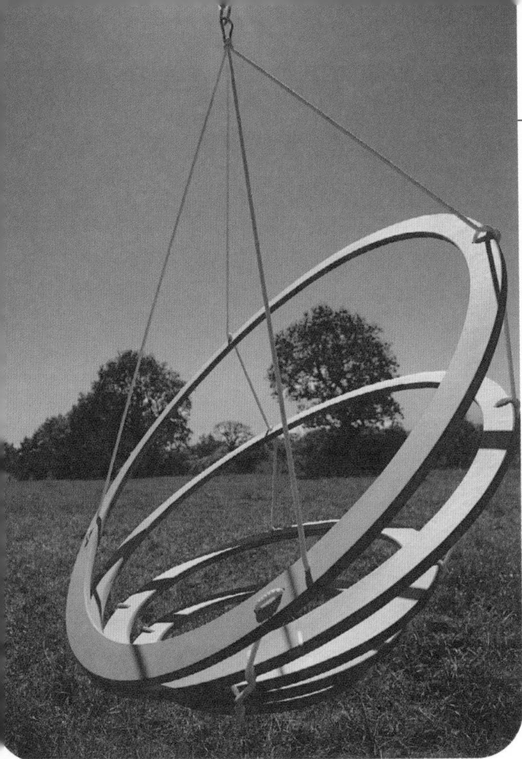

Circa and Double

The hanging chairs of the Swinging Sixties find a sustainable cousin in this flat-pack, fold away series of designs for solo (Circa) or paired (Double) seating. Each chair is cut by machine to minimize wastage of birch plywood. Suitable for indoor or outdoor use, the simplicity of the geometry provides a visual focus and tempts us back to childlike pleasures.

✎	rawstudio, UK	309
⚙	Small batch production	
▬	FSC exterior grade birch plywood, rope	
♺	• Renewable materials from managed sources • Minimal wastage	325 327

Chair (Lusty's Lloyd Loom)

Lusty's Lloyd Loom Company engaged Gschwendtner, a former student of the Royal College of Art, London, to create these occasional/dining chairs. Her designs build on traditional designs, so strongly associated with tearooms and Empire, and open more contemporary modern markets to the manufacturer. Economical use of materials combines with durability to produce a serviceable dining/occasional chair.

✎	Gitta Gschwendtner	306
⚙	Lusty's Lloyd Loom, UK	318
▬	Lloyd Loom woven paper & wire, steel	
♺	• Reduction in materials used • Renewable materials	325 325

Eco

These stackable chairs are cut from a single piece of veneer-faced ply and follow in the Scandinavian tradition of working with bent ply, such as the designs of Gerald Summers for the firm Makers of Simple Furniture based in London in the late 1930s. Simplicity, economy and functionality meet in this award-winning design.

	Peter Karpf, Denmark	307
	Iform, Sweden	316
	Plywood	
	• Renewable materials *• Reduction in materials used* *• Low-energy production*	325 325 326
	iF Ecology Design Award, 2000	332

LoveNet

During the 1950s Europe and North America witnessed a boom in woven 'capsule' chairs on spindly metal frames, appearing on the glossy pages of interior design and architecture magazines. This legacy lives on in Lovegrove's homage to economical usage of technosphere materials, the LoveNet. Opting for lightweight woven polyethylene (PE), rather than a natural material, such as rattan, increases the embodied energy of the materials for the seating but if the strength of the PE permits greater longevity then the embodied energy per lifespan year of both options may not be so different. At the end of its life the galvanized and powder-coated steel under the webbing and stainless-steel frame can be separated for recycling, although questions remain regarding the feasibility of recycling the PE.

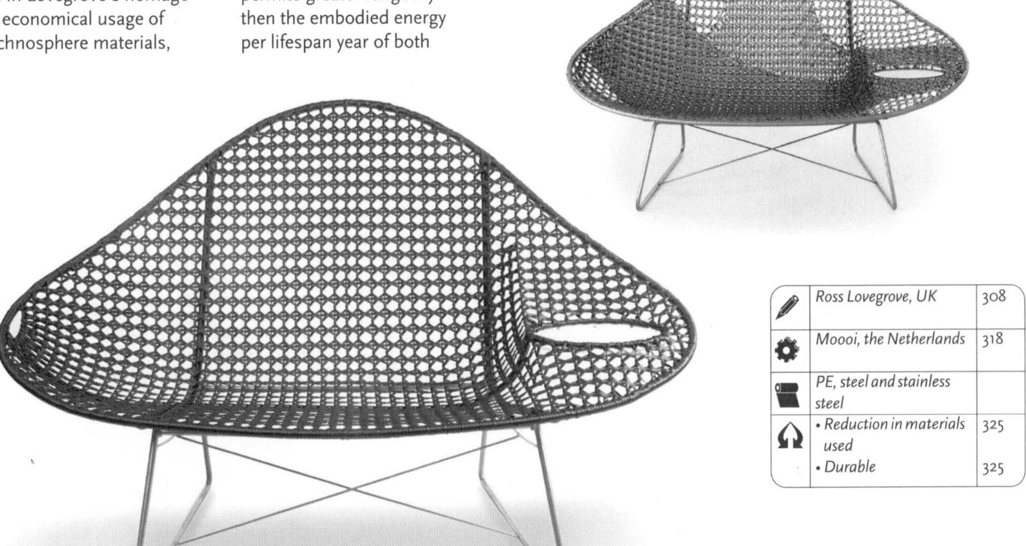

✎	*Ross Lovegrove, UK*	308
⚙	*Moooi, the Netherlands*	318
	PE, steel and stainless steel	
	• Reduction in materials used	325
	• Durable	325

Wiggle series

Originally designed in 1972 as economical furniture and manufactured by Jack Brocan in the USA, the Wiggle side chair has been reproduced by Vitra since 1992. Each layer of corrugated cardboard is placed at an angle to the next layer to provide significantly increased durability compared with the folding cardboard chairs by the likes of Peter Raacke and Peter Murdoch in the 1960s.

✎	*Frank O. Gehry, USA*	306
⚙	*Vitra, Switzerland*	322
	Cardboard, glue	
	• Renewable materials	325
	• Low-energy production	326

B. M. Horse Chair

B. M. refers to the Bell Metal Project, 1998–2000, for which Pakhalé applied and refined the process of the *cire perdue* or lost-wax casting technique popular in central India for hundreds of years. Building on his experiences creating an eclectic range of tableware (see B. M. vase, fruitbowl and spoon; p. 170), this chair is laden with primal and symbolic forms bridging the past and present. The alchemy extends beyond the mixed recycled brass, bronze and tin to the combination of hand, computer and technical skills used to make this piece. Sandblasting the surface is the last technique applied to realize the final form. A blurring of high-craft with high industrial design shows fresh opportunities for the future.

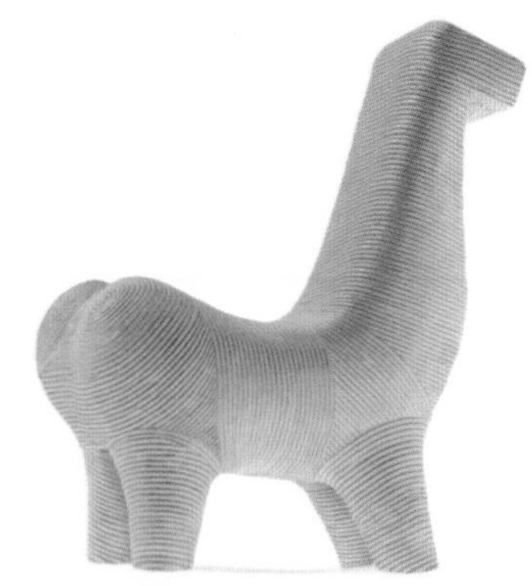

	Satyendra Pakhalé, the Netherlands and India	309
	Atelier Satyendra Pakhalé, the Netherlands	309
	Brass, bronze, copper scrap	
	• Recycled metals *• Low-energy production* *• Innovation of traditional craft skills*	325 326 326

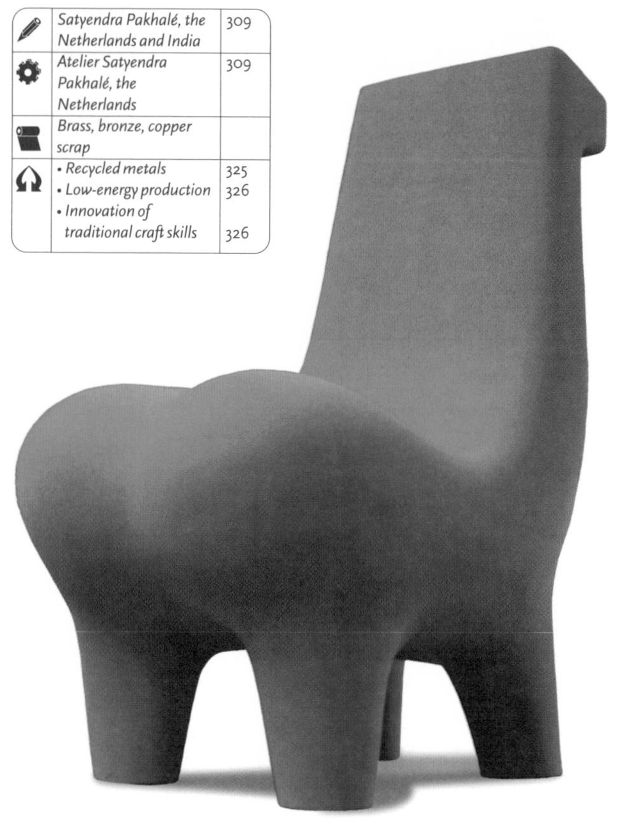

Blähtonhocker

Clay is an elemental material indelibly bound with mankind's history and creativity. Robert Wettstein's expanded clay stool is proof that the relationship continues to evolve. Working in that energy charged zone between art, craft and design, this object challenges our perceptions and stimulates our senses.

	Robert A. Wettstein, Switzerland	311
	One-offs and small batch production	
	Clay	
	• Abundant geosphere material	325

Ghost

Purity of form and function can often be achieved by focusing on the properties of one particular material. Cini Boeri and Tomu Katayanagi have taken a single piece of 12 mm- (½ in-) thick toughened glass and cut and moulded it into an extraordinary object. They juxtapose the contradictory characteristics of the material – its fragility and toughness – and create a durable, rather timeless design. Ghost provides food for thought on how other familiar materials can be modified or mutated to fit new forms and functions. Being composed of a single material facilitates recycling at the end of the product's life and encourages closed-loop recycling, where the manufacturer uses its own recycled materials to produce new goods.

✎	*Cini Boeri and Tomu Katayanagi, Italy and Japan*	304
⚙	*Fiam Italia SpA, Italy*	315
	Glass	
	• Recyclable *• Single material* *• Durable*	325 325 325

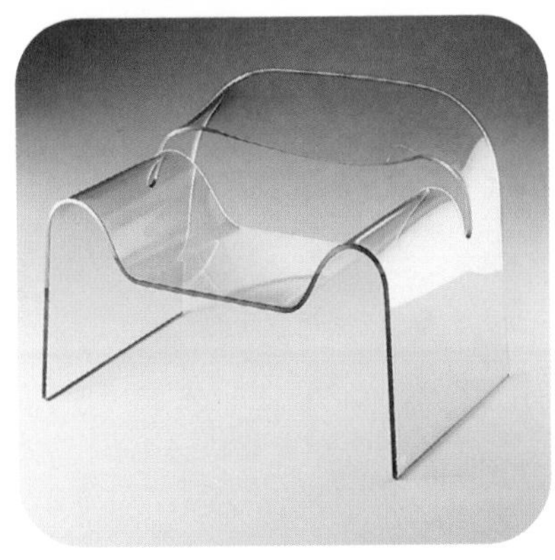

SE 18

It is difficult to pinpoint exactly how or when a product becomes a 'classic' design. Classic implies that the design is respected, valued and maintains its socio-cultural relevance. Such a design is also likely to be well made, durable and repairable. Originally designed in 1952 by Professor Egon Eiermann, the SE 18 received the 'Good Design' Award 1953 and was immediately acquired for the Museum of Modern Art's collection in New York. Fifty-two years after the first chairs left the factory, the efficient production system, with its emphasis on hand-finishing for quality, has hardly changed. Here is a philosophy of manufacturing that is inherently more sustainable than the fashion output of numerous furniture manufacturers. There is no need to keep spending on a continuous cycle of re-tooling and marketing. This chair virtually markets itself by appealing to consumers' design instinct for knowing a classic when they see one.

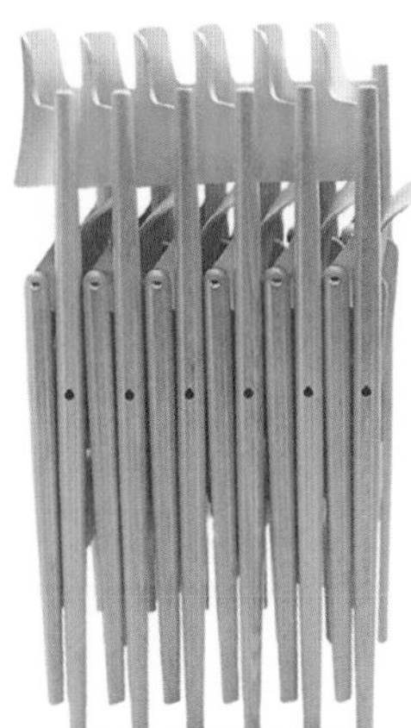

✎	*Prof. Egon Eiermann, Germany*	323
⚙	*Wilde & Spieth GmbH, Germany*	323
	Beech, metal	
	• Low-energy production *• Classic design*	326 324

Trinidad No.3298

'Industrial craft' production will undoubtedly prosper in the 21st century if the workmanship and graphical form of this ash chair are a measure of the output of today's furniture manufacturers.

Designer	*Nanna Ditzel, Denmark*	305
Manufacturer	*Fredericia Furniture A/S, Denmark*	315
Materials	*Ash, metal*	
Eco	*• Renewable material*	325

Corks

Tactile, warm, giving and durable all characterize cork, from *Quercus suber*, cork oak. Morrison approaches this material with his eye for economy and subtlety to produce a versatile object that encourages a little leisure – for sitting, leaning or for resting your coffee cups. In a gesture the embattled cork industry of the Iberian peninsula might appreciate, here's a great use for cork to offset recent gains made by plastic corks in the global wine industry.

Congregate Communal Seating

Wood from reclaimed pews from Gilmerton Church in Edinburgh finds a new life in this seating area whose aim is to encourage coming together, interaction and exchange. The original formal linear arrangement of the pews has been quietly re-directed into more informal circles of communal activity that would have characterized pre-Reformation worship. This seating encourages sharing and the notion of collective experience.

Designer	*John D. O'Leary, BlueGreen&Co., UK*	308
Manufacturer	*One-offs*	
Materials	*Reclaimed furniture, textiles, padding*	
Eco	*• Reclaimed materials* *• Social design*	325 325
Award	*Scottish Furniture Makers Association Morgan Furniture Award 2008*	

Designer	*Jasper Morrison, Germany and UK*	308
Manufacturer	*Moooi, the Netherlands*	318
Materials	*Cork*	
Eco	*• Renewable material* *• Low-energy material* *• Reusable or recyclable*	325 325 329

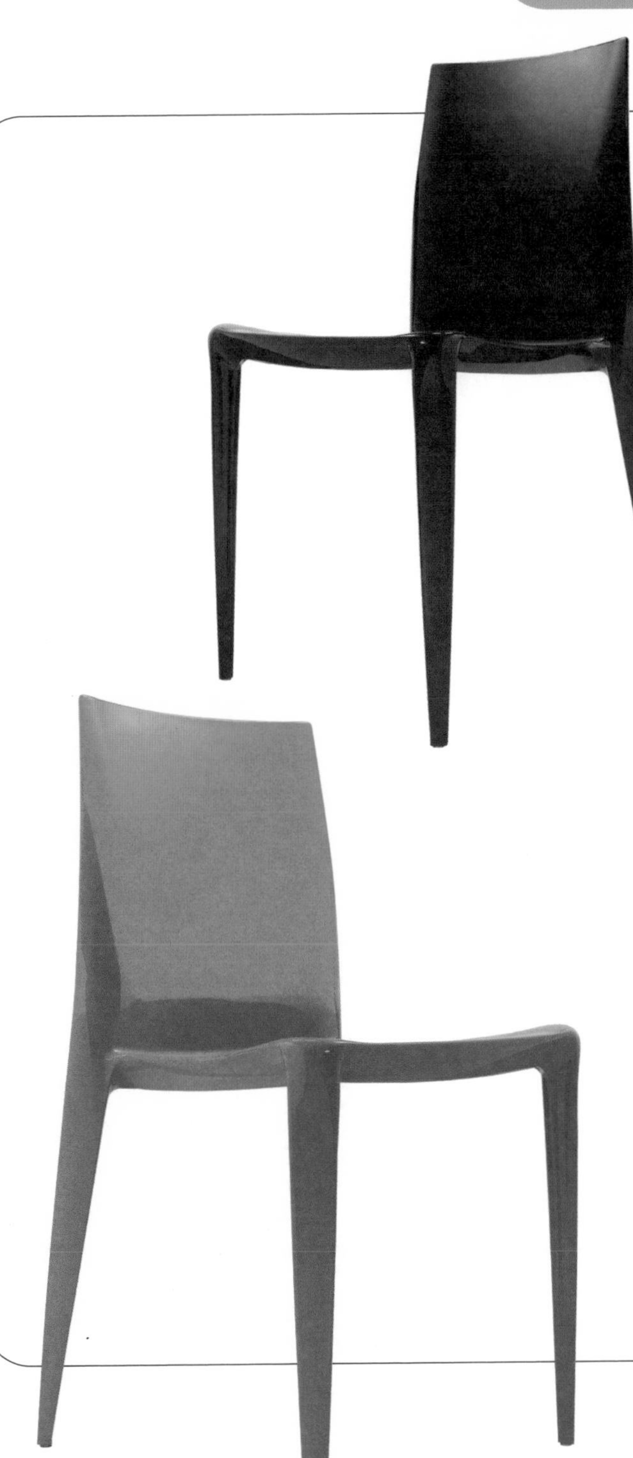

UltraBellini Chair®

Nylon, an industrial plastic with a reputation for toughness, fine engineering and strength, can be used in diverse applications from stockings to gear wheels, consumer products to furniture. Mario Bellini celebrates the material's character in pure lines, while tacitly acknowledging an Italian design lineage going back to Gio Ponti's 1957 Superleggera Model 699 chair, and the Italian industry's undoubted masterly use of synthetic polymer production techniques since the 1960s. There is a potential irony that this nylon chair, made from a by-product of oil derivatives, might outlast the very oil resources from which it came as we move into a post-peak oil economy in the later part of the 21st century.

✎	*Mario Bellini, Italy*	308
⚙	*Heller, Italy*	316
	Nylon polymer	
	• *Single material* • *Recyclable* • *Durable* • *Simplicity, maybe a future 'classic'*	325 325 325 324

✎	*Kenneth Cobonpue, the Philippines*	305
⚙	*Kenneth Cobonpue, the Philippines*	305
▬	*Rattan cane/slats, abaca rope, steel, nylon, stains & dyes*	
♺	*• Renewable materials from local sources* *• Recycled content* *• Low-energy production* *• Reduction in materials used* *• Retention of traditional craft skills*	325 325 325 325 326

Yoda, Pigalle, Croissant and Lolah

Natural materials (rattan, abaca rope) are combined with tradition, ingenuity and grace to create diverse furniture forms in Kenneth Cobonpue's Philippine studio. Unexpected juxtapositions of formal lines and wilder elements imbue the range with a liveliness and lightness of structure permitted by natural, minimal materials. Cobonpue constructs his pieces using a variety of traditional skills, such as those used in weaving baskets, making fish traps, and building boats, while the combination of recycled steel and tough nylon invests the furniture with durability and strength without compromising the overall aesthetic sensibilities. The furniture is animated and inviting: the Croissant sofa and Lolah capsule enfold their occupants, while the waveforms of the Pigalle armchair and the naturally sprung back of the Yoda range offer theatrical seating options.

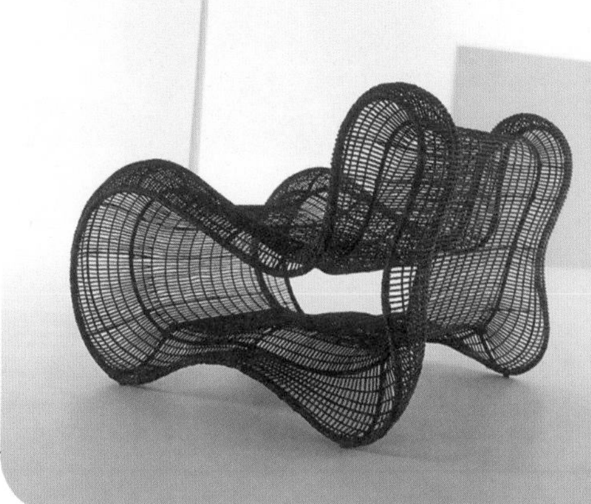

Lloyd Loom chair

Responding to market needs is always challenging for manufacturers using traditional materials. This easy chair reflects the renewed interest in organic and biomorphic forms while showing off the raw materials – Lloyd Loom fabric of wire and twisted paper – to the maximum. The durability of these materials is well documented – pre-World War II chairs are still going strong.

These new designs allow younger generations to acquire seating that will survive to the next generation and beyond.

Designer	*Jane Dillon and Tom Grieves*	305
Manufacturer	*Lloyd Loom of Spalding, UK*	318
Materials	*Lloyd Loom woven paper & wire, steel*	
Eco	• *Reduction in materials used* • *Renewable materials*	325 325

Raita Bench

This eclectic bench is made from DURAT®, a polyester-based plastic that contains 50% recycled material and is itself fully recyclable. Warm and smooth to the touch, the colourful stripes indicate the range of standard DURAT® colours. Two versions are available, 120 cm or 160 cm (47 in or 63 in) long, both standing 45 cm (17¾ in) high and 40 cm (15¾ in) deep.

These dimensions ensure it can double up as a coffee table too. Durable and easy to clean, this bench should give many years of service.

Designer	*Tonester, Finland*	322
Manufacturer	*Tonester, Finland*	322
Materials	*Recycled and virgin polyester-based plastics*	
Eco	• *Recycled content* • *Recyclable*	325 325

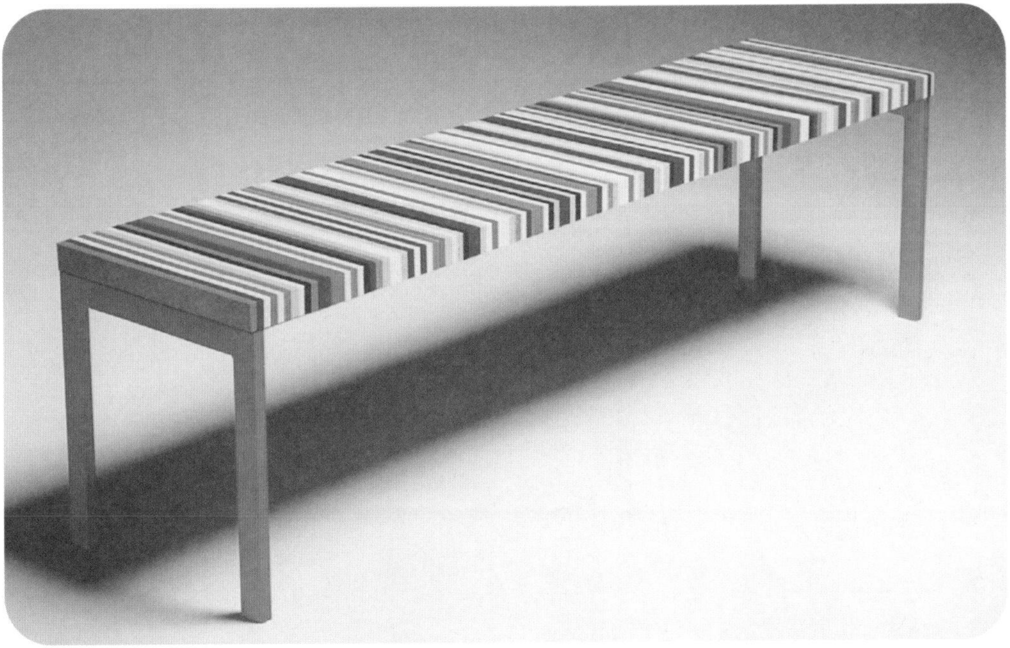

Imprint Chair

The moulded seat shell of the Imprint Chair is constructed of Cellupress™, a material with a distinctive surface formed by compressing plant fibres. There is a range of options to allow different fibre specifications – such as the use of spruce or bark. The shell sits on a solid ash wood frame or, alternatively, on stackable steel rod legs. This is not intended to be a 'fashionable' design but one that stands the test of time.

✎	*Johannes Foersom and Peter Hiort-Lorenzen, Denmark*	306
⚙	*Lammhults, Denmark*	317
▬	*Cellupress™, solid ash wood, steel tubing, rubber glide*	
♺	*• Cellupress™ reduces environmental impacts compared to conventional material options*	326
	• The company is certified to ISO 14001 standard	330
✪	*Best of NeoCon, Innovation Award, 2005*	

Pancras

Without sacrificing the comfort levels associated with a traditional armchair, the Pancras collection is a slimmed-down version to reduce material consumption. Oak and beech veneer for the laminate is rotary cut from Swedish trees to minimize waste and transport energy.

The laminate is compression-moulded to form the support and attachment points for the supporting polished, rather than chromed, stainless-steel frame. A removable Polytex or leather cushion completes the construction.

An ottoman is also available with a detachable cushion, enabling it to be converted into an occasional table. Swedish traditions, contemporary detail and care for the environment meet in these designs.

✎	*Tore Borgersen and Espen Voll, Sweden*	304
⚙	*Iform, Sweden*	316
▬	*Oak or beech veneer laminate, Polytex or leather*	
♺	*• Renewable materials from managed sources*	325
	• Reduction in materials used	325
	• Low-energy production	325

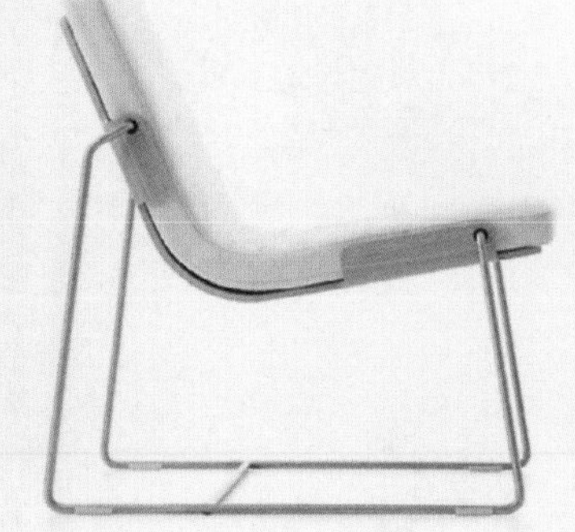

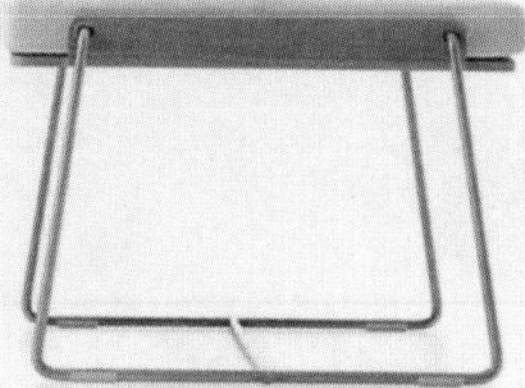

Your Stool 2

Your Stool 2 is adaptable and multifunctional, and it acts as a companion artefact, growing with you from childhood to the twilight years. This versatility is achieved through a sophisticated design of plywood elements that can be easily re-assembled again and again at home as needs shift from rocking horses to high-backed rocking chairs. Precision cutting of the plywood through the latest computer numeric control (CNC) machines enables bespoke production for each user.

✎	*Ryuichi Tabu, UK*	310
⚙	*One-offs*	
▬	*Plywood*	
♻	• *Personalization for better emotional bonds* • *Bespoke production for each use*	328 326

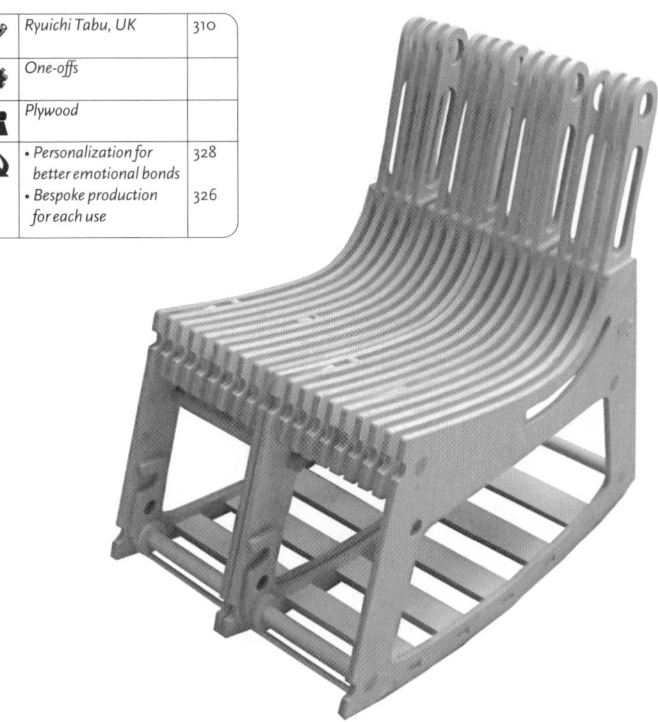

Louis chair

Expect to see more and more imaginative reuse of reclaimed furniture in the future. This designer-maker studio offers ready-made revitalized chairs as well as accepting commissions to revive an old tired chair. This particular example has been reupholstered with a Christian Dior scarf with a new silk back. Restoration work is undertaken by qualified seamstresses in the UK and local suppliers are used wherever possible.

A percentage of the profits go to help Tree Aid, a charity that supports reforestation projects in Africa.

✎	*The Baobab Tree, UK*	304
⚙	*One-offs*	
▬	*Reclaimed wooden chairs, recycled silk scarves, Duppion silk, eco-paint*	
♻	• *Recycled and reclaimed materials* • *Local suppliers and tradespeople* • *Percentage of profits to a reforestation charity*	325 324 327

C10 Springback chair

In this intriguing design by David Colwell, a staunch advocate of sustainable furniture manufacturing practices since the 1980s, native fast-grown ash is steam bent and tensioned. Using forestry thinnings of small roundwood trees provides income for the foresters, and steam bending is ideal as it actually strengthens the wood. The result is a comfortable, individualistic dining/everyday chair that will endure.

✎	*David Colwell Design, UK*	305
⚙	*David Colwell Design, UK*	305
▬	*Ash wood, stainless steel, organic oil finish*	
♻	*• Renewable materials from local sources* *• Recyclable material* *• Durable* *• Organic finishes*	325 325 325 326

Grownup stool

This is a stool for DIY enthusiasts and gardeners with patience. You can't buy Christopher Cattle's stool, but you can buy the jigs or the cutting plan for the jigs to grow your own. Plant three saplings of your choice in your garden (sycamore and maple are good fast growers), train and graft the trees using the jig, wait about five years, cut your stool base, air dry for a few weeks, then add a seat of your choice. Voila: your very own 'harvested' stool and an unforgettable lesson from nature.

✎	*Christopher Cattle, Grown Furniture, UK*	305
⚙	*One-offs grown by user*	
▬	*Wood*	
♻	*• DIY* *• Deepens eco-awareness* *• Renewable materials from local sources*	326 324 325

Toy Chair

Tough, durable and colourful, polypropylene has been a favoured material with designers for nearly half a century. Toy Chair is a wonderful celebration of technological progress in single-piece injection moulding, with its hollow arms and 'angle-iron' legs. Starck brings his usual wit and economy of line to this chair, which is equally happy in a garden, an urban loft or a café. As it is fabricated entirely from PP with a small, easily removable stainless-steel plug around the drainage hole in the seat, it is easy to recycle the materials at the end of the item's life. By 2030 manufacturers may even be requesting that their products be returned by the current custodian for dismantling and recycling of components and materials. The material content of the Toy Chair will then be valued as much as the comfort and pleasure given throughout its lifetime.

Philippe Starck, France	309
Driade SpA, Italy	314
Polypropylene	
• *Recyclable single material* • *Reduction in materials used*	325 325

Thinking Man's Chair

Tubular and flat steel are combined in a deliberately 'engineered' look, further enhanced by the red oxide-type finish complete with written dimensions.
A durable design for indoor or outdoor use, which, being made from a single material, is easily recycled.

Jasper Morrison, UK	308
Cappellini SpA, Italy	313
Steel	
• *Recyclable single material*	325

Ceramic chair project, 2001–2002

It is rare that traditional materials, ubiquitous today and throughout history, cause as much excitement or attract the design profession in quite the same way as seductive techno-materials such as high-strength composites, shape-memory metals and synthetic plastics. Yet Pakhalé suffuses new life and brings out exciting qualities from an abundant, but highly variable geosphere material, clay. 'Playing with clay' was a collaboration with EKWC (the European Ceramics Centre) in the Netherlands, drawing on the skills of craftsmen and -women and technical knowledge about the properties of different clays, especially in relation to consistent crack-free firing. Handmade prototypes paved the way to developing pressure-cast moulds for industrial production. Where uprights and legs required joining to other parts, a special glued joint was designed using computer modelling. Technical achievements aside, each chair (Ceramic Pottery Chair, 2001; Flower Offering Ceramic Chair, 2001; Roll-Ceramic Chair, 2000–2001) possesses its own spirit and presence, reflecting new possibilities where craft is in symbiosis with industry, perhaps encouraging others to re-examine the potential in this symbiotic relationship. Conversations between highly skilled computer technicians, artisans and designers seem to offer fresh hope in revitalizing socio-cultural value and meaning in our industrial production. The Ceramic chair project is proof of a subtle re-balancing and suggests myriad opportunities.

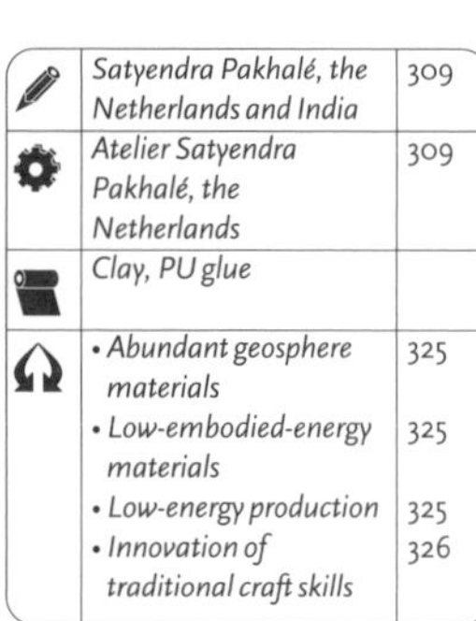

✎	*Satyendra Pakhalé, the Netherlands and India*	309
⚙	*Atelier Satyendra Pakhalé, the Netherlands*	309
	Clay, PU glue	
	• *Abundant geosphere materials*	325
	• *Low-embodied-energy materials*	325
	• *Low-energy production*	325
	• *Innovation of traditional craft skills*	326

Hudson 3, Navy 1006 2 and Emeco stools

Founded after World War II, Emeco harnessed the skills of immigrant German craftsmen to sculpt aluminium into the 1006 chair. This quickly became a classic with the US government and military and exists today as the Navy 1006 2. In 2000 the French designer Philippe Starck was involved in a commission for the new Hudson Hotel in New York and began collaborating with Emeco to evolve the 1006 into a pure, reductive form. Reducing weight but not sacrificing strength, and using 60–70% post-consumer aluminium, the latest incarnation, the Hudson 3, celebrates the raw material, and the combination of production and craft skills while exuding durability. Emeco stools reflect Starck's economical eye but his latest addition, the Kong bar stool, reverts to more traditional post-modern whims and fancies. Is this luxury design or a 'yours eternally' approach? Or is it simply providing a temporary resting place for high-embodied-energy materials patiently waiting until the next reincarnation?

✎	*Philippe Starck, France, and Emeco, USA*	309
⚙	*Emeco, USA*	315
▆	*Recycled and virgin aluminium*	
♺	*• Single recyclable material* *• Recycled content* *• Reduction in materials used*	325 325 325

Pm

This solid natural beech, cherry or maple tabletop is extendable from 180 cm to 230 cm (71 to 90½ in) by the insertion of a lightweight core tamburato mat extension that sits underneath the original top. A detachable square section steel frame finished in 'silver' or 'titanium' epoxy powder completes the minimalist purity of this design.

✎	*Pallucco Italia, Italy*	319
⚙	*Pallucco Italia, Italy*	319
▤	*Solid wood, steel, tamburato mat*	
♺	• *Multifunctional* • *Durable*	328 325

✎	*Setsu Ito, Italy*	307
⚙	*Front Corporation, Japan*	315
▤	*Steel*	
♺	• *Single material* • *Durable*	325 325

Saita

Almost 90% of the steel in circulation has been recycled at some time, so it is refreshing to see steel being used with great sculptural panache in this table design. Long the preserve of architectural and structural engineers, steel offers fresh perspectives for furniture designers.

Take away

Laptop computers expanded the concept of work to the commuter train, aeroplane or café. This idea has not been wholly successful, with its blurring of boundaries between different activities. Nor does a laptop constitute a desk – where do you put the coffee? Enter the portable one-person office: a lightweight plywood shell and desktop sit on foldable aluminium legs, with neat stowaway compartments in the hinge-up cover. Take away is not the exclusive domain of the laptop, rather it feels receptive to recalcitrant pencil-and-paper creatives, or to covert watercolour painters. It is ideal for setting up and folding away in small urban apartments or homes and yet small enough to take with you on a working holiday.

Designer	*Beat Karrer, Switzerland*	307
Manufacturer	*Prototype*	
Materials	*Plywood, aluminium*	
Eco	• *Multifunctional* • *Lightweight*	328 325

Lost & Found Tables

Originally commissioned for a café interior at the East London Design Show, these distinctive tables are a combination of sustainably harvested wood and re-combined elements from discarded furniture. Each piece bears a brass plaque indicating where the different elements were 'lost' and 'found', hinting at their history in a previous incarnation. The Lost & Found collection continues to grow with new additions, such as a range of stools, emerging from the hybridization of defunct forms.

Designer	*&made, UK*	304
Manufacturer	*One-offs*	
Materials	*Wood, reclaimed furniture, brass*	
Eco	• *Local sustainably sourced solid timber* • *Reclaimed furniture* • *'Second-life' objects*	325 325 325

Handy

Portability and indoor/ outdoor use were the priorities for these strong, lightweight polypropylene tables. The double slots accommodate the underside supports for a tray or cushions, depending on the user's needs. Flexible, unobtrusive and durable, these tables provide functional service yet permit customization.

✎	*Luisa Bocchietto, Italy*	304
⚙	*Serralunga, Italy*	321
▮	*Polyethylene*	
♻	• *Multifunctional* • *Single material* • *Recyclable*	328 325 325

Ash round table

Combining excellent rigidity and ample leg-room, the simplicity of this design relies on the strength of the solid ash, which comes from local English woodlands. A range of table sizes to seat three to ten people is manufactured to the same basic design. David Colwell successfully blends traditional furniture-making techniques with a modern aesthetic to produce durable, quality seating, tables and shelving.

✎	*David Colwell and Roy Tam, UK*	305
⚙	*David Colwell Design, UK*	305
▮	*Solid ash wood*	
♻	• *Renewable materials with stewardship sourcing* • *Low-energy construction techniques*	325/ 326 326

Tischbockisch

Minimal machining is required to create the solid, untreated ash wood horizontal ribs and legs, fitted with non-slip rubber stops, for each trestle frame. Pairs of ribs are available in different lengths so users can define and make their own table tops, although the manufacturer will supply laminated MDF tops to order if required. This pared-down design enables customers to source appropriate local materials, helping to reduce the transport energy per table compared with centralized manufacturing and distribution. This is a rich seam of design opportunities that helps users to become part of the design process to realize the full function of the manufactured components.

	Jakob Timpe, Germany	310
	Nils Holger Moormann GmbH, Germany	308
	Untreated ash wood, rubber	
	• Low-energy manufacturing	326
	• Reduced transportation energy	327
	• User involvement	328

Raw

Here is a robust, honest table that reminds us of the inherent beauty of its materials, oak and steel. The first component is from one of nature's most revered trees and the second is still the backbone of industrial development. Their embrace is complimentary, their durability undoubted. The graphite coloured steel frame holds 42 mm-thick (1⅔ in) staves of tough solid oak wood, left with a natural finish, that can be aligned or staggered according to the wishes of the user.

	Garth Roberts for Zanotta, Italy	309
	Zanotta, Italy	323
	Varnished steel frame, solid oak	
	• Sustainably sourced oak wood	326
	• Simple, solid, durable construction	325

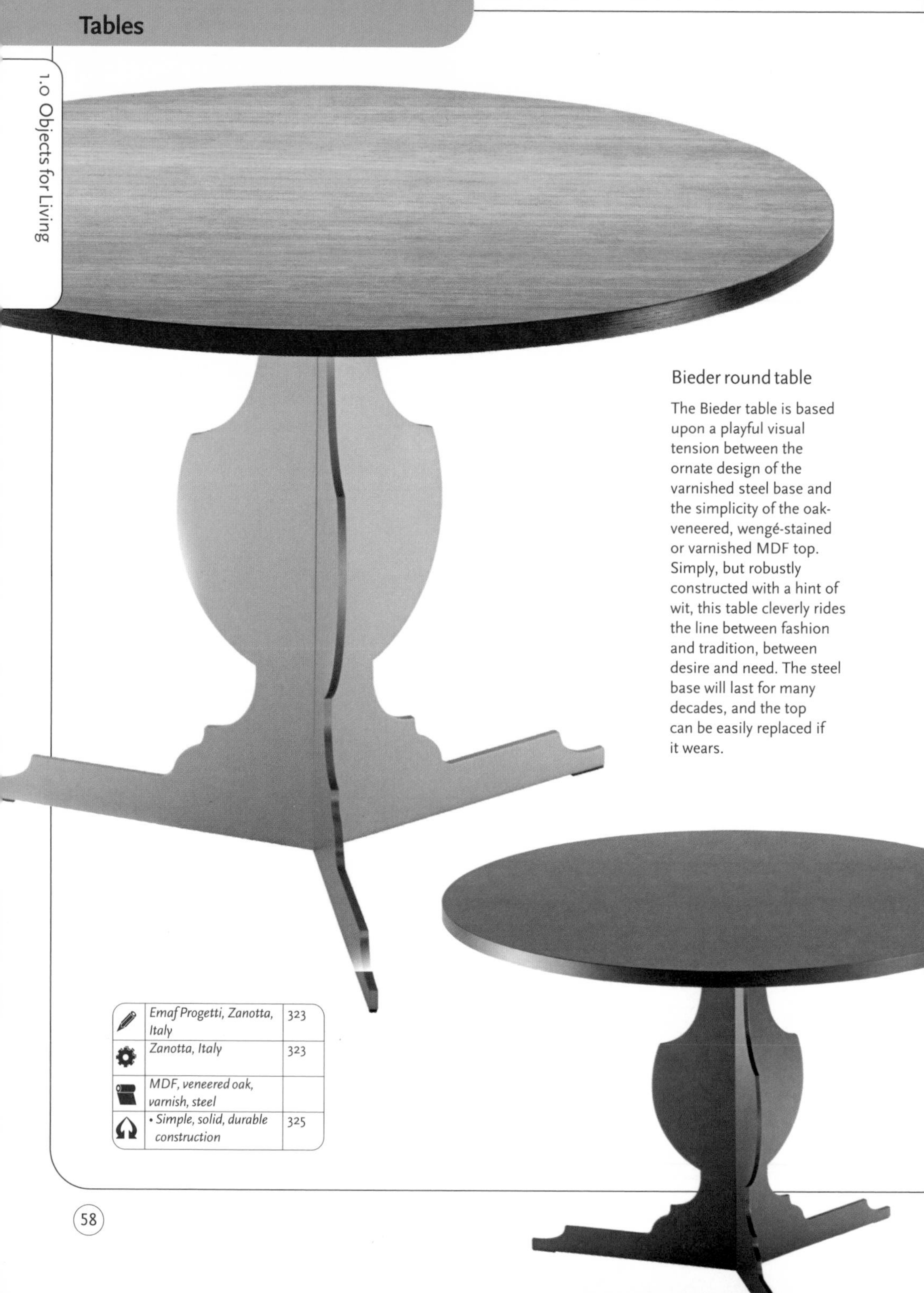

Bieder round table

The Bieder table is based upon a playful visual tension between the ornate design of the varnished steel base and the simplicity of the oak-veneered, wengé-stained or varnished MDF top. Simply, but robustly constructed with a hint of wit, this table cleverly rides the line between fashion and tradition, between desire and need. The steel base will last for many decades, and the top can be easily replaced if it wears.

✎	*Emaf Progetti, Zanotta, Italy*	323
⚙	*Zanotta, Italy*	323
	MDF, veneered oak, varnish, steel	
	• Simple, solid, durable construction	325

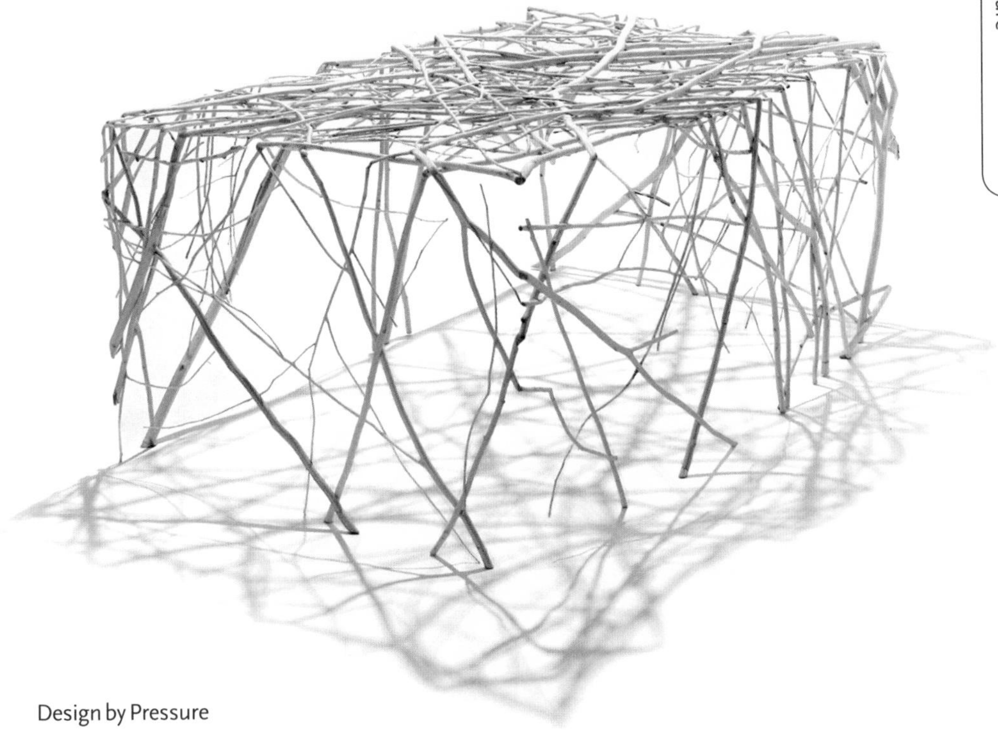

Design by Pressure

This table is part of a larger project by Front called 'Design By' where part of the design process is guided by external factors and random events. Design by Pressure is the outcome of subjecting harvested branches and twigs to high pressure. Design by Sunlight, Gravity, Temperature, Scale and so on, lead to creative interpretations and forms in other media. This typifies the imaginative output from Front, a design collective whose search for new meaning in objects, and the processes by which they originate, challenges our preconceptions.

	Front, Sweden	315
	One-offs	
	Wood	
	• *Simplicity and truth to materials* • *Light-weighting* • *Single material*	325 325 325

Re-imagined sideboards

Lucy Turner gives well-constructed sideboards (usually made of tough teak) from the 1960s and 1970s a new lease of life by carefully restoring and reinvigorating them with beautiful laser-cut colour laminate designs. This is more than just a makeover as each piece takes on a new character, cheekily announcing its renewed presence.

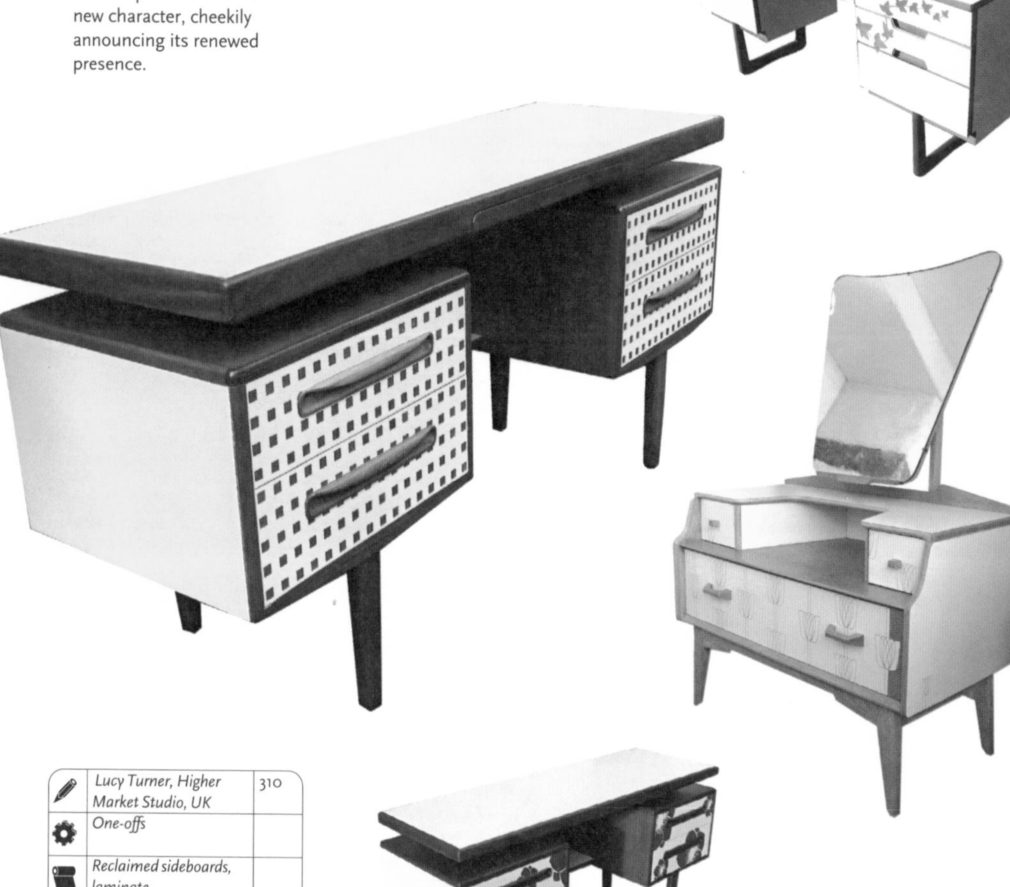

✎	*Lucy Turner, Higher Market Studio, UK*	310
⚙	*One-offs*	
	Reclaimed sideboards, laminate	
	• Reclaimed furniture *• Existing components*	325 326

Hanging Shelves

Young ash thinnings from forest management operations are ideally suited to this solid wood shelving that demonstrates flexibility and grace. Durability is guaranteed by the hardness of the ash and reliable stainless-steel fixings, and the shelves are easily suspended from two wall screws. The overall effect offers a subtle tension often missing from the (predictable) offerings of large furniture superstores.

✎	*Roy Tam Design, UK*	309
⚙	*Roy Tam Design, UK*	309
▮	*Ash wood, stainless steel*	
♻	• *Locally sourced wood from managed forests* • *Simplicity of construction* • *Organic finishes*	325 337 326

Brosse

Our curiosity is aroused by Sempé's intriguing storage unit based on a metal frame whose contents are hidden by a curtain of industrial brushes. It is the unexpected application of these ready-made brushes that challenges our reactions. While the use of ready-mades does not confer instant low environmental impact it is an interesting route for designers to explore. Existing manufacturing capacity can be more efficiently deployed if its output has suitable applications in other sectors, although this concept needs careful application.

✎	*Inga Sempé, France*	309
⚙	*edra SpA, Italy*	314
▮	*Lacquered aluminium, propylene industrial brushes, metal*	
♻	• *Use of existing manufacturing capacity*	325

Plus Unit

Flexibility and easy assembly and disassembly are provided by the simple, extruded polished aluminium 'X' connectors that lock each ABS injection-moulded drawer or carcass unit. Heavyweight castors attached to the base create a mobile storage unit. Users or owners can upgrade and expand their system as they can afford, or as new needs arise. The solid, well-made and elegant design encourages consumers to take a longer view. This is design as an investment, design for a lifetime.

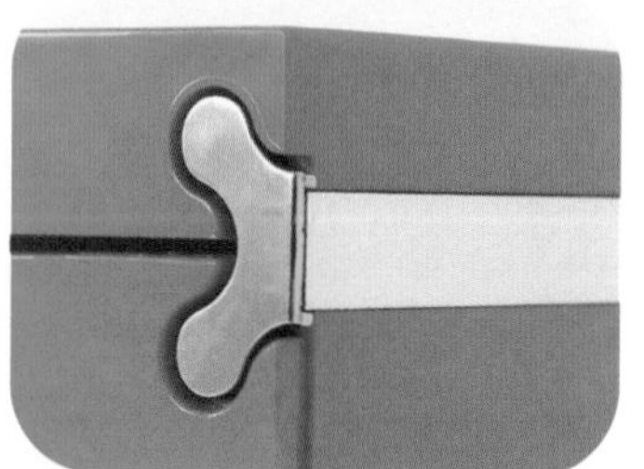

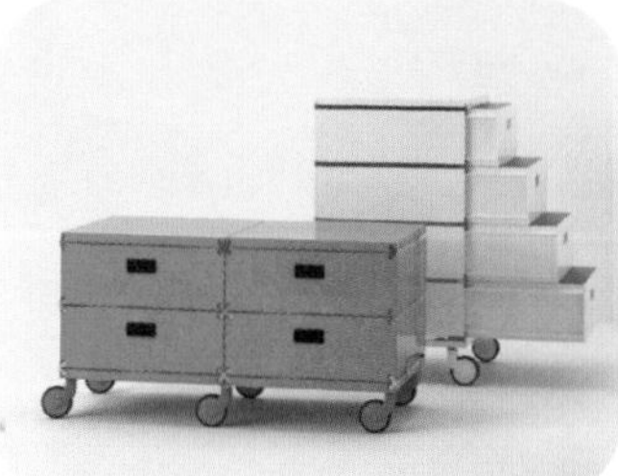

✎	*Werner Aisslinger, Germany*	304
⚙	*Magis SpA, Italy*	318
▮	*ABS, Aluminium*	
♻	*• Recyclable, single material components* *• Modular* *• Customizable* *• Durable*	325 328 328 325

Aluregal

This lightweight high-strength modular stacking shelving system is 3D-formed from 2 mm anodized aluminium sheet, each shelving unit being tensioned with galvanized steel cross-struts. Each module is 1.6 m (5 ft 3 in) long, 33 cm (13 in) deep and 36 cm (14 in) tall, with a maximum stacking height of five modules. Components promise long life but are easily separated for removal, maintenance or recycling at the end of their life.

✎	*Atelier Alinea, Switzerland*	312
⚙	*Atelier Alinea, Switzerland*	312
▮	*Aluminium, steel*	
♻	*• Reduction in materials used*	325

Wandregal

This is an exercise in maximum functionality with minimum wastage. Rectangular cut-outs from one standard birch multiplex or white laminate-faced MDF sheet serve as the shelves while lateral sections reinforce the frame. Steel struts take the load on the shelves. Simply lean the shelving unit against a wall and load with weight to achieve full stability. Transportation is facilitated by the fact that the shelving units can be virtually flat-packed.

Atelier Alinea, Switzerland	312
Atelier Alinea, Switzerland	312
Birch multiplex or laminate-faced MDF, steel	
• Reduction in materials used	325

Celia

All too often the glitzy world of contemporary design feeds off a limited palette of seductive technological materials. It takes the vision of the Campana Brothers to realize exciting tactile and aesthetic pleasures from humbler man-made materials. The Celia range uses the ubiquitous strandboard found on building sites the world over for temporary, disposable screens or shuttering. In the strandboard, nature is roughly thrown together and bonded with glue, heat and pressure. The Celia sideboard reminds us of the precious and unique nature of these resources and lends a quiet dignity to the material. Modest and non-elitist, this furniture hints at a quality that all members of society could aspire to and afford.

Fernando and Humberto Campana, Brazil	304
Campana Studio, Brazil Habitart, Brazil	304
Strandboard	
• Single material *• Renewable materials* *• Recyclable materials*	325 325 325

Pocket

Most paperback books' dimensions follow international standards. This observation served as the stimulus to design minimalist shelving scaled to the depth set by these standards. Avid collectors of specific publishing brands, such as Penguin, can delight in arranging broad swathes of orange and turquoise blue spines in abstract patterns. In fact, Pocket allows users to create their very own wall art feature using lovingly thumbed copies of their favourite paperbacks. This is a minimalist altar to the joy of reading.

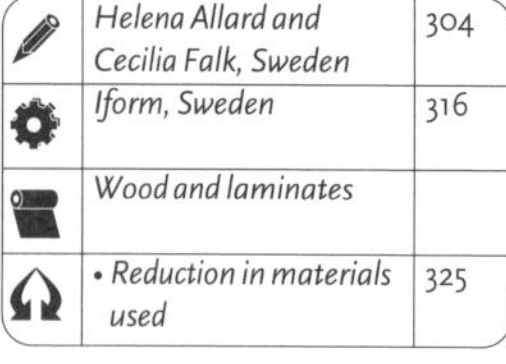

✎	*Helena Allard and Cecilia Falk, Sweden*	304
⚙	*Iform, Sweden*	316
	Wood and laminates	
	• Reduction in materials used	325

Es

Nine beech wood rods are inserted into four plywood panels and locked into place using plastic rings. Grcic tests the boundaries of stability with a design that wobbles yet doesn't fall over. His design appears to fly in the face of man's desire to remove nature from the process of manufacturing, being deliberately made to look naïve and in a DIY style. The rods permit the shelving to double as a coat rack and clothes stand.

✎	*Konstantin Grcic, Germany*	306
⚙	*Nils Holger Moormann GmbH, Germany*	308
	Beech wood, plywood, plastic	
	• Renewable materials, economically applied *• Design for assembly/ disassembly*	325 326

zehn hoch

Experimenting with Finnish plywood led to the creation of the 'zehn hoch' drawer modules. Each module is finished in melamine laminate, which ensures toughness and creates an easy-glide surface for each drawer. Modules can be arranged as steps, a chest of drawers or a sideboard, according to one's needs.

✎	*Häberli/Huwiler/ Marchand, Switzerland*	306
⚙	*Röthlisberger, Switzerland*	320
▭	*Plywood, melamine*	
♻	• *Multifunctional*	328

Shell

Moulded aircraft-grade plywood, just 3 mm (about ⅛ in) thick, is fixed with 3D metal corner fixings to create a basic shell that can be fitted out internally as required with shelving and a clothes rail. A longitudinal hinge permits much better access than conventional wardrobes.

✎	*Ubald Klug, France*	307
⚙	*Röthlisberger, Switzerland*	320
▭	*Plywood, metal*	
♻	• *Reduction in materials used* • *Low-energy manufacturing*	325 326

Continua

Continua is an appropriate name for this evolving modular shelving and storage system; the design originated in 1994 but has undergone continuous evolution. The latest addition is the 'Continua glass' and 'Continua glass console' 12 mm (½ in) hardened glass shelving with underside screen prints. Core to the system concept is a series of powder-coated aluminium vertical wall rails and horizontal shelf supports. These can accommodate a diverse range of shelving, clothes hangers and storage units. Shelves are available in steel, wood and glass, according to personal taste and needs. A variety of steel drawers on wheels interlock with the fixed wall system. Continua is suitable for domestic and contract markets. Eco-efficiency is built into the concept as the original fixings can accommodate upgrades, newly designed accessories or extensions. Underpinning Continua is the pure minimalist approach of a manufacturer that understands its customers well, knowing that they want a delicate balance between delivering predictable quality and offering new aesthetic and practical choices.

✎	*Pallucco Italia, Italy*	319
⚙	*Pallucco Italia, Italy*	319
	Aluminium, steel, wood, glass	
	• *Modular* • *Upgradable* • *Multifunctional*	328 328 328

Nomadic lifestyle

As geographic boundaries fade in a greater European Union, and citizens gain rights to working and living in EU states, there is an expectation that people movement will increase in the next decade. Recognizing that there is a new generation of cultural and economic migrants, the Nomadic lifestyle project suggests an economical solution to moving and storage. Boxes used for removal purposes can easily be converted into shelving units by the addition of internal shelves and strengthened components. Regular movers will know that half-a-dozen moves is all you can expect from conventional boxes. A little attention to detail can extend the life of this specialist corrugated cardboard box.

✎	*Willem Jansen, graduate student 2003, Design Academy Eindhoven, the Netherlands*	307
⚙	*Prototype*	
▬	*Corrugated cardboard*	
♺	*• Lightweight strong materials* *• Recyclable materials* *• Multifunctional*	325 325 328

China cabinet

Turning the concept of a cabinet to hold china inside-out, China cabinet is made from porcelain. Each piece is constructed in the same fashion but deforms during the firing to produce a series of one-off modular shelf units. The unexpected forms and unusual application of the material combine to subvert expectations of precision manufacturing. Applying traditional materials to new forms is an exciting arena for experimentation in sustainable design and is ideal for exploring the creation of local sustainable enterprises.

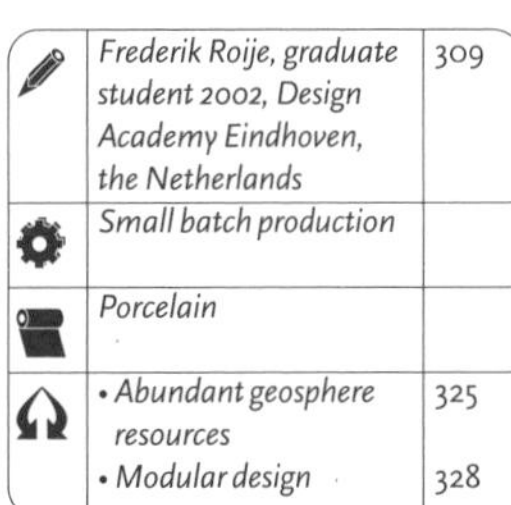

✎	*Frederik Roije, graduate student 2002, Design Academy Eindhoven, the Netherlands*	309
⚙	*Small batch production*	
▬	*Porcelain*	
♺	*• Abundant geosphere resources* *• Modular design*	325 328

Dia

Adaptability and durability are the two primary prerequisites for furniture that is intended to survive the elements and robust use in the garden. This range offers a high degree of flexibility – the chair has an upright and a low position, the table height is adjustable and the sunbed has eight possible permutations. Polished stainless steel and strong fabric, impregnated with waterproofing and UV-stabilized, ensure a long life. Thanks to these high-quality materials, this range of furniture is also suitable for indoor use and thus offers flexibility and dual-functionality.

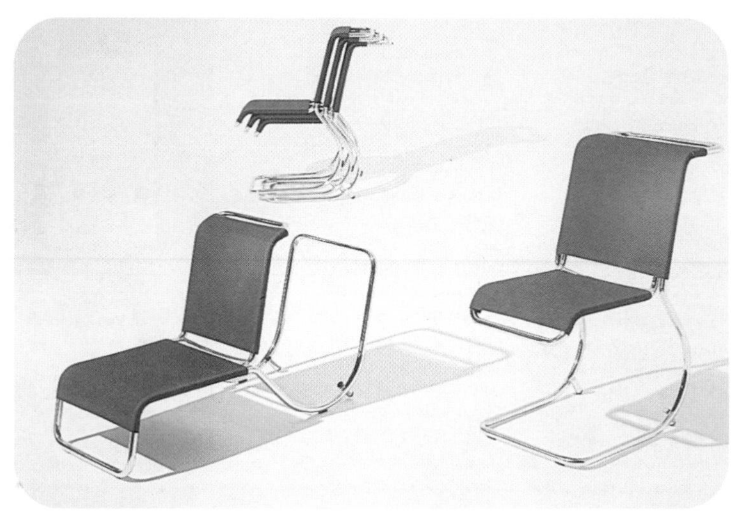

✎	*Gioia Meller Marcovicz, UK*	308
⚙	*ClassiCon, Germany*	313
	Stainless steel, waterproofed/UV-stabilized fabric	
	• Durable *• Multifunctional*	328 328

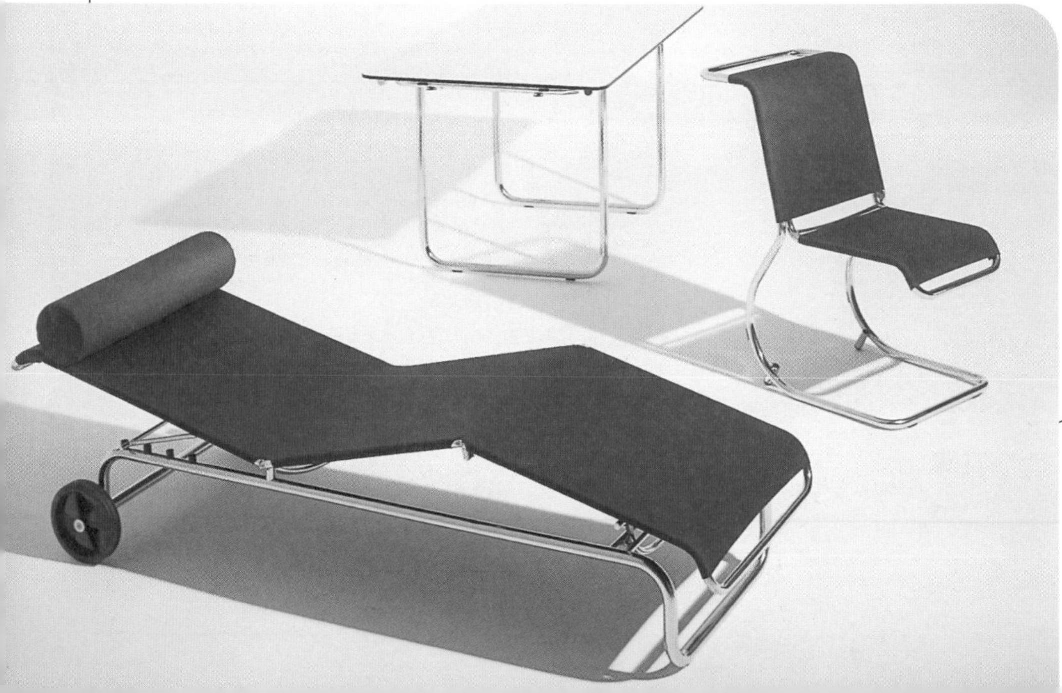

eo

Colour is integral to our emotional well-being, yet most furniture designs are available only in hues that designers and manufacturers feel are in tune with overall fashion trends. This modular storage system incorporates low voltage LEDs and some clever electronics to allow users to mix blue, red and green light using a remote control. Light is projected onto the back panel of each module and reflected light is 'captured' by the matt glass panels. Individuals can create the exact hue and tone to match their emotional needs – set the modules to bright orange to pump up the (visual) volume or switch to meditation mode with calm greens if you want to chill out. Aside from eo's function as an emotional barometer, its basic modules can be fitted with internal glass or steel shelving to suit a variety of artefacts from CD players to tableware or a mini wine cellar. With such intelligent application of electronics and lighting to benefit our sense of well-being, it is only fitting that the system won the prestigious Design Award of the Federal Republic of Germany in 2002.

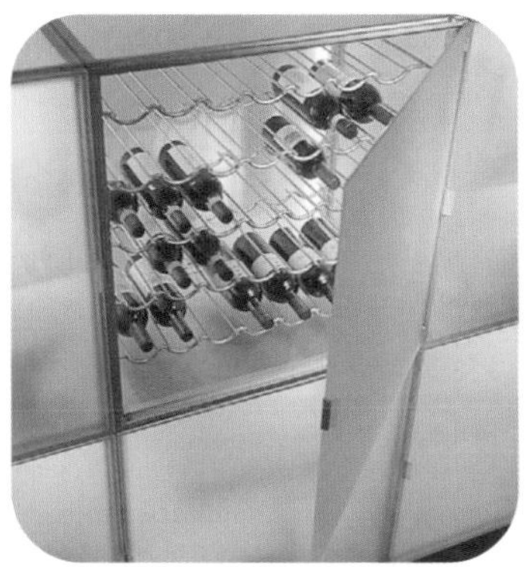

	Professor Wulf Schneider and Partners with Stephan Veit, Germany	309
	Interlübke, Germany	317
	Aluminium frame and rails, glass, LEDs and electronics	
	• *Low voltage* • *Modular* • *Multifunctional* • *Improved well-being*	329 328 328 328

Kokon

Old wooden furniture is revived by covering it with a PVC-based coating. The opportunities to create quirky new custom furniture are legion but the technique needs further refinement to find a substitute for PVC, whose environmental track record is poor. How to isolate the timber of the reclaimed furniture from intimate contact with the PVC and what to do with the items at the end of their lives are unanswered questions.

✎	*Jurgen Bey, Droog Design, the Netherlands*	304
⚙	*Limited batch production, Droog Design, the Netherlands*	314
▬	*Reclaimed furniture, PVC coating*	
♺	*• Use of ready-made components*	327

Téo from 2 to 3. Snoozing Stool

Téo challenges the hyper-productivity of the modern era. Originating in 1998 but still produced today, Téo suggests that time out for doing nothing in particular – reading a book, taking a nap – might make us more productive when we become active again. 'Do not disturb' signals the unequivocal intention of the user and causes the viewer to think and react.

Crasset suggests that taking a nap on the Téo satisfies our need to step aside from the non-stop world. While setting a challenge for our daily work practices, Téo is also a useful multifunctional object in the home – it can also serve as an overnight guest bed or side table.

✎	*Matali Crasset, France*	305
⚙	*Domeau & Pérès, France*	314
▬	*Wood, double density and high resilience foam, coated cloth*	
♺	*• Multifunctional • Encourages slowing down*	328

Kauna-phok mats & cushions

In the swamps of Manipur, India, there is a valuable local resource, the tough, hard-wearing kauna-phok rush, which is harvested to make this range of cushions, pads and mats. Once worked, the material retains its inherent sponginess, making it ideal to provide a soft but firm surface. Each design is handmade and is traded according to Fair Trade principles.

✎	*Indigenous designers, India*	
⚙	*Natural Collection, UK*	318
▮	*Kauna-phok plant materials*	
♻	*• Renewable and sustainably harvested materials* *• Fair Trade products*	325/ 326 325

Table #2

This low side/coffee table finds new life for older birchwood trees which tend to suffer in later life from fungal attack, and are normally burnt releasing CO_2. Table #2 locks up the carbon in a useful, durable object that doubles as a place to sit. The distinctive bark of the birch logs provides a striking aesthetic in harmony with the table's simple construction method.

✎	*Patrick Fredrikson and Ian Stallard, Fredrikson Stallard, UK*	306
⚙	*Fredrikson Stallard/ David Gill Galleries, UK*	306
▮	*Birch wood, steel*	
♻	*• Locally sourced timber* *• Design for assembly & disassembly*	325 326

Hülsta furniture

Germany is undoubtedly one of the 'greenest' consumer markets in the European Union and Hülsta is a significant manufacturer of domestic and office furniture with a proven commitment to environmental performance. It was one of the first companies to reach the quality assurance standard ISO 9001, and its entire production is certified under the Blue Angel eco-label scheme. In collaboration with Danzer, a leading veneer company, Hülsta initiated the 'veneer passport' guaranteeing that it does not originate from a tropical rainforest. Only four of their current ranges of furniture use solid wood, again not sourced from rainforests. Particleboard or MDF is the primary material. In-house designers apply lifecycle analysis to extend the projected lifespan of products, of which most are already expected to last between thirty and forty years.

✎	*Hülsta, Germany*	316
⚙	*Hülsta, Germany*	316
▮	*Veneers, solid wood, particleboard*	
♺	• *Renewable materials* • *Blue Angel eco-label* • *Corporate environmental vision and policy*	325 330 330

Spiga

Mimicking an ear of corn, this lightweight coat rack, made of seven thin, wave-shaped, plywood cut-outs attached to a metal-rod frame, is an ideal resting place for coats, hats, umbrellas, bags, newspapers and more throughout its entire length.

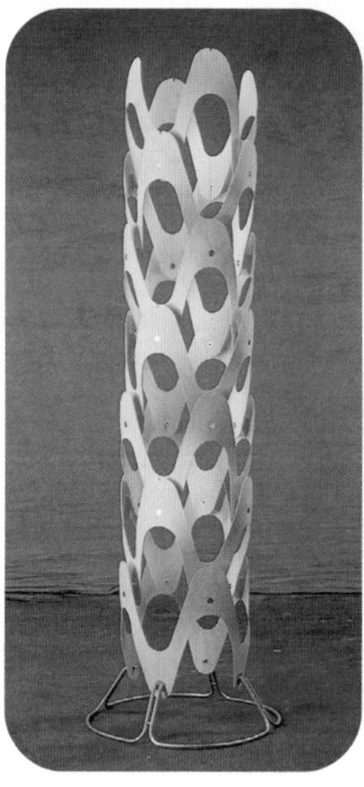

✎	*Ubald Klug, France*	307
⚙	*Röthlisberger, Switzerland*	320
▮	*Plywood, metal*	
♺	• *Reduction in materials used* • *Multifunctional*	325 328

black 90, black 99

Black bamboo is suspended in an American black walnut frame to neatly counterbalance the delicacy of the bamboo with the solidity of the frame. In another screen, interlocking bamboo forms foldable concertina-like sections. The modernist mantra of 'truth to materials' is eloquently demonstrated here. These natural materials have an extensive history and unique colour signatures.

Designer	*Paola Navone, Italy*	308
Manufacturer	*Gervasoni SpA, Italy*	316
Materials	*Bamboo, black walnut wood*	
Sustainability	*• Renewable materials*	325

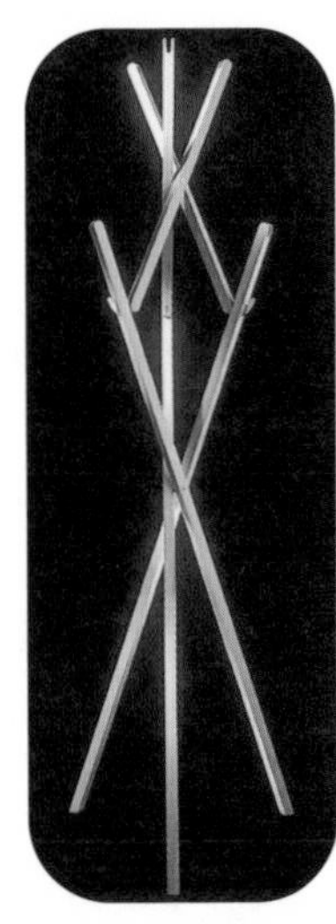

Hut Ab

Aluminium fixings allow simple machined pieces of ash wood to articulate around a pivot to provide a multifunctional clothes and hat stand, drying rack or structure for suspending house plants. Low-energy requirements during production make this an efficient, low cost, design.

Designer	*Konstantin Grcic, Germany*	306
Manufacturer	*Nils Holger Moormann GmbH, Germany*	308
Materials	*Ash wood, aluminium*	
Sustainability	*• Recyclable and renewable materials* *• Low-energy manufacturing* *• Multifunctional*	325 326 328

Lattenbet

Anyone who has moved house knows that the most cumbersome item is the double bed. Not so for this superb example, an entire double bed that can be neatly carried in its own suitcase. Bucking the trend for self-assembly furniture to be flat, stylistically drab and infuriatingly difficult to assemble, Steinmann and Schmid have devised a construction system that is not only rapid to assemble but also visually appealing.

Designer	*Peter Steinmann and Herbert Schmid, Switzerland*	310
Manufacturer	*Atelier Alinea, Switzerland*	312
Materials	*Beech wood, plywood, steel, rubber*	
Sustainability	*• Portable, self-assembly furniture* *• Reduction in materials used*	328/ 327 325

do Create

Droog Design are renowned for their wry humour ('droog' translates as 'dry'), which is apparent in the project 'do Create' in 2000. Kesselskramer, a Dutch publicity company, set up a fictional experimental brand called 'do'. Ten designers were invited to participate in the Droog Design project, the outcomes of which were presented at the Milan Salone. The 'do' brand is incomplete without continuation of the design process by the consumer, involving a personalization or customization of the original artefact. Marijn van der Poll's 'do hit' armchair is a cube of sheet steel before being transformed into its final form. More modest physical activity is required for Martí Guixé's 'do scratch', a light box that comes to life as the owner scratches a message or drawing into the black paint. Guixé's 'do reincarnate' is a loose recipe for reviving lifeless, familiar objects by magically suspending them by an 'invisible' thread and by an extension of the wiring from a ceiling light rose. This new lighting appliance parades in surreal splendour. The designer is facilitator to the consumer's imagination, resulting in whimsical designs that affirm the personality and imagination of the consumer and are a world away from the predictability of mass-produced design. This concept of designer and consumer as co-designers provides fertile ground for cementing a long-lasting relationship with the brand, an emotional mortar to create a treasured object.

✎	*Marijn van der Poll and Martí Guixé, Droog, the Netherlands*	310
⚙	*One-offs*	
	Various materials	
	• *Personalization and customization* • *Emotional interaction*	328 328

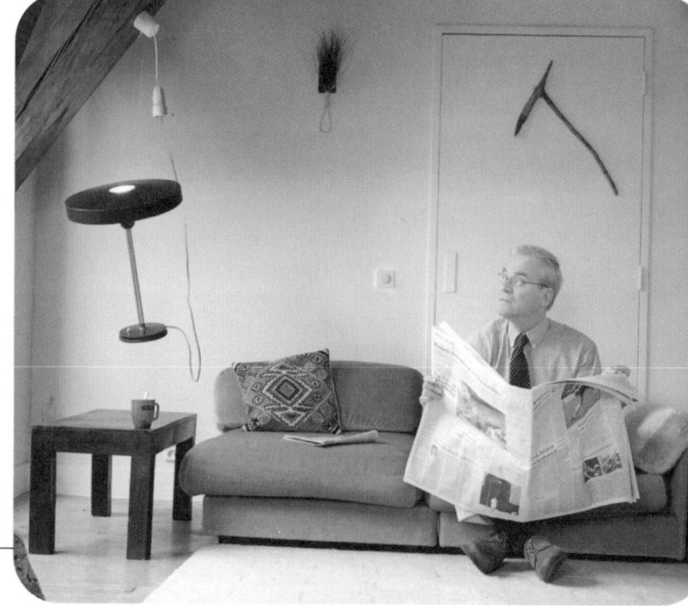

Carta

Shigeru Ban creates structures with cardboard and succeeds in elevating this humble material to a new aesthetic level. His use of cardboard tubes in projects as diverse as furniture, temporary housing for refugees and buildings for communities reveals superb understanding of the capabilities of the material. Carta is a range of furniture that makes minimal use of low-embodied-energy materials.

Designer	*Shigeru Ban, Japan*	309
Manufacturer	*Cappellini SpA, Italy*	313
Materials	*Cardboard*	
Eco-features	*• Renewable materials* *• Reduction in materials used* *• Recyclable*	325 325 325

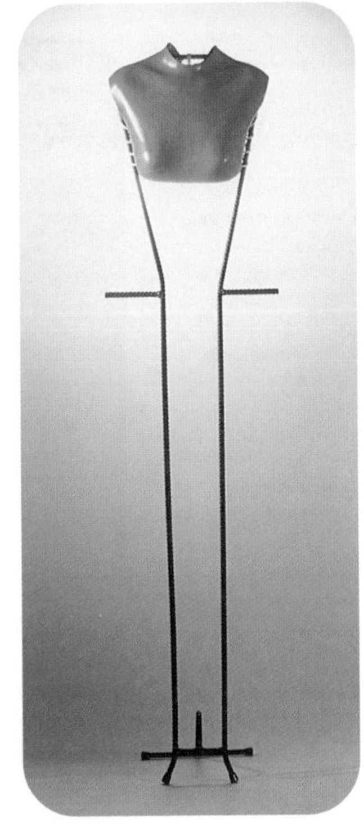

Herz

Steel reinforcing rods, similar to those used in construction with concrete, are welded and bolted into a simple frame to which a moulded leather breastplate is attached, providing a functional, minimalist coat stand. The materials used are easily recycled and the design is both aesthetically pleasing and durable.

Designer	*Robert A. Wettstein, Germany*	311
Manufacturer	*Anthologie Quartett, Germany*	312
Materials	*Leather, steel*	
Eco-features	*• Recyclable and compostable materials* *• Reduction in materials used* *• Low-energy manufacturing*	325 325 326

3Star

Initial experiments with paper and scissors led to the development of this dynamic sculptural form, die-cut from one triangular sheet of polycarbonate. The flat-pack design is simply bent into shape requiring no adhesives or clips. It represents an ethical stance where care is given to tread lightly, through minimizing materials, energy and costs. Simultaneously there is a desire to enhance the perceived and actual quality of light, in this case a compact fluorescent lamp, by offering a range of colours depending upon the functionality required – ambience, task or other lighting.

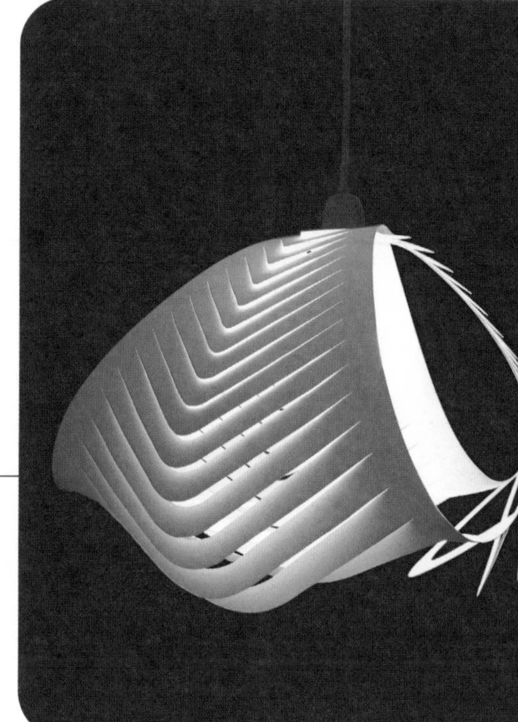

✎	*David Henrichs, DH Product Design, UK & Germany*	307
⚙	*One-offs and small batch production*	
▮	*Polycarbonate sheet*	
♻	*• Single material, recyclable at end of life* *• Light weighting*	325 325

Horeta

Polycarbonate is a durable, versatile polymer that can be easily recycled and is available in translucent or opaque forms. It is ideal for lightweight flat-pack, self-assembly light shades, as demonstrated here by Setsu Ito. This multilayered pendant shade lets through varying amounts of light depending on the angle of view.

✎	*Setsu Ito, Italy*	307
⚙	*One-offs*	
▮	*Polycarbonate*	
♻	*• Single material* *• Recyclable* *• Self-assembly* *• Reduction in energy of transport*	325 325 327 327

Ash Pendant No. 1

The exuberant liveliness of this lightshade hints at the fast-growing ash trees that provided the raw material. The shades are crafted out of thin strips of green ash timber sanded to equal thickness, which are then steam-bent into coils and assembled onto a spherical frame. The simple, low-energy production system means that it is possible to make a range of diameters from 55 cm (21⅔ in) up to 5 m (16.4 ft), allowing bespoke designs. This offers a viable, renewable, local model for future production.

✎	*Tom Raffield, UK*	309
⚙	*One-offs and small batch production*	
▬	*Ash wood, 20W energy saving cathode lightbulb*	
♻	• *FSC-certified wood* • *Bespoke designs ensure long life through uniqueness* • *Low-energy manufacturing*	326 324 325

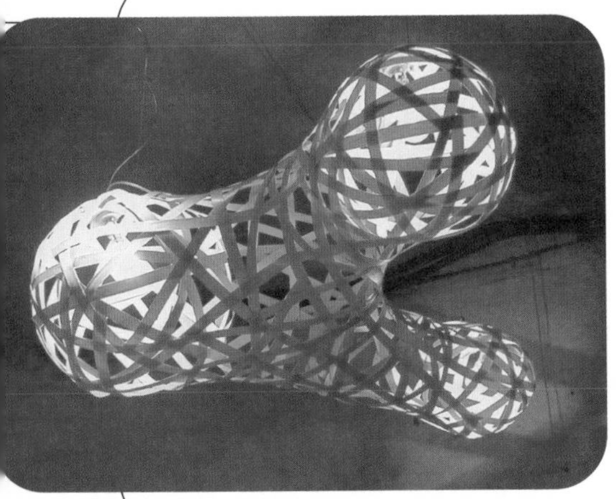

FluidSphere

This pendant light shade made from criss-crossed maple veneer strips explores an organic geometry. Beams of light escape from a cell-like structure that floats in space. Wood, that most malleable of materials, is perfectly adapted to creating new aesthetic pleasures with light.

✎	*Leo Scarff, Ireland*	309
⚙	*One-offs and small batch production*	
▬	*Maple veneer, steel suspension fittings*	
♻	• *Reduction in materials used* • *Renewable materials*	325 325

Agave

Initial public reaction to compact fluorescent lightbulbs was as muted as the luminosity of the bulbs themselves. Happily, compact fluorescent technology has substantially improved, leading to their adoption by most major light manufacturing companies to enhance their product range. Luceplan has experimented with transparency, reflection, refraction and diffusion in its latest range of light shades suitable for pendant, wall and floor lighting. Transparent methacrylate is injection-moulded into a radial structure comprising a series of arc-shaped ribs that simultaneously diffuse and 'conduct' the light. This produces a complex interplay of light quality and colour that can be further enhanced and personalized by fitting graded coloured filters (yellow, red and blue) to the shade. Spherical, elliptical and double-bodied shades add further variations to the range. Methacrylate is renowned for its distinctive optical clarity and impact resistance, which is why it is used for illuminated signs and motor vehicle lights. Unfortunately it requires acetone and cyanohydrin in its manufacture, so the best way to eco-redesign this light would be to examine how to reduce environmental impacts while retaining the brilliance of the luminosity from the compact fluorescent bulbs. Reduction in weight and alternative materials are feasible.

	Diego Rossi and Raffaele Tedesco, Italy	309
	Luceplan, Italy	318
	Methacrylate, compact fluorescent bulbs	
	• Reduced energy consumption	325
	• Customization	324

Bloom pendant

When the lightbulb is switched on in this aero-plywood lampshade, two complementary patterns are revealed: the laser-cut holes, through which light streams, and the subtler backlit surface pattern on the plywood. In addition to this lovely juxtaposition of patterns, the Bloom features technical achievements in folding of plywood and finding new explorations for this renewable resource. The shade can be flat-packed and delivered in a standard size envelope.

✎	*Rentaro Nishimura, UK*	308
⚙	*One-offs and small batch production*	
▬	*Lightweight plywood*	
♻	• *Lightweight, renewable materials*	327

Lamp shade

This reversible shade provides a choice of two strong lighting directions depending upon whether the reflector is uppermost (for down-lighting) or on the underside (for up-lighting). This eloquent design embodies principles of minimalism and dual-functionality, both of which are very relevant to designs with reduced environmental loads.

✎	*Sebastian Bergne, UK*	304
⚙	*Radius GmbH, Germany*	320
▬	*Steel*	
♻	• *Dual-function design* • *Reduction in materials used*	324 325

Flute Pendant

In Giles Miller's imaginative lamp design, corrugated cardboard is transformed from the banal into a beautiful surface through which light can play. Strips of recycled or FSC-certified source cardboard are cut by a CNC machine, and the floral pattern cut-outs are made by fret saw. A cut-out from one shade is inserted in the gap in another shade, with the corrugations aligned in opposite directions to create the beguiling surface texture. These simple techniques reveal the designer's ability to make basic materials sing.

	Giles Miller, UK	308
	One-offs and small batch production	
	Recycled cardboard	
	• *Recycled and/or certified materials* • *Recyclable*	325 325

Flexlamp

Playful interaction is an innate human behaviour that can be encouraged by intelligent design. Sam Hecht utilizes the flexible properties of silicone to permit users to bend the Flexlamp shade to meet their requirements. In encouraging the user to play with the aesthetic form of the product, and actively engage in designing his/her surroundings, the designer helps bond the user with the product in more meaningful ways which hopefully means it will be cherished for longer.

	Sam Hecht for Droog, the Netherlands	306
	Droog, the Netherlands	314
	Silicone, 11W energy efficient lightbulb	
	• *Invites interaction and encourages personalization* • *Single material for easy recycling*	324 325

MultiSheer collection

Lifecycle thinking is about adding new value in each cycle of a material's life, ensuring that it can be continually reincarnated for a new existence. Kate Goldsworthy seeks to do that by making lightweight textile screens and lighting products from recycled or reclaimed polyester, then adding laser-cut or etched floral elements derived from reused PET plastic packaging. This combination of discarded materials is effectively re-animated in a fresh design.

✎	*Kate Goldsworthy, UK*	306
⚙	*One-offs*	
▮	*Recycled and/or reclaimed polyester and PET*	
♻	*• Reused, recycled and recyclable materials* *• Design for disassembly*	325 326

Liese Lotte

This one piece polycarbonate (PC) shade is a beautifully simple product that can be snapped over a bare lightbulb: simply ease the shade apart along the slit to place it over the light fitting of the bulb, then click back together. To clean the Liese Lotte shade, just undo it, wipe clean and replace.

✎	*Constantin Wortmann, Büro für Form, Germany*	311
⚙	*Next Home Collection, Germany*	318
▮	*Polycarbonate*	
♻	*• Single material easily recycled at end of life* *• User-friendly* *• Easy to clean*	325 328 329

See-Through Light

Two products with short consumer lives – discarded polyethylene terephthalate (PET) plastic water bottles and laddered stretch nylon stockings or tights – are brought together in a new symbiotic relationship in this lamp. A little cutting and stitching transforms these abandoned artefacts of consumerism into new desirables. In doing so it encourages self-exploration through contemplating the detritus resulting from our contemporary way of life. It signals that all materials have value if we care to look, and legitimizes the importance of thinking 'zero waste' as a viable future strategy.

✎	*Tea Un Kim, UK/South Korea*	310
⚙	*One-offs and small batch production*	
▬	*Post-consumer PET water bottles and old stockings/tights*	
♻	*• Reused materials* *• Cold, low-energy construction* *• Potential DIY design*	325 326 326

Shahin

Contemporary compact fluorescent bulbs have migrated towards the ubiquitous Edison lightbulb form, facilitating uptake of the lower energy demanding bulbs. This classic form was the inspiration for two clip-together polypropylene moulds that embed the bulb and the lamp cable, the latter forming a hook enabling this light to be hung anywhere. There is minimal waste as the mouldings are cut by a CNC machine, then vacuum formed. Being a mono-material design it is easily recycled at the end of its life.

✎	*Alan Crummey & Artein Hossein, Artal Designs, UK*	305
⚙	*Small batch production*	
▬	*Polypropylene*	
♻	*• Single material* *• Zero packaging* *• Affordability*	325 329 327

Punga Light

The design on these pendant light shades is inspired by growth patterns found on the native Punga tree in New Zealand. Through the use of a 'pseudo' dove-tail joint, thin laser-cut strips of hoop pine plywood are assembled, without fasteners or adhesives, to create a mesmerizing repeat pattern that blends the natural and the modern.

Designer	*Christopher Metcalfe, New Zealand*	308
Manufacturer	*Small batch production*	
Materials	*Lightweight plywood, steel ring*	
Environmental	*• Lightweight, flat pack* *• Design for assembly/ disassembly*	327 327

Bendant Lamp

This lightweight chandelier, made of powder-coated laser-cut steel, arrives as a flat-pack design that the user can customize by bending the cut-out sections to suit personal preferences, or the specifics of the lighting environment. The steel used to make the lamp contains recycled content and is locally sourced.

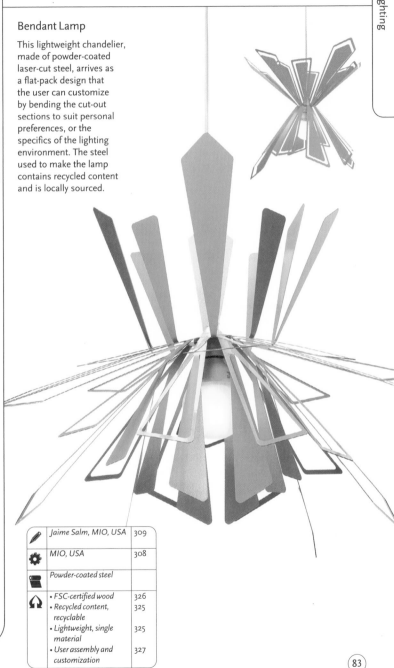

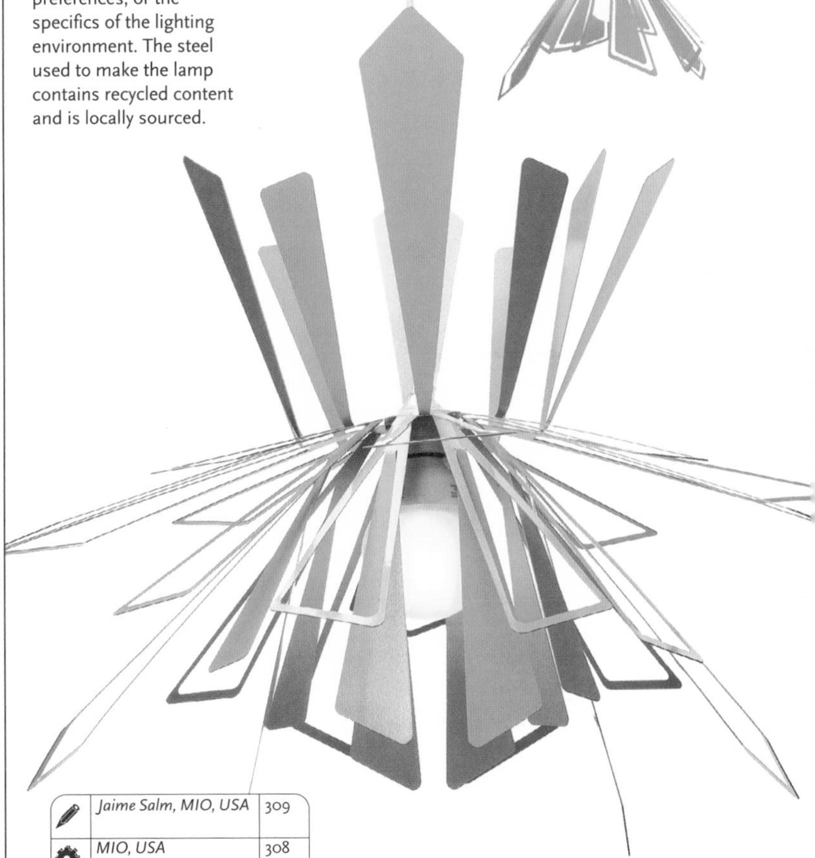

Designer	*Jaime Salm, MIO, USA*	309
Manufacturer	*MIO, USA*	308
Materials	*Powder-coated steel*	
Environmental	*• FSC-certified wood* *• Recycled content, recyclable* *• Lightweight, single material* *• User assembly and customization*	326 325 325 327

Flamp

This wooden-based table lamp is dipped in phosphorescent coating so that it absorbs the energy from sunlight and re-radiates it for up to 20 minutes. An ideal 'emergency' light after sunset.

Designer	*Martí Guixé, Spain*	306
Manufacturer	*Small batch production*	
Materials	*Phosphorescent paint, wood*	
Sustainability	*• Solar powered non-electric light*	329

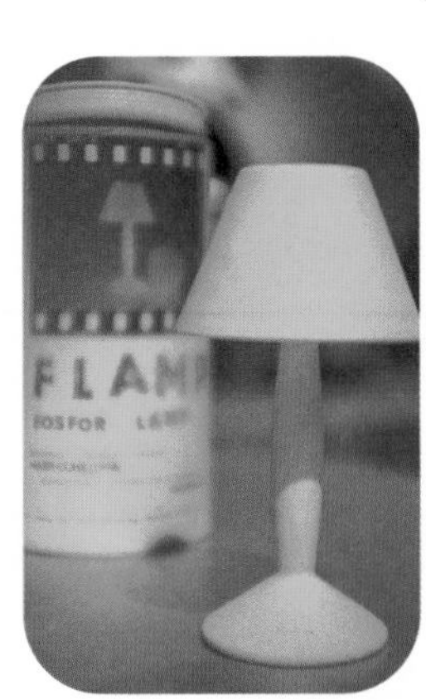

Clips

A simple stainless-steel frame clips over a discarded drinks can and supports a polypropylene shade. As your favourite brand of drink changes you can dispose of the old one (at a can bank) and insert a can that held the flavour of the month.

Designer	*Bernard Vuarnesson, Sculptures-Jeux, France*	310
Manufacturer	*Bernard Vuarnesson, Sculptures-Jeux, France*	310
Materials	*Polypropylene, stainless steel*	
Sustainability	*• Encourages reuse of ready-mades* *• Reduction in materials used*	325 325

Fake Lamp

Guaranteed to add a surprise dimension to any room, the Fake Lamp is a large 171 cm- (67½ in-) wide sheet of white MDF that challenges our perceptions of domestic lighting.

Designer	*Sophie Krier, the Netherlands*	307
Manufacturer	*Moooi, the Netherlands*	318
Materials	*MDF*	
Sustainability	*• Zero energy use* *• Improved well-being*	329 328

Buttoned Up

This free-standing floor lamp, with an ash frame and aero plywood shade, is a mellow mixture of wood grain whose form is reminiscent of the spun lunar tripod lamps of the 1950s. Buttoned Up re-contemporizes its predecessor in an engaging way: when it is switched on, light issues through the perforated pattern in the ply bringing it to life by mimicking the grow rings and grain of the tree. All materials are sourced locally in Cornwall, UK, including the buttons from a haberdashery store, which are used ingeniously to lock the cylinder shade of plywood together.

Designer	*Simon White Design, UK*	309
Manufacturer	*Small batch production*	
Materials	*Ash wood, aero plywood, buttons, Danish oil*	
Eco	*• Renewable, local materials*	325

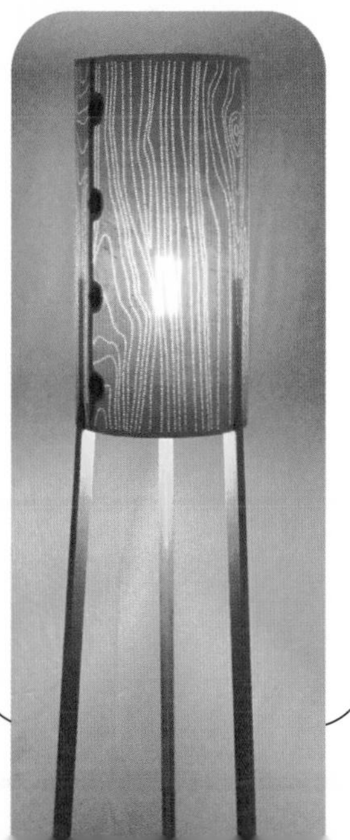

Airswitch Az Baby

Mathmos continue to provide magical moments from their lighting range in the form of the Airswitch Az Baby, activated, dimmed, brightened or switched off by simply moving the hand over the light. A design derived from the innovative Airswitch 1, the latest version is made of hand-blown glass on a metal base. This design can accommodate a low-energy CFL bulb. It is impossible not to love the technology that makes a humble table lamp part of everyday/every night theatre in our lives. No more fumbling in the dark for the switch!

Designer	*Mathmos Design with Shin and Tomoko Azumi, UK*	318
Manufacturer	*Mathmos, UK*	318
Materials	*Glass, metal base, electronics, compact CFL*	
Eco	*• Non-mechanical, electronic, switches for long life* *• Low-energy bulb fitment* *• Emotionally durable product*	328 329 328

Bamboo lamp

A beautifully lacquered stem of bamboo dovetails with a shiny metal base topped with a bamboo/cotton textile shade and an idiosyncratic toggle for the pull-cord switch. The contrast in materials is well balanced and the quirky pull-cord adds an element of surprise – and each lamp has a unique pull-cord. While this is a modest application of the material, we can expect to see much more of bamboo in the 21st century: it is a material with an illustrious history and a bright future.

Designer	*Committee for Moooi, the Netherlands*	305
Manufacturer	*Moooi, the Netherlands*	318
Materials	*Bamboo, cotton, metal*	
Eco	*• Renewable materials* *• Simplicity of construction*	325 327

bolla S, M, L, XL

Mimicking seed pods hanging from a tropical tree, these biomorphic rattan floor lamps fulfil many functions and needs. The designs embed what Christopher Day, the British architect and writer, called 'spiritual functionalism'. He was referring to the ability of buildings to lift the human spirit by their form, spatial dynamics and sensory palette but the term can also be applied to products. Sodeau's lyrical use of woven rattan makes this much more than just a light. It evokes many senses, from an appreciation of the aesthetic form to the tactility of the material and, in doing so, it generates emotional and mental reactions that aim to improve our well-being.

Designer	*Michael Sodeau, UK*	309
Manufacturer	*Gervasoni SpA, Italy*	316
Materials	*Rattan, MDF*	
Sustainability	• *Renewable materials*	325
	• *Retention of local craft skills*	325
	• *Improved sense of well-being*	328

	Artemide SpA, Italy	312
	Artemide SpA, Italy	312
	Synthetic polymers, low-voltage lamp	
	• Reduction in consumables used *• Energy conservation* *• Easily repaired and disassembled*	325 329 329
	Design Sense awards, shortlist, 1999	

e-light

The e-light integrates a number of technological improvements over conventional desk lamps. The lifetime of the lighting filament is 20 times greater than that of an incandescent bulb and two to three times that of a fluorescent bulb, and it uses one-fifteenth as much mercury as the latter. Creating a light spectrum similar to daylight, it is five times brighter than a tungsten bulb. As the e-light produces negligible thermal emissions, the need for heat-resistant materials is significantly reduced. Components can easily be separated, facilitating recycling and reuse. Reversible joints and compact design provide flexible lighting configurations and a small footprint.

Lagoon Light

Industrial rubber silicone destined for the landfill is rescued, cut, reassembled and reborn as a sensuous bespoke floor lighting system, with an internal LED system of low-energy bulbs. Re-utilizing materials will be a pre-requisite in the coming decades, and if done with sensitivity as in the Lagoon Light, the recycled materials come to life again.

✎	*Lucy Fergus, Re-silicone, UK*	306
⚙	*One-offs and small batch production*	
▬	*Reused rubber silicone, LEDs*	
♺	*• Reused materials* *• Low-energy bulbs*	325 329

Pixelate

An MDF boxed frame encloses three fluorescent 10w low-energy tube lights hidden by a card insert, which is laser cut with a pattern of 'pixels' or squares, protected by a transparent polymethacrylate sheet. Five pixelated card inserts are provided. The user simply takes one and pushes out the perforated pixels to create a pattern, message or intervention to their liking. The more pixels removed, the more light, enabling the user to control the final result – ambience mixed with personal symbolism and creativity.

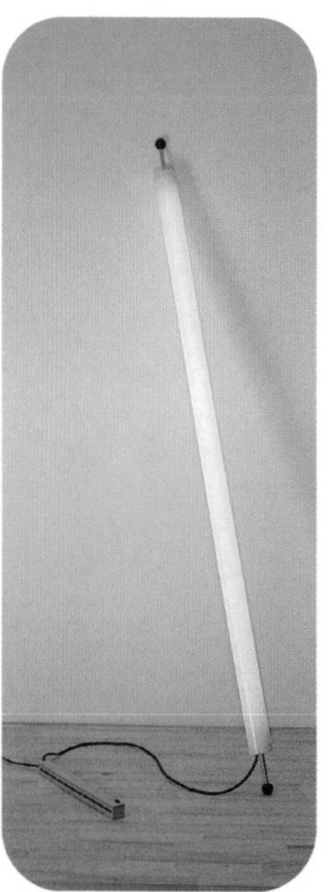

Tube

The familiar fluorescent light gets the minimalist treatment from Christian Deuber. A slender synthetic tube protects the light with steel- and rubber-footed closures, allowing the light to be placed wherever it is required. The use of fluorescent bulbs, in this case 58W, which are much more efficient users of energy than incandescent sources, adds to the versatility of this product.

✎	*Christian Deuber, N2, Switzerland*	305
⚙	*Pallucco Italia, Italy*	319
▬	*Fluorescent light, steel, rubber*	
♺	*• Multifunctional* *• Reduction in materials used* *• Low energy consumption*	328 325 329

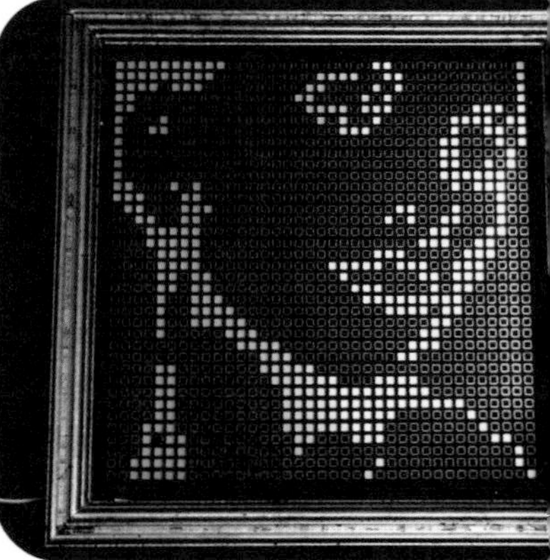

✎	*Alison Edwards, UK*	305
⚙	*One-offs*	
▬	*MDF, card, polymethacrylate sheet, fluorescent tube (low energy) strip lights*	
♺	*• Low-energy lighting* *• Personalization to create enduring relationships with products*	329 328

PO98 10/10C, 11/11C, 12/12C

In a clever extrapolation of scale, the table lamp becomes a floor or standard lamp. These lightweight constructions combine visual stimulation and humour in an economical design.

✎	*Marcel Wanders, the Netherlands*	310
⚙	*Cappellini SpA, Italy*	313
▬	*Wire, polymer*	
♻	*• Reduction in materials used*	325

Crush

Jamsheed Todiwala has taken his inspiration for this lamp from the zero-waste culture of the slum dwellers in Mumbai, India, where everything has a use. Crush combines finely crushed glass with a bio-resin extracted from sunflowers, and is cast by mixing the two constituents in a mould, itself made from recycled plastic.

The cured shade is finished by polishing with a glass sanding machine. This imparts a glow reflected in every particle of the composite and hints at other potential uses for this 'second-time around' material. End-of-life recycling is accomplished by heating to 340° C (644 °F) to separate the two component materials.

✎	*Jamsheed Todiwala, UK*	310
⚙	*One-offs and small batch production*	
▬	*Glass, bio-resin*	
♻	*• Recycled materials* *• Bio-/renewable materials* *• Cold manufacturing process*	325 325 326

Sinus

Polypropylene is a versatile, lightweight material favoured by lighting designers. This particular engineered solution deploys simple, yet sophisticated, interlocking snap joints to hold the shade and bulb fixture. The unusual spine-like joint provides a dramatic aesthetic whether the lamp is lit or unlit, as well as enabling the product to be sold as a flat-pack.

✎	*Andreas Nydahl, UK*	308
⚙	*One-offs and small batch production*	
▮	*Recycled and recyclable cardboard*	
♻	*• Lightweight single material* *• Design for easy assembly/disassembly*	325 326

Gloworm

✎	*Draigo, UK*	305
⚙	*One-offs and small batch production*	
▮	*Reused plastic bottle tops, LEDs*	
♻	*• Reused materials* *• Low-energy bulbs*	325 329

The Gloworm ambient light deals with the challenge of finding a use for plastic bottle tops that are often not wanted by local recyclers. A waterproof line of 36 LEDs is inserted into the 100 post-consumer bottle tops to create a sinuous eclectic coloured worm that instantly adds dynamic energy to any space. Consuming just 3 watts this uplifting light can literally brighten your life.

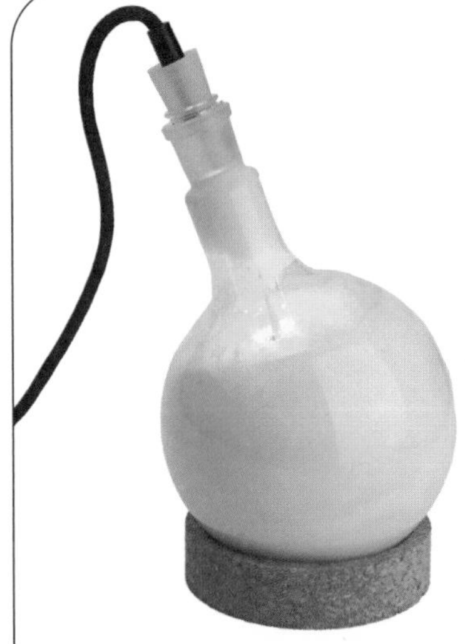

Slow Glow Lamp

In addition to being an extremely unusual light to cuddle and a reminder that energy transforms materials, this intriguing lamp provides a conversation starter! When the lightbulb buried within a glass container of solid vegetable fat heats up, the fat gradually liquefies, and the resulting light grows brighter and warmer. Beautifully conceived and constructed, this light can become a cherished lifelong object that gives meaning and character to our lives.

✎	*NEXT Architects & Aura Luz Melis for Droog, the Netherlands*	308
⚙	*Droog, the Netherlands*	314
	Glass, vegetable fat, cork, 25W G9 bulb	
	• *Abundant renewable and lithosphere materials*	325
	• *Encouraging deeper emotional bonding*	328

Sun Jar

A traditional Mason jar, used for preserving and storing food, has been fitted with a solar cell, rechargeable battery and warm-coloured LEDs. Pop it on a window, or even out in your garden (it is waterproof) in the morning and it will harvest then store the energy of the sunlight. A light sensitive sensor automatically switches it on when darkness falls to provide up to five hours of illumination. A soft, diffused light emanates from the frosted glass – coloured blue or yellow to give a cool or warm radiance. Simple, effective and durable.

✎	*Suck UK, UK*	321
⚙	*Suck UK, UK*	321
	Glass, solar cell, rechargeable battery, LED bulbs	
	• *Uses existing manufactured product*	327
	• *Renewable energy*	329
	• *Design for easy assembly/disassembly*	326

Stitch One

Designer	*Louisa Cranmore, UK*	305
Production	*One-offs*	
Materials	*Reused glass jars and reclaimed wool*	
Eco	• *Reused and recycled materials* • *Encouraging DIY using 'at hand' materials*	329 326

The beauty of compact, low-energy lights deploying fluorescent technology is that little or no heat is produced (unlike tungsten incandescent lighting), so the bulbs are suitable for enclosing. Louisa Cranmore takes advantage of this by fitting a small CFL bulb and inside a conventional screw top jam jar. This is then covered with a crocheted sleeve using reclaimed wool (undo any jumpers or hats you don't like anymore). Patterns can be made to accommodate round, square, tall, short or unusual-shaped jars. Her message is 'why buy, when you can make?'

Tibia floor lamp

The nonchalant posture of the Tibia floor lamp is achieved by the structure being casually lent against the wall. A square frame of Highwood synthetic timber profile of 100% recycled polystyrene, made of post-consumer drink containers and packaging, hides the electrics and supports a plain cotton shade. The specification of the synthetic timber can be varied to meet different grades of impact, UV stability and colour, requiring no further painting or surface treatment.

Designer	*Vanessa Battaglia & Brendan Young, Studiomold, UK*	304
Production	*Small batch production*	
Materials	*Recycled polystyrene, cotton, steel*	
Eco	• *Recycled materials* • *Recyclable*	325 329

Anglepoise® Original 1227 and Giant Model 1227

This fifth generation family company has been making the sturdy, yet functional, Anglepoise® lamps for over 75 years. Model 1227 is one of the most popular in their range with its strong aluminium chassis, 3-springs and classic 1934 looks. Many of these lamps are passed down from generation to generation, parts being replaced as required, testimony to the Modern movement's belief in form, function and durability. The rotation, stability and flexibility of this table/floor lamp is renowned, and has been appreciated by many, including Roald Dahl, the famous children's author. Dahl worked with an Anglepoise® lamp for years, so to honour this relationship – and to celebrate the 70th anniversary of the Original 1227 – the company has built the Giant Model 1227, a version three times life size, for the Roald Dahl story centre. The Giant Model is handmade in England.

Originally by George Carwardine, UK	312
Anglepoise, UK	312
Aluminium, steel, electrics	
• Robust, durable engineering *• Repairable* *• Dependable*	325 329 327

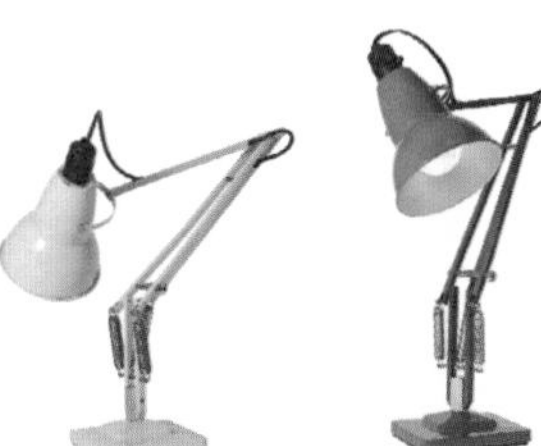

Birzi

The Birzi table lamp is made of soft flexible silicone that can be manipulated and moulded into a variety of shapes to control the dispersion and mood of the lighting. This positive interaction engages the user with the light source enabling a high degree of personalization. Silicone is a tough and resistant yet tactile material that protects the lamp fitting while encouraging long-term use.

✎	*Carlo Forcolini and Giancarlo Fassina for Luceplan, Italy*	306
⚙	*Luceplan, Italy*	318
▬	*Silicone, electrics, fittings*	
♺	• *Strong user interaction*	328

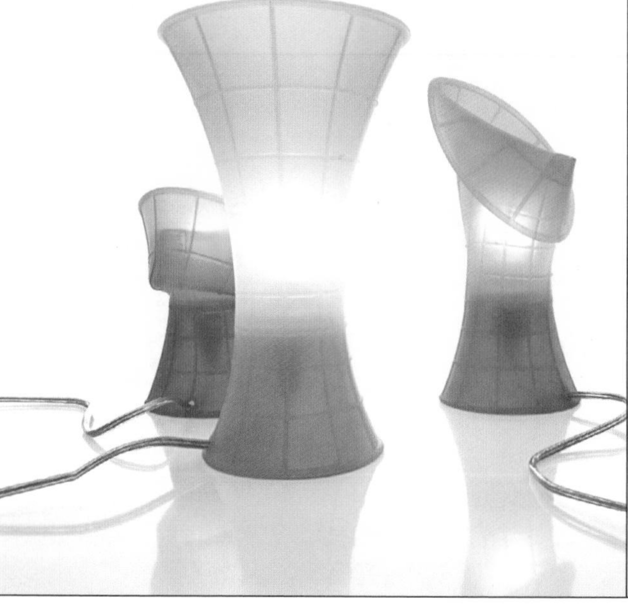

Block Light

By carefully balancing the properties of the materials with the function they are intended for, Nina Tolstrup strips this portable light – that doubles as both hanging lamp and table lamp – down to the essentials. Its simplicity and ease of assembly/ disassembly offers directionality for manufacturers in a marketplace filled with 'over designed' and 'over specified' products. If you need a light to move around your personal or work space, the Block Light does all that is required.

✎	*Nina Tolstrup, Studiomama, UK*	310
⚙	*One-offs and small batch production*	
▬	*FSC-certified pine wood, aluminium tubing, paint, 11W CFL bulb, electric cable*	
♺	• *Simplicity of construction and ease of assembly* • *Minimal utility*	326 324

power glass®

A sandwich of conductive material, which is completely transparent, is embedded between layers of ordinary glass. Single or multi-laminate conductive glass affords different power-carrying capacities, so this patented technology can be used in a range of applications for lighting, switches, electronic displays and so on, especially for low-voltage applications.

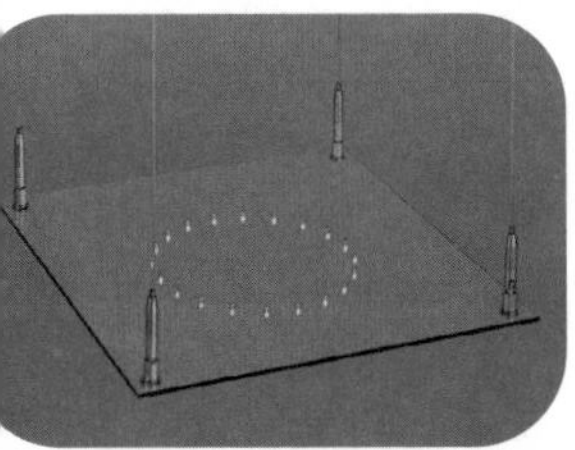

✎	Glas Platz, Germany	316
⚙	Glas Platz, Germany	316
▬	Transparent conductive material, glass	
♲	• Multifunctional	325

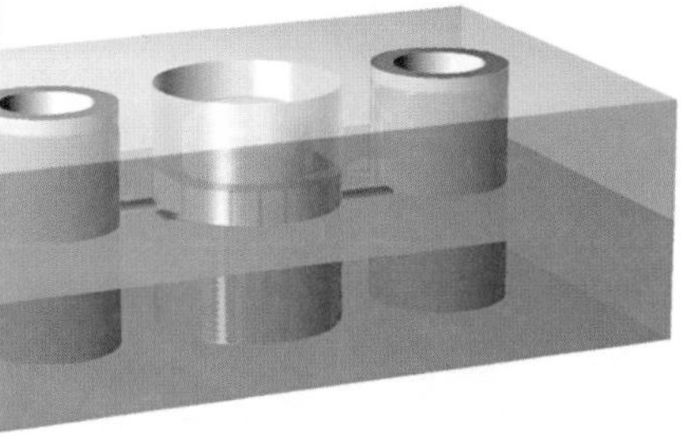

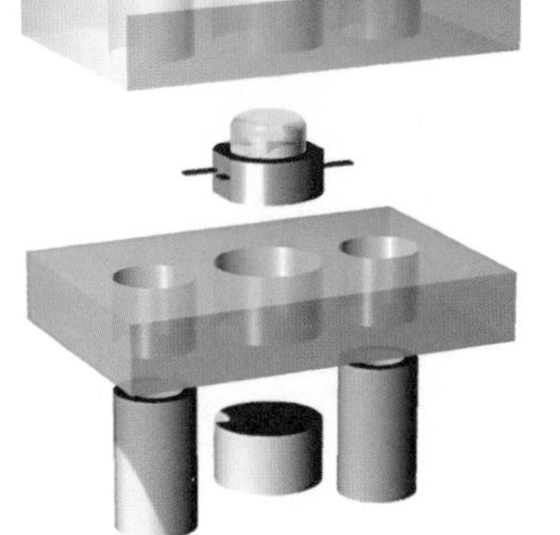

SugaCube

Individual LEDs are encased in click-together two-tone acrylic blocks, enabling consumers to create complex patterns of blocks for bespoke lighting. This playful concept combines the technological flexibility of LEDs with the fun of children's Lego™ building bricks. Users continue the design process by customizing the final lighting product.

✎	Studio Jacob de Baan and Frank de Ruwe, the Netherlands	310
⚙	Conceptual prototype	
▬	Polymer, single LED	
♲	• Low-energy lighting • Modular • Multifunctional	329 328 328

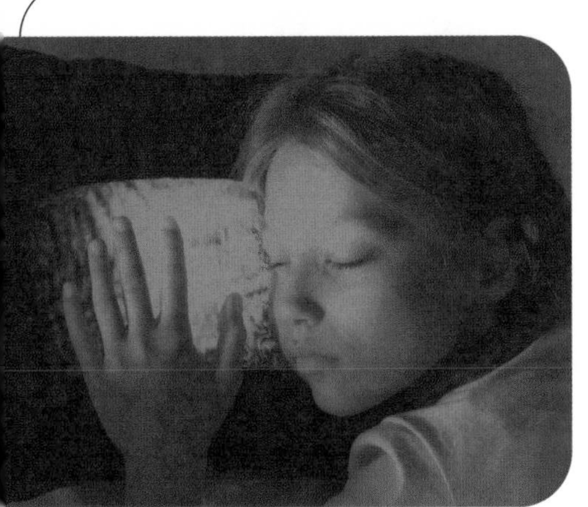

Pillow Light

Switch off your room lights! There are times when a soft glow in the dark is a comfort. This black hemp fabric cushion contains internal illumination creating an interference-light effect by using red LEDs operated by a 6V battery on a 20-minute timer switch. The low-voltage LEDs ensure long battery life but maybe a future model could include a recharge or wind-up battery facility.

✎	Stiletto DESIGN VERTReiB, Germany	310
⚙	Stiletto DESIGN VERTReiB, Germany	310
▬	Hemp fabric, LEDs, battery and switch	
♲	• Low-energy light source • Renewable fabric	329 325

Firewinder

This outdoor light has an innovative double-helix wing that captures wind energy and converts it into a unique light show. In gentle breezes a spiral of light is created, while strong winds generate a brighter pillar of light that pulses with energy. The Firewinder requires no batteries or wiring, and is easily installed. It consists of a smart combination of existing technology, modern computer-aided design (CAD), and manufacturing techniques that facilitate end-of-life disassembly for future recycling. Additionally, the technology offers a platform from which to develop other applications.

✎	*Tom Lawton, The Firewinder Company, UK*	308
⚙	*The Firewinder Company, UK*	308
▰	*ABS plastic, acrylic, stainless and mild steel, aluminium, copper wire, neodymium iron boron rare earth magnets, recycled cardboard, polypropylene (packaging), electronic components including LEDs*	
♺	*• Renewable energy source*	329
	• Restriction of Hazardous Substances (RoHS) compliant electronic components	326
	• Design for disassembly	326

Pod Lens

Most lighting is static, irredeemably rooted to a building's electric cabling. Pod Lens is a modular system of a polycarbonate pod unit with bulb and flex, and a series of bases for standard or floor lighting. For indoor or outdoor use, the pods provide flexible and decorative lighting at the whim of the user.

✎	*Ross Lovegrove, UK*	308
⚙	*Luceplan, Italy*	318
▬	*Polycarbonates, electrical components*	
♻	• *Multifunctional lighting system* • *Upgradable and repairable*	328 328

Moonlight MFL

A robust, weatherproof, semi-translucent, polyethylene material is moulded in four sizes and fitted with different sockets to enable the low-wattage lamps (5–23 watts) to be fixed into the earth or used on hard surfaces. Feel a mood swing coming on? Simply change the coloured bulb filter, choosing from up to 250 colours. Moonlight MFL is a versatile, low-energy, 'mood and colour', indoor/outdoor lighting system.

✎	*Moonlight, Switzerland*	318
⚙	*Moonlight, Switzerland*	318
▬	*Polyethylene*	
♻	• *Multifunctional* • *Low energy consumption*	328 329

Tsola

Most outdoor solar-powered lights are above-ground installations, which makes them vulnerable to the elements, accidental damage and vandalism. Tsola is designed to be installed flush with the ground and can be walked or driven upon without damage. This low-maintenance light is equipped with a timer that automatically switches the light off in extended hours of darkness to conserve the stored energy in the battery.

✎	*Sutton Vane Associates, UK*	310
⚙	*Light Projects Ltd, UK*	317
▬	*Photovoltaics, heavy-duty glass, stainless steel, battery*	
♻	*• Solar power*	329

Solar Bud

A photovoltaic panel generates energy from sunlight, stores it in a battery and releases it to three low-voltage, red LEDs, all in a self-contained unit that is placed in the desired position by pushing it into the soil or other suitable medium. Ideal for garden decorative or safety lighting, the Solar Bud would also be at home in the window box of an urban bedsitte

✎	*Ross Lovegrove, UK*	
⚙	*Luceplan, Italy*	
▬	*Metals, photovoltaics and light emitting diodes (LEDs)*	
♻	*• Solar-powered lighting* *• Very low-energy LED bulbs*	

Sherpa/Sentinel

Just 30 seconds' winding on the handle linked to the AC alternator gives 30 minutes of light from the 3.3V 1565mA high-efficiency Xenon dual-filament bulb on normal beam setting. An LED charge level indicator tells you the optimal winding speed. Alternatively plug it into the mains supply using the AC/DC adaptor. A fully charged NiMH 1000mAh battery gives five hours of light on full beam. Ideal for everyday or emergency use, the Sherpa is available in the USA under the Coleman Sentinel brand.

✎	*Freeplay Energy, South Africa/Europe*	315
⚙	*Freeplay Energy, South Africa/Europe*	315
▣	*Various materials*	
♻	*• Renewable energy options*	327

StarLed® Light

Candles remain a potent sign of human faith. Here, perhaps, if we aspire to a more sustainable future, is the 21st-century technological equivalent. StarLed® is a portable lamp using a single bright LED given jewel-like prominence by its transparent or metalized methacrylate body and prismatic head. A dedicated electronic circuit and three rechargeable AA nickel metal hydride (NiMH) batteries sit in the base plate of the candle. A single charge generates four hours of light and it is recharged by placing the candle on a mains recharger that fits to the base. This design does much to focus on issues of lighting energy, and creates a special mood lighting, but only a full lifecycle analysis would reveal how many candles you could burn to equate to the embodied energy of manufacturing, distributing and retailing each StarLed®. However, over time, as the embodied energy is negated by recharging the product, it will become a 'zero energy' device, providing that the mains electricity emanates from renewable sources. Future evolution of this product may consider an alternative to the methacrylate. This is a high-embodied-energy polymer with excellent optical qualities but the manufacturing process is chemically complex and involves hydrocyanins. Glass, glass composites and other translucent or transparent materials may provide acceptable substitutes.

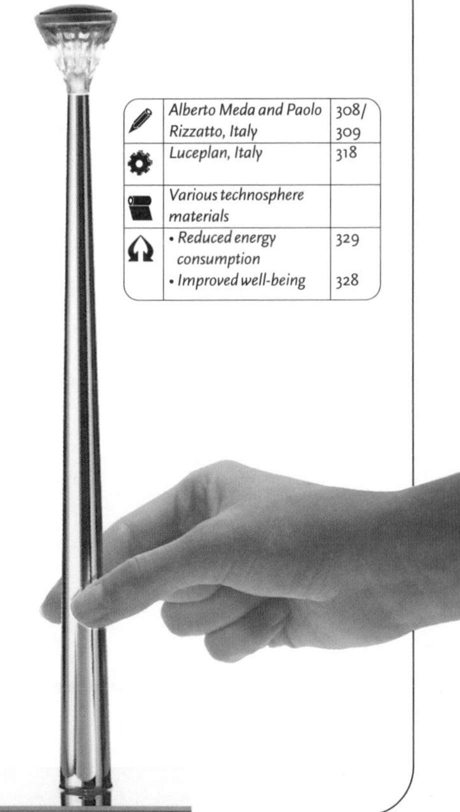

✎	*Alberto Meda and Paolo Rizzatto, Italy*	308/309
⚙	*Luceplan, Italy*	318
▣	*Various technosphere materials*	
♻	*• Reduced energy consumption* *• Improved well-being*	329 328

Jonta Flashlight

Freeplay Energy has consistently improved their wind-up technology over the last decade to produce reliable, durable, human-powered devices. The Jonta utilizes the latest in LED technology, a 1 Watt Star LED designed to last 100,000 hours, to produce a beam capable of lighting up to 50 metres (164 ft) away. It also has an integrated circuit that enables three lighting modes – full, intermediate, and emergency. Over 74% of the kinetic energy produced by winding up the battery goes into storing electricity in the rechargeable built-in Ni-MH battery, with a 60-second wind giving 10 minutes of light in energy saving mode. Occasionally you can perk up and improve battery life with a mains recharge which offers 12 hours of continuous lighting.

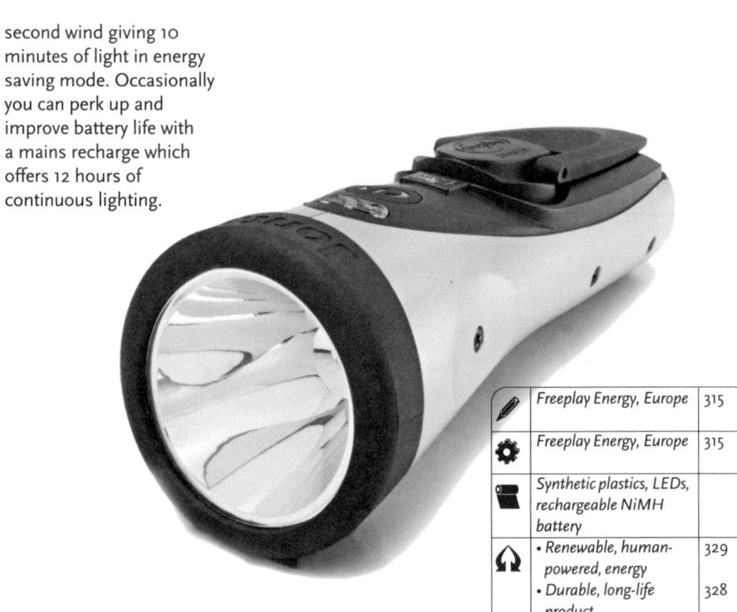

	Freeplay Energy, Europe	315
	Freeplay Energy, Europe	315
	Synthetic plastics, LEDs, rechargeable NiMH battery	
	• *Renewable, human-powered, energy*	329
	• *Durable, long-life product*	328

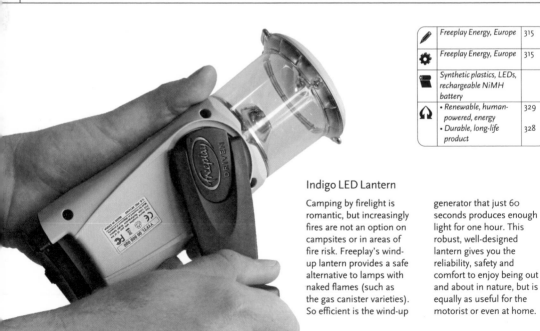

	Freeplay Energy, Europe	315
	Freeplay Energy, Europe	315
	Synthetic plastics, LEDs, rechargeable NiMH battery	
	• *Renewable, human-powered, energy*	329
	• *Durable, long-life product*	328

Indigo LED Lantern

Camping by firelight is romantic, but increasingly fires are not an option on campsites or in areas of fire risk. Freeplay's wind-up lantern provides a safe alternative to lamps with naked flames (such as the gas canister varieties). So efficient is the wind-up generator that just 60 seconds produces enough light for one hour. This robust, well-designed lantern gives you the reliability, safety and comfort to enjoy being out and about in nature, but is equally as useful for the motorist or even at home.

Jigsaw

Prior to the arrival of electricity and the tungsten bulb, oil lamps were a primary source of light. Today they are rarely used in Western Europe despite widespread availability of suitable fuel. The Jigsaw oil-burner is a beautiful glass receptacle with cotton wick. It is functional, aesthetically pleasing and economical to operate. Grouped together or as solitary lamps they cast a warm comforting light and are suitable for mood lighting in bars, restaurants and the dining table at home.

✎	*K. C. Lo, UK*	308
⚙	*Small batch production*	
▬	*Glass, cotton*	
♻	*• Low-energy lighting*	329

SL-Torch

An 80% reduction in materials used is achieved by making the battery into the handle in this neat torch design. Insert the battery into a housing, which holds the bulb, and twist to turn on the torch.

✎	*Antoine Cahen, Les Ateliers du Nord, Switzerland*	304
⚙	*Leclanché, Switzerland*	317
▬	*Battery, bulb, plastic*	
♻	*• Reduction in materials used*	325

Solaris™ lantern

Two hours of sun provide one hour of light for this lantern, which is capable of functioning at -30°C (-20°F) and altitudes in excess of 7,000 m (23,000 ft). Fully charged, the NiMH battery, which is free of mercury, cadmium and lead, will provide light for six hours, but if the battery discharges 90% of its capacity a low-voltage disconnect is automatically triggered. This saves battery life and ensures it will last for up to a thousand recharges.

✎	*Light Corporation, USA*	317
⚙	*Light Corporation, USA*	317
▬	*Photovoltaics, plastics, NiMH battery*	
♻	*• Renewable power source* *• Avoidance of hazardous substances in the battery*	329 328

Eco Kettle

The UK Energy Saving Trust estimates that this 3kW kettle uses, on average, 31% less energy than normal kettles. This is achieved through a dual chamber design. The top chamber acts as a reservoir from which the exact amount of water needed can be released through a valve into the boiling chamber below, assuring that only water needed just then is boiled. This simple mechanism, facilitated by a highly visible transparent calibrated chamber, helps control the amount of water and electricity used.

Designer	*Eco Kettle, UK*	314
Manufacturer	*Eco Kettle, UK*	314
Materials	*Various synthetic polymers, heating element, electrics*	
Eco	• *Reduced energy and water consumption*	329

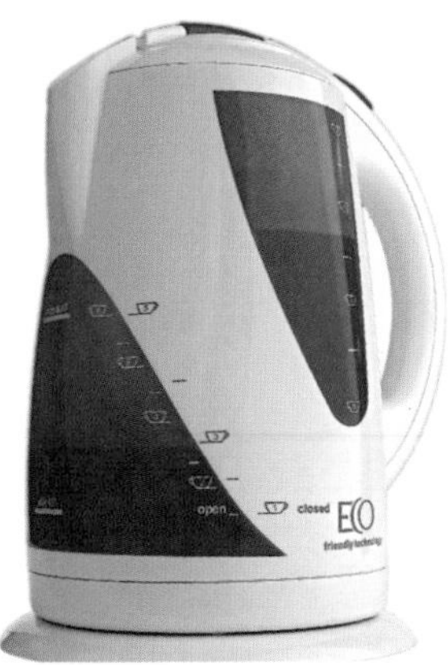

CS3000T Induction Hob

This 3kW single tabletop hob uses induction technology, where magnetic fields transfer heat from an electric wire coil directly to a cooking vessel. This is very energy efficient technology: 86% more efficient than gas cookers and 50% more efficient than halogen cookers. Glass is used to separate the wire coil from the cooking pan or kettle but, as all the heat is transferred to the vessel, the glass top remains cool, providing safety and cleaning advantages over conventional cookers. Precise temperatures, with power settings from 1–99, can be selected by using a digital display.

Designer	*Induced Energy, UK*	317
Manufacturer	*Induced Energy, UK*	317
Materials	*Metal, glass*	
Eco	• *Energy efficient* • *Improved safety*	329 328

Bel-Air

Indoor air quality can sometimes be quite poor, especially during warm and humid periods when synthetic plastics used for furniture, and other goods emit various air-borne pollutants (benzene, formaldehyde and trichloroethylene). This off-gassing phenomenon was noted as far back as the 1980s by NASA who discovered toxic levels of such pollutants in the astronauts who lived and worked in the spacecraft. NASA identified several plants/greenery for their ability to absorb toxic gases: the gerbera, the philodendron, the spathiphyllum, the pathos and the chlorophytum are most effective. This phenomenon affects our everyday lives as well as those of astronauts, and we are familiar with topics such as 'sick building syndrome' and the increase in allergy and asthma sufferers. Bel-Air aims to optimize elimination of these airborne pollutants by harnessing the power of plants to filter the toxins out. The natural filter system has three components in a patented design system – the leaves, the roots and a humid bath. It is conceived as a new guardian angel, where plants are released from purely ornamental, decorative tasks and become a true object of service in a vegetal cranial box of aluminium and Pyrex.

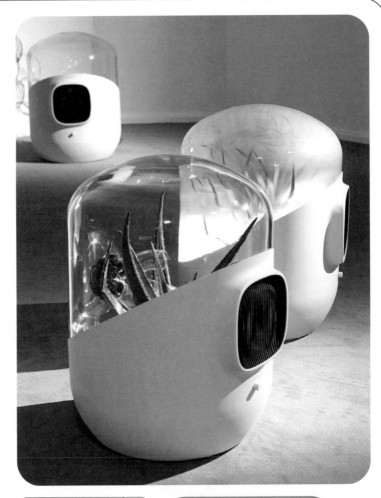

	Mathieu Lehanneur, France, with David Edwards, USA for Le Laboratoire, France	307
	Prototype	
	Fans, Pyrex, aluminium, plants	
	• Improved health environment	328

POLTI Ecologico AS810

For those who suffer from asthma this vacuum cleaner is a boon. Its 8-stage filtering process – including EcoActive and HEPA filters and a water-based filtration system – removes 99.99% of dust, up to 0.3 microns including pollen, dust-mite faeces and cigarette smoke. Paper bags are replaced with a removable water filter, which is emptied after use and has a lifetime of approximately six months. The appliance has a range of action up to 8.84 metres (29 ft) and comes with a wide selection of accessories for soft and hard floors.

	Polti, Italy	320
	Polti, Italy	320
	Various plastics, metals	
	• *Reduction in use of consumables* • *Improved health environment*	329 328

	Dyson Appliances, UK	314
	Dyson Appliances, UK	314
	Various plastics, rubbers, motors, metals	
	• *Reduction of materials, light weighting*	326
	• *Design for easy maintenance*	328
	• *Five-year guarantee with full labour and parts*	328
	• *Avoids use of consumables (paper bags)*	329
	• *Special user considerations for allergy sufferers and pet owners*	328

The Dyson DC24 All Floors

More than 15 years after the DC02 Dual Cyclone technology vacuum took the world by storm, Dyson Appliances is still creating new models with leading-edge technologies. The Dyson DC24 All Floors is part of the upright range of vacuums, with a full-size pick-up head but a body one-third smaller than the rest of the range, making it ideal for compact living spaces or difficult corners, and the lightest upright at 5.4 kg (11.65 lbs). Aside from the patented Root Cyclone™ technology that prevents loss of suction power, the DC24 can easily be turned with the flick of a wrist. Fitted with a washable HEPA filter, approved by the Asthma and Allergy Foundation of America, and with a motorized brush bar to assist with dirt and pet-hair removal, the DC24 ensures very high standards of cleaning, and so assists with overall environmental health in the home. Dyson Appliances strives for more and more efficient engineering solutions, and supports aspirant young engineers and designers through the James Dyson Foundation and award schemes. As a contribution to closing the materials loop, the company will even recycle your old vacuum if you buy a new Dyson.

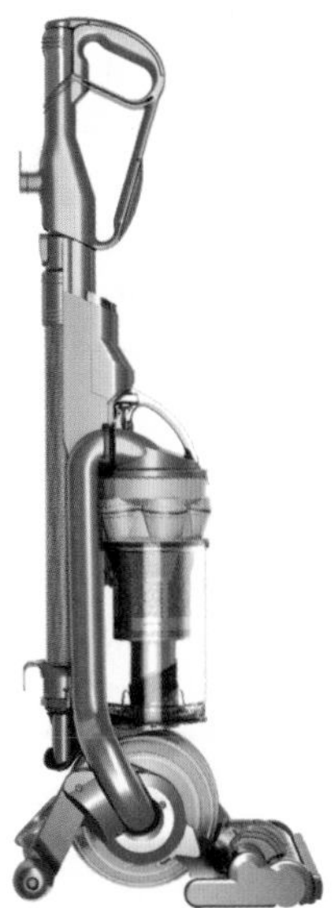

POLTI Vaporetto ECO Pro 3000

The ECO Pro uses super-heated steam, an effective agent in cleaning and sterilizing carpets, mattresses and upholstery that obviates the need for strong, toxic chemical cleaners. Steam is also safer than insecticides for killing dust mites and other insects. The outlet hose can be fitted with a range of brushes, and the ECO Pro has an operational range of 8.14 metres (26.7 ft).

✎	*Polti, Italy*	320
⚙	*Polti, Italy*	320
▮	*Various plastics, stainless steel*	
♺	*• Reduction in use of consumables* *• Improved health environment*	329 328

Staber Washer

Unlike horizontal-axis-driven front-loading washing machines, the Staber Washer offers a top-loading machine into which the stainless-steel basket of laundry is loaded. Energy-saving features include the use of a variable-speed motor. Easy access to the internal components can be gained by lifting the front panel and fitting a self-cleaning filter, thus facilitating maintenance. The manufacturers claim reduced energy, water and detergent consumption.

✎	*Staber Industries, USA*	321
⚙	*Staber Industries, USA*	321
▮	*Stainless steel, steel, resin and various other materials*	
♺	*• Energy and water conservation* *• Energy efficient to Energy Star guidelines*	329 329

Lavamat 86850

This is one of the best washing machines in the AEG-Electrolux range with a water consumption of just 6.4 litres of water per kilogram of wash load (7 kg with just 45 litres per wash, 1.19 kWh per wash). Under the European system of energy efficiency it is rated 'A' class on both wash and spin-drying performance up to 1600 rpm. Measuring 800 mm (31½ in) wide, 600 mm (23½ in) deep and 850 mm (33½ in) high, this machine easily slips under the kitchen work surfaces or fits in a utility area. Careful planning of washing routines (not under-loading) and the eco-efficiency reductions achieved above, are all significant in reducing the CO_2 emissions of appliances for Europe, which currently amount to 4% of total emissions.

✎	*AEG-Electrolux, Europe*	312
⚙	*AEG-Electrolux, Europe*	312
▬	*Various metals, synthetic polymers, motor, electronics*	
♻	*• Reduction of water and energy consumption*	329

Santo 70170-38 TK

AEG-Electrolux offer this 152-litre under-counter refrigerator with a low energy consumption of 120kWh per year, and compliance with the latest CFC- and HFC-free refrigerants. It has been 'energy efficiency recommended' by the Energy Saving Trust in the UK and is 'A' class rated by the EU's Energy Labelling scheme.

With an external LED digital temperature indicator it is easy to keep an eye on what's happening, and the Coolmatic system ensures rapid cooling if warm goods are placed inside. Of course the user can help by minimizing opening the fridge door.

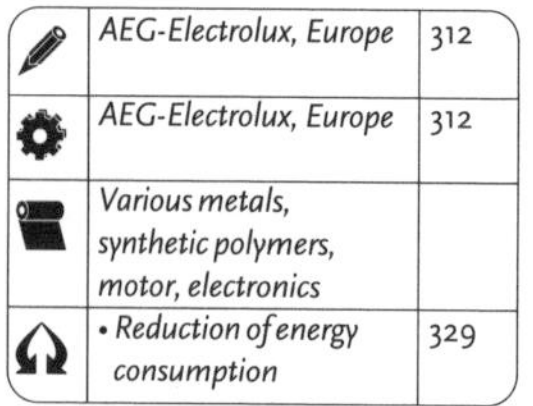

✎	*AEG-Electrolux, Europe*	312
⚙	*AEG-Electrolux, Europe*	312
▮	*Various metals, synthetic polymers, motor, electronics*	
♲	*• Reduction of energy consumption*	329

Wind

In the Industrial Revolution iron and steel usurped natural materials, so it is refreshing to see the process cleverly reversed in the housing of this electric fan, in which woven rattan replaces the conventional pressed sheet steel or plastic. At the same time the fan is transformed from an object of cold functionalism to one of playful character. Most of the materials can be recycled or composted.

✎	*Jasper Startup, Startup Design, UK*	310
⚙	*Gervasoni SpA, Italy*	316
▮	*Rattan, steel, electrics*	
♲	*• Recyclable and compostable materials*	325

Vestfrost SW 350 M

Vestfrost is one of the world's largest manufacturers of refrigerators and freezers and took an early lead in showing environmental responsibility by removing all CFCs and HFCs from its range in 1993. Using the alternative 'Greenfreeze' refrigerants, Vestfrost was the first manufacturer in Europe to hold the EU Eco-label for this category of appliances. The Vestfrost SW 350 M continues to uphold these high standards: this 249-litre refrigerator has a 123-litre capacity freezer with an automatic defrost facility, and consumes just 0.99 kWh per day, making it energy efficiency class A.

✎	*David Lewis, UK*	308
⚙	*Vestfrost, Denmark*	322
▮	*Metal, plastics, rubber, electric motor and compressor*	
♺	• *Low energy consumption* • *Clean production*	329 326

Felt 12 x 12

Be your own fashion designer using Fortunecookies' felt squares backed with Velcro: assemble a jacket, trousers, wedding dress or any other garment in your own personalized style. Bored with the look? Deconstruct your design and start again. Fashion is placed back in the hands of the consumer.

	Fortunecookies, Denmark	306
	One-offs	
	Felt, Velcro	
	• Modular system for reuse of components *• Renewable material (felt)*	328 325

Infiniti

Often all that is needed to reinvigorate an item in our wardrobe is a facelift, or the addition of a new bit of trim, or by changing the cut or length. Recognizing that we all like change, Benjamin Shine has conceived a system where this is facilitated. With the Infiniti you can play with a huge number of permutations to zip and press together a variety of components to make your 'ideal' garment for each situation. The 14 pieces permit hundreds of combinations enabling daytime or nightime outfits.

✎	*Benjamin Shine, UK*	309
⚙	*Prototype*	
▬	*Cotton jersey fabric, zips*	
♲	*• Multifunctional* *• Adaptable* *• Repairable*	328 324 329

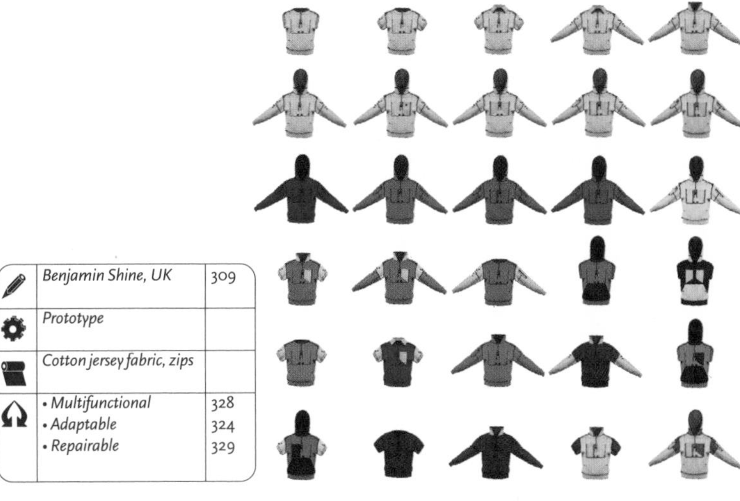

Uptown Hoody

Since 1994 Indigenous Designs has encouraged the development of a global network of artisan workers and co-operatives to create garments using natural fibres and renewable synthetic fibres. This means sourcing fibres from specialist farmers and communities. Alpaca comes from the Puno and Cusco areas of Peru with a range of up to 52 natural colours available; organic cotton is sourced again in Peru mainly from the Trujillo and Amazonas; merino wool from the southern Andes in Argentina and Tencel, a fibre derived of cellulose extracted from wood pulp originating from sustainably managed forests. Each garment benefits from the characteristics of natural fibres to breathe and wick moisture away from the skin. The Uptown Hoody offers a good combination of functionality, style and detailing that ensures a long-lasting garment.

✎	*Indigenous Designs, USA*	317
⚙	*Indigenous Designs, USA*	317
▬	*Tencel, alpaca and wool, non-toxic natural dyes*	
♲	*• Renewable materials* *• Fair Trade scheme & support to local agricultural communities*	325 324/ 325

Eco-Boudoir

Established in 2005 by Jenny White, Eco-Boudoir brings an ethical consciousness to the most intimate of garments and challenges the notions of the luxury end of the market. The collection, inspired by the colours of the forest, includes organic silks and the first printed organic cotton and Lenpur (cellulose fabric made from the wood pulp of fir trees using 'closed-loop' production). Silks are digitally printed using natural dyes and fabrics certified by Eco-Tex, Skal and IMO, three leading sustainable textile labelling systems. The company's environmental and ethical policy ensures that its carbon emissions are offset by supporting the Iracambi rainforest in Brazil and their garments are manufactured locally in the UK. So, while you cocoon yourself in sumptuous fabrics you can relax knowing that nobody is being harmed elsewhere by your indulgence.

✎	*Eco-Boudoir, UK*	314
⚙	*Eco-Boudoir, UK*	314
▤	*Organic silks and cottons, Lenpur, natural dyes*	
♺	*• Renewable, organic materials* *• Certified labelling schemes* *• Supply chain management* *• Reduced water & energy in manufacturing*	325 330 326 326
✪	*International Design Awards 2008, category Fashion, Land & Sea*	

Solar Vintage

These two experimental collections contest the aesthetics and functionality of smart clothing by exploring the inherent beauty and melancholy of the hand-crafted artefact. Corchero's wearable solar technologies bring fresh elements of surprise and a distinctive performative quality. The hand-held fan embedded with solar panels gathers sunlight, stores it in a small battery then re-radiates it through LEDs integrated into the fan. The collections explore the emotional value of garments, jewelry and keepsakes by humanizing the inherent coldness of the technologies they harness.

These designs strive for a harmony between the natural, human and technological; a challenge that is supreme in the 21st century.

✎	*Elena Corchero, UK*	305
⚙	*One-offs*	
	Various textiles, hand embroidery, electronic circuitry, solar cells, LEDs, battery	
	• *Re-humanizing technology*	326
	• *Re-vitalizing the value of the handmade*	326

whiSpiral

This shawl is embedded with electronic circuitry that enables messages from loved ones and friends to be 'released' when the garment is wrapped around the user. Sensors in each miniature audio module are activated by a soft caress or wrapping movement to activate personalized pre-recorded messages. Messages are recorded in each module by connecting a microphone to a yellow leaf, where the message is stored in persistent memory. Intended as a gift for a lover, family member, or friend when they are far away, the whiSpiral offers a tenuous but positive emotional thread as we navigate our highly mobile contemporary societies where long-distance relationships have become an established norm.

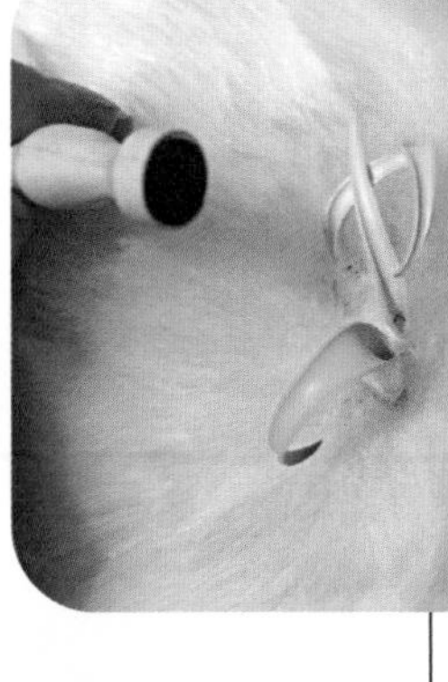

✎	*Elena Corchero and Stefan Agamanolis, UK*	305
⚙	*One-offs*	
▬	*Synthetic fur, electronic circuitry and components, battery*	
♻	*• Re-humanizing human-to-human connections*	328

Hug Shirt™

CuteCircuit works in the area of wearable technology by developing wearables that make technology more usable, emotionally fulfilling and fun. Their Hug Shirt™ utilizes Bluetooth technology (sensors and actuators) for Java-enabled mobile phones allowing electronic 'hugs' to be sent by SMS messages. HugMe™ Java software communicates to sensors that monitor the strength of the touch, skin warmth and heartbeat of the sender, and actuators that recreate this 'hug' in the recipient's Hug Shirt. The red parts of the shirt contain the active sensors and actuators in a smart technology pad that can be removed so the shirt can be washed. All components are lead-free and non-toxic, and the system operates on rechargeable batteries. User-centred and participatory design methods during prototyping ensured that the humanistic dimensions of the project were always kept in sharp focus: during tests, 'hugs' received generated the release of positive natur[al] chemicals in the recipie[nt]. In our highly mobilized cultures these shirts see[m] to be able to deliver emotional healthcare.

✎	*CuteCircuit, USA*	314
⚙	*CuteCircuit, USA*	314
▬	*Various fabrics, electronic components, rechargeable battery*	
♻	*• Improve the emotional balance of the wearer* *• Electronic components comply with the European RoHS (Restriction of Hazardous Substances) regulations*	328 326

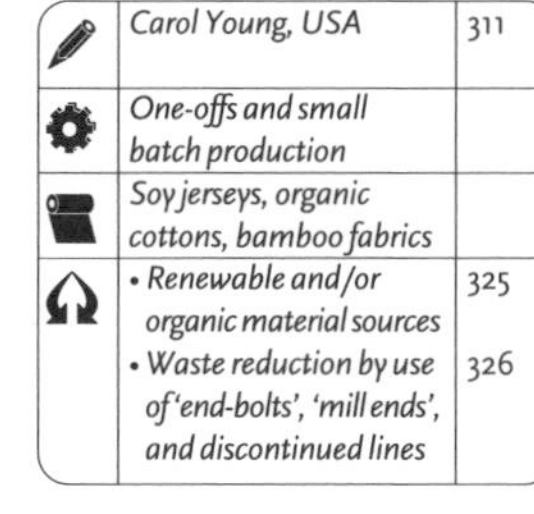

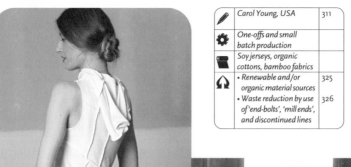

	Carol Young, USA	311
	One-offs and small batch production	
	Soy jerseys, organic cottons, bamboo fabrics	
	• *Renewable and/or organic material sources* • *Waste reduction by use of 'end-bolts', 'mill ends', and discontinued lines*	325 326

Undesigned

Attention to detail is the hallmark of an accomplished fashion designer and today this must include the source of materials and an examination of the supply chain. Carol Young is an independent fashion designer whose company reflects a growing trend for flexible, small batch production using natural, organic textiles. Her signature style of seasonless clothing, combined with a focus on stitching, texture and cut, accentuate these natural fabrics and ensure a distinctive garment that is intended to enjoy a longer life. Each piece is hand-crafted and individually made, with a focus upon the ability to mix and match pieces from the collections. No20 and No17 from her collection are classic lines that have emerged as bestsellers, made of bamboo jersey (93% bamboo, 5% Lycra) using velvet denim cotton blend from designer surplus stock, and bamboo denim (28% bamboo, 70% organic cotton and 2% lycra) with a 100% silk trim, respectively. Her latest collection features organic cotton twill (100% organic cotton) with a lining of mill-end stock. All fabric scraps not used in the clothing are donated to local children's art programmes.

John Patrick ORGANIC

While women can choose from a diverse range of organic clothing labels, men's options have been rather limited to outdoor, recreational or sportswear. John Patrick changes all that with his urban-chic menswear collections using organic and recycled cottons and wools in eclectic or subtle combinations to suit a broad range of tastes.

Much of the clothing is made from sustainable materials in the US, and the organic knitwear is produced by a cottage-industry of artisans in Peru. Now it is possible for men to build up a palette of well-cut tailoring and one-offs for their very own organic wardrobe.

	John Patrick ORGANIC, USA	307
	John Patrick ORGANIC, USA	307
	Organic and recycled cottons, recycled wool, reused garments and accessories	
	• *Recycled and reused materials* • *Organic renewable sourced textiles*	325 325

Keep & Share

Defining itself as an 'alternative luxury label', Keep & Share creates knitwear that transcends short-lived trends and becomes your 'best friend'. The emphasis is on handmade durability for a line of knitwear ranging from cardigans, wraps and jumpers to dresses. All garments use high quality and naturally coloured organic and recycled yarns, and often feature fine alpaca and cashmere. In a commitment to the emerging concept of 'slow fashion', Keep & Share build up close relationships with their customers by a free try-before-you-buy-scheme, aftercare wash and repair services, and even encourage wearer involvement by running knitting workshops using Keep & Share's own patterns. Owners are encouraged to use low-impact environmental care methods, such as old-fashioned hand washing and air drying.

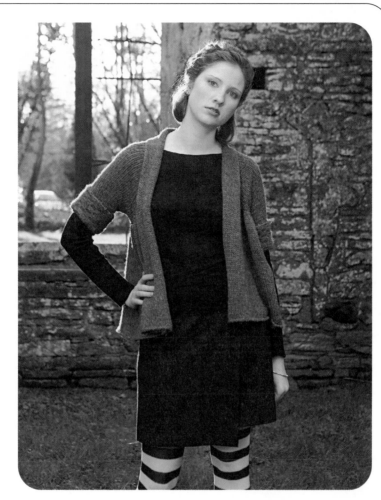

	Amy Twigger Holroyd, Keep & Share, UK	307
	Keep & Share, UK	307
	Organic and recycled yarns	
	• *'Slow fashion'* • *Improved emotional bond to artefact* • *Renewable and recycled materials* • *Good customer aftercare*	324 328 325 328

El Naturalista

The rapid global expansion of the El Naturalista brand attests to this innovative manufacturer's strong environmental and social consciousness alongside comfortable and ergonomic designs. The range of shoes represents a mix of contemporary and traditional designs, with attention to detail a consistent hallmark. The women's shoes include Ambar – with natural/recycled rubber soles – that subtly reference the ankle ties of espadrilles and Spanish dancing shoes, as well as Ikebana, with aesthetics derived from a Japanese/Spanish hybrid, and soles made of recycled polyurethane. The company's eco-policy is clearly articulated, as are its humanitarian goals such as fair wages, and the establishment of The Atauchi Project, which provides shelter and education for children in Peru.

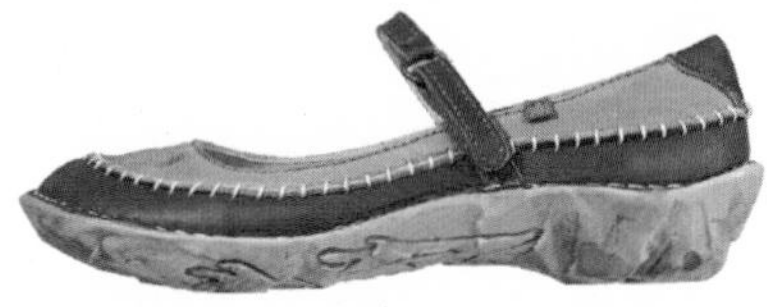

✎	*El Naturalista, Spain*	315
⚙	*El Naturalista, Spain*	315
▬	*Naturally-dyed leather, natural and recycled rubber*	
♻	*• Renewable and recycled materials* *• Biodegradable materials* *• Non-toxic materials* *• Traditional forms of production with skilled labour*	325 325 325 324

Felt Slippers

These elegant contemporary felt slippers build on a traditional Finnish design using 100% wool felt. Lightweight, comfortable and easily stowed away when travelling, these slippers offer physical, emotional and mental comfort at the end of a busy day. The fabrication of these slippers also keeps alive the skills and traditions that have evolved since the factory opened in 1921.

✎	*Huopaliike Lahtinen, Finland*	316
⚙	*Huopaliike Lahtinen, Finland, sold via Matteria, Spain*	316
▬	*Wool*	
♻	*• Renewable fibre materials* *• Innovation of traditional craft skills*	325 324

Worn Again

Proving that the second time around the materials loop can be just as exciting as the first, Terra Plana are continually extending the boundaries of shoe design with their eclectic Worn Again range. Utilizing recycled and/or reused materials – from parachutes, T-shirts, leather scraps, motorbike tyres and ex-military jackets – challenges the designer to find practical yet attractive ways of giving these materials a second life. The company donates 5% of sales to a UK organization called Anti-Apathy that promotes awareness for positive social change.

✎	*Worn Again, UK*	309
⚙	*Worn Again, UK*	322
▬	*Wide variety of recycled and reused materials*	
♺	• *Recycled and/or reused materials* • *Ethical and socially responsible production*	325 326

Veja Tauá

Harking back to the days of simple canvas shoes with rubber soles, Veja has updated the concept by operating a responsible system of manufacturing for these contemporary casual recreation shoes. Organic cotton comes from the northern Brazilian town of Tauá, and the soles are from individually tapped wild Amazonian rubber trees, rather than plantation rubber, helping support indigenous communities and protect areas of virgin rainforest. Both raw materials are backed by foundations to help sustain local communities and co-operatives with Fair Trade practices, and Veja buys in organic cotton at twice the world market rate to help encourage the growers.

✎	*'agnès b' with Veja, France*	304
⚙	*Veja, France*	322
▬	*Organic cotton, wild latex rubber*	
♺	• *Renewable materials sourced from organic or wild production systems* • *Ethical production practices* • *Social responsible production*	325 326 326

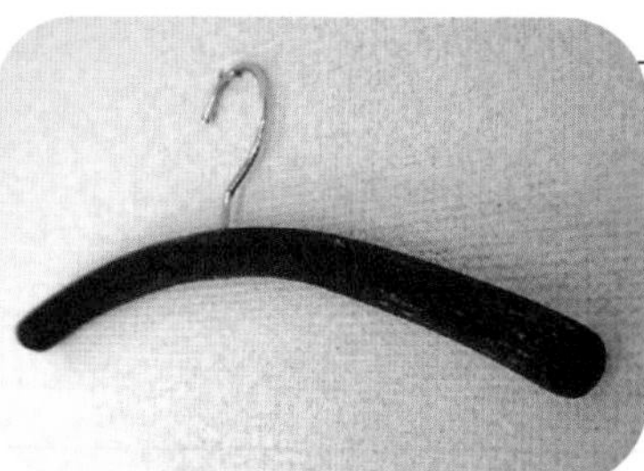

Charcoal hanger

Waste pieces of wood from sustainably harvested timber are bent and moulded to shape then carbonized by firing in a furnace. Charcoal dust is washed off the surface to prevent any damage to clothing. Carbonization of the wood ensures that excess humidity and odours are absorbed, giving the hanger additional clothing care qualities. Beyond the useful life of the hanger the charcoal can be returned to the soil as a nutrient or used for water filtering to complete the lifecycle of the product.

	Koji Takahashi, Japan	310
	Small batch production	
	Charcoal	
	• Renewable and sustainably harvested material	325
	IDRA award, 2002–2003	

Treetap® Rubber Bag

If you want to make an ethical statement with your handbag, then look no further that the Treetap® Rubber Bag. The leather look and appeal is achieved by a fabric rubberized with natural latex extracted from wild rubber-plant forests in the Amazon. The vegetal leather was developed by the Treetap® Project, established in 1991 to improve the lives of indigenous Indian and Amazon rubber tappers and their families. Workers are paid four times more than the normal market price for wild rubber to support and valorize the cultures, traditions and biodiversity of the Amazon region.

	Bags of Change, UK	312
	Bags of Change, UK	312
	Natural latex, fabric	
	• Renewable, sustainable material *• Supports socio-economic benefits to local communities* *• Ethically driven design*	325 324 324
	Green England Awards, 2008: Most Innovative Green Product	

A-button

Designers often neglect important aspects of everyday living. Not so Antonia Roth, whose elegant button design takes a fresh look at something that people with limited dexterity struggle with daily. The elongated shape makes it easier to hold and to pull through the button-hole, so it offers improved independence for children, the elderly or anyone with arthritis or limited hand mobility.

	Antonia Roth, Fachhochschule Hannover, Germany	309
	Prototype	
	Polymer	
	• Universal design	325

Um bags

✎	*Josh Jakus, USA*	307
⚙	*Small batch production*	
▬	*Waste and reused industrial wool felt*	
♲	*• Renewable materials* *• Reused and recycled materials* *• Lightweight, flat-pack, durable design*	325 325 327

The sculptural aesthetic for these pressed wool felt bags is achieved by transforming flat pieces of felt through the simple action of zipping them up. This enables the bags, derived from factory excess of approximately 85% wool and 15% other fibres, to be easily stored, cleaned and re-zipped ready for use. Sizes for these simple, beautiful, functional and ecological bags range from a small change purse to a shoulder bag.

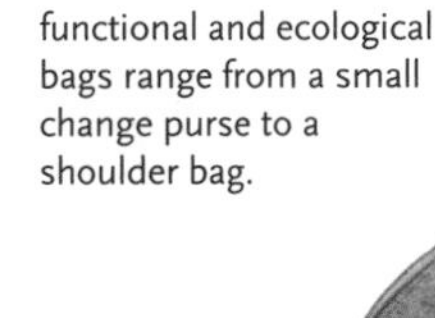

Morsbags

A new phenomenon has emerged via the Internet – the concept of downloadable designs – and it is finding support at the grass roots level. Morsbags offer a cutting plan and description of how to make your very own bag from old curtains, bed sheets or other unwanted fabrics. The driving ambition of Morsbags is to relegate the plastic shopping bag to history, claiming that one lovingly handmade bag replaces 80 of the ephemeral plastic ones. Perhaps another key benefit is the conviviality that is apparent at a local Morsbags stitching session. We will definitely see more downloadable design as people place renewed emphasis upon hand-crafted items and self-sufficiency.

✎	*Morsbags and thousands of DIY designers, UK and anywhere*	308
⚙	*Individual designer-makers, anywhere*	
▬	*Various reused textiles*	
♲	*• Downloadable Internet designs for everyone* *• Reduce waste and reuse materials* *• Encourages positive social networking and well-being*	326 326 328

Yogi Family low chair, bench and low table

Michael Young, UK	311
Magis SpA, Italy	318
Plastic	
• Single material *• Recyclable* *• Durable*	325 329 328

Young plays with his audience, mixing the aesthetics of children's play-things with the dimensions of usable adult furniture. This blurring of ergonomic boundaries creates a fresh appeal to this family of indoor/outdoor moulded plastic furniture and challenges the need to produce separate furniture ranges for children and adults. The solid mono-material is not easily damaged, which ensures longevity and also means that the plastic can be recycled when the object no longer fulfils its promised function.

Stokke™ Tripp Trapp®

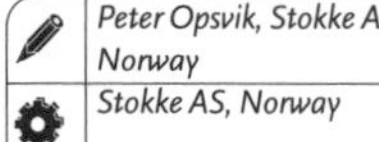 *Peter Opsvik, Stokke AS, Norway*
Stokke AS, Norway
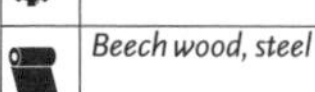 *Beech wood, steel*
• Multifunctional *• Universal design* *• Durable*

This modern classic emerged from the design studio some 30 years ago, but still fulfils its original purpose to provide a chair that grows with the child. Made from solid beech wood from cultivated forests, the Tripp Trapp can accommodate six-month-old babies up to teenagers or even adults by moving the relative height of the seat and footrest components in the frame. The chair encourages good posture by allowing the sitting position to be adjusted to the table height. This arrangement also permits freedom of movement. Wooden components are guaranteed for seven years and a range of safety rails and cushions enable further customization. Here's one chair you can raise your entire family in.

IoLine Crib

This sturdy baby crib/cot is precision manufactured using CNC technology, milling bamboo laminate to produce an intricate pattern of enclosure, light and space that references Arabesque windows. The vertical orientation of the design is enhanced by the natural striations of the bamboo, creating an interesting light environment while simultaneously providing a feeling of security. The laminate surfaces are protected with a non-toxic finish of linseed oil (flax oil), orange peel citrus extract and carnauba wax. The crib is designed to fit a standard mattress, and a kit enables the crib to be easily converted into a toddler's bed, extending the functional life of the base. Lovingly conceived and well made, this is a piece of children's furniture that can become a cherished heirloom.

	Michaele Simmering and Johannes Pauwen, Kalon Studios, USA and Germany	309
	One-offs, made in USA	
	Bamboo laminate, nickel-plated and stainless-steel hardware, non-toxic glues, plant based oil/wax finishes	
	• Fast-growing renewable material	325
	• Durable, adaptable, long-life product	328
	• Recyclable and/or biodegradable	329

Stokke Xplory™

After the success of Stokke's Tripp Trapp adjustable children's chair (p. 122) it was only a matter of time before the company tackled new design challenges in their range of products. Xplory™ includes an adjustable footrest, maintains a low centre of gravity, and has four wheels to ensure stability at all times. The central spine is the key to the innovative range of positions in which the child seat can be fixed, so the angle of recline can be chosen to suit the child's requirements. Air bubbles inside the 'solid' rubber tyres and a shock absorber between the back axle and the frame provide a comfortable ride. Accessories are also attached to the spine and include a changing mat. Stokke has also thought long and hard about the ergonomics in relation to the adult carer.

Handle height is adjustable from 67–116 cm (26–46 in) above ground and the U-shaped rear axle means that a relaxed walking posture is easily maintained.

Designer	*Bjørn Refsum, Stokke AS, Norway*	309
Manufacturer	*Stokke AS, Norway*	321
Materials	*Aluminium profiles, polymers, polyester fabric*	
Function	*• Multifunctional* *• Improved ergonomics*	328 328

Leggero Twist

Safety features of this bicycle trailer for children include a low centre of gravity, seat belts, a protective plastic shell and a warning flag. All-weather protection allows flexibility of use and ensures that the children have a good view and can feel the breeze.

Designer	*Christophe Apotheloz, Switzerland*	304
Manufacturer	*Brüggli Produktion & Dienstleistung, Switzerland*	313
Materials	*Various*	
Function	*• Human-powered transport for the family*	329
Award	*iF Design Award 2003*	332

The Kangaroo Bike

This Danish tricycle provides a dedicated vehicle for moving a young family around in safety and comfort. Strong, yet lightweight, 7005 aluminium framing means that it weighs only 39 kg (86 lbs) fully equipped. Impact-resistant polyethylene for the cabin protects its occupants, while the three-point steering system permits easy manoeuvring, and the seat and handlebar positions can both be easily adjusted to maximize comfort. Here's a genuine performer for short-haul urban travel or longer countryside excursions.

✎	*A. Winther A/S, Denmark*	312
⚙	*A. Winther A/S, Denmark*	312
	Aluminium, polyethylene, various synthetics	
	• *Toughness, durability, long life* • *Lightweight* • *Multifunctional*	328 325 328

Viking and Viking Explorer

If future generations are to revive the art of cycling in our post-peak oil world, then encouraging them to learn and develop their mobility when very young seems a great ambition. This is the mission of the Danish manufacturers A. Winther, who make a diverse range of tricycles, bicycles, scooters and other accessories for the nursery or school playground. Their aim is to encourage 'learning by moving'. Typical of the range is the no-nonsense traditional Viking with a strong oval-shaped downtube, durable weather-resistant rubber and steel components, all compliant with the latest safety regulations. For the boy and girl racers there's the Viking Explorer with its horizontal frame and lever steering system. Ease of assembly and repair are crucial, so there is an extensive parts supply department that will keep the Vikings in service and extend their lifespan.

✎	*A. Winther A/S, Denmark*	312
⚙	*A. Winther A/S, Denmark*	312
	Steel, aluminium, weatherproof rubber, various synthetics	
	• *Lightweight but tough* • *Encourages healthy development of children*	325 328

XO Laptop

This is an ambitious project to enable children in the developing world to gain access to personal computers and the Internet. In conjunction with the One Laptop Per Child (OLPC) movement, the XO Laptop is a remarkable design collaboration between designers, product engineers, interface designers and software engineers. Capable of delivering and recording still and video photography and audio files, or playing e-books and games, the XO Laptop comes with a unique child-friendly interface, and has a rotating screen that is both a full-colour image screen and a high-contrast black-and-white screen so that it can be read under any light condition. The Wi-Fi antennae 'ears' can be articulated to maximize wireless signal reception, and a dual mode touchpad supports finger and stylus inputs. Built with a minimum of toxic materials, the XO is entirely recyclable, and uses a fraction of the power required by ordinary laptops. For the relatively small donation of $199 an XO Laptop will be given to a child, helping to empower young people by giving them access to an incredible learning environment.

✎	*Yves Behar, fuseproject, USA*	304
⚙	*Quanta Computer Inc. and OLPC, USA*	320
	Polycarbonate/ABS, rubber, computer electronics and screen	
	• Recyclable materials *• Affordable, low-energy electronic equipment* *• Open source software*	329 329 324
	• IDEA Gold Award 2008 *• Red Dot Design Award 2008* *• London Design Museum's Brit Insurance Design Award 2008* *• INDEX Community Award 2007* *• I.D. Magazine Concept Design 2007*	

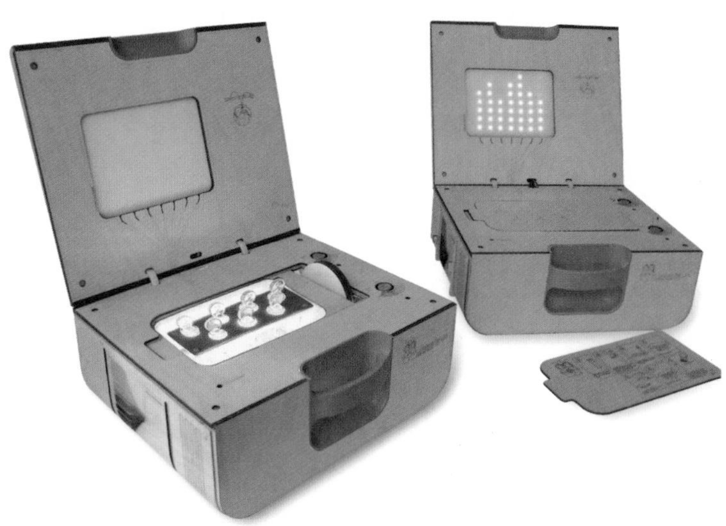

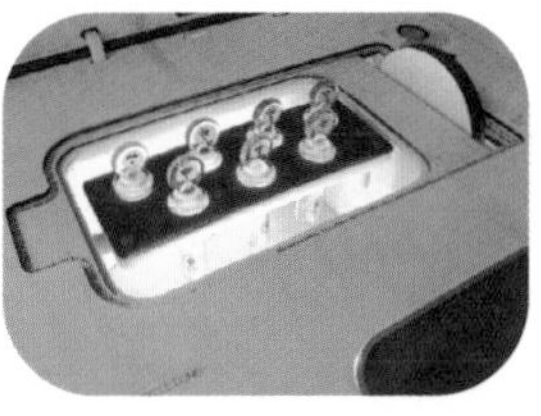

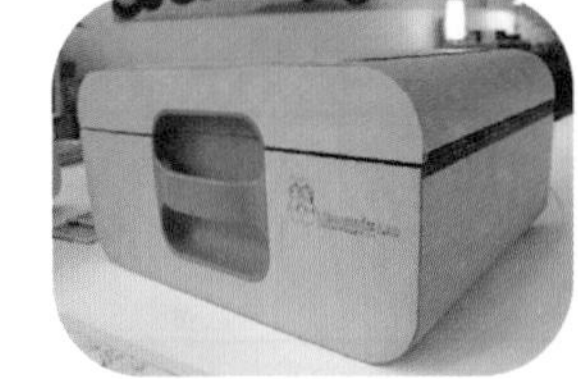

Lunchbox Laboratory

Futurefarmers are a design collective in California whose ambition is to be active in 'cultivating consciousness' by creating design interventions and projects that focus on important issues from food production to water conservation and energy generation. The Lunchbox Laboratory is a collaborative project between Futurefarmers and the Biological Sciences Team at the US National Renewable Energy Lab. It is a toolbox that can be sent to schools to encourage students to help collect and screen algae strains that are most efficient for the production of hydrogen, a renewable energy form. Schoolchildren can collect algae in their local area using the portable lab, test the productivity of these strains, and help rule out the non-productive strains. By sharing their research data with the network of students and scientists affiliated with the Lunchbox Laboratory, young scientists play a valuable role in speeding up the research and development process.

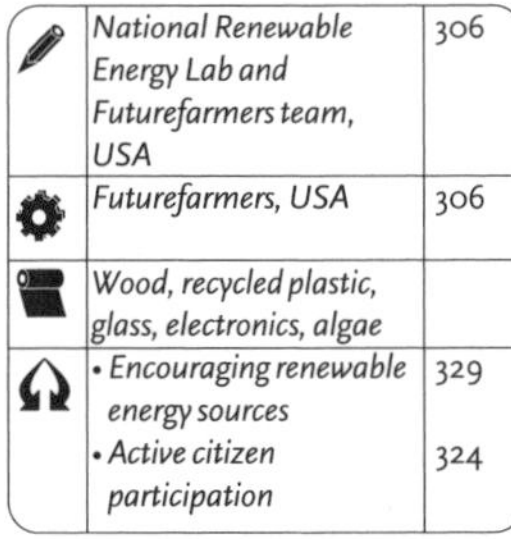

	National Renewable Energy Lab and Futurefarmers team, USA	306
	Futurefarmers, USA	306
	Wood, recycled plastic, glass, electronics, algae	
	• Encouraging renewable energy sources *• Active citizen participation*	329 324

Virtual Water

Designer Timm Kekeritz has turned data from a 2004 UNESCO-IHE report about the 'Water Footprint of Nations' into a visually appealing poster that demonstrates the relationship between what we consume and the amount of 'virtual water' used in the production, distribution and consumption cycles. One side of the poster illustrates the water footprint of selected nations, showing their imports and exports of virtual water, and the other side shows the water content of selected foods and commodities.

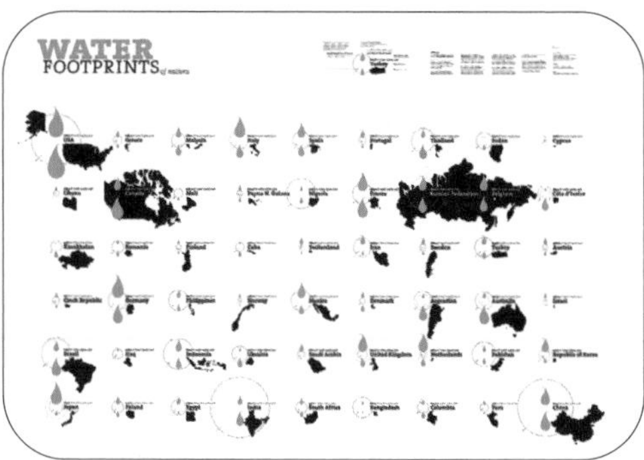

Vegetarians will be happy to note that their diet tends to consume much less virtual water than the omnivores, although even a humble loaf of bread takes 650 litres (171.7 gallons) of virtual water.

Thankfully, apples, oranges and eggs are much more water efficient, although the consumer will also have to factor in where they were produced, the water sources used for production and, of course, the air miles.

✎	*Timm Kekeritz, Germany/USA*	307
⚙	*Timm Kekeritz, Germany/USA*	307
▮	*150g/m² paper, cyan and black printing inks*	
♻	*• Eco-awareness* *• Behavioural change agent*	324 328

Play Rethink

The aim of this eco-design game is to encourage players to creatively rethink how to make everyday objects and services more positive from a social and environmental standpoint. On each turn, a player takes a rethink card, each card offering a specific 'eco-strategy' challenge, such as how to make a toaster that is recyclable or multifunctional, or that is inspired by biomimicry (mimicking nature's designs), or that provides a service to the community. There are blank cards so that players can generate their own challenges. The game creates fertile conversation and elicits individual and collective creativity. You can even upload the results of your game to the Play Rethink website, and share your genius for the greater good. As expected for such a game, the production and printing of Play Rethink also treads lightly.

✎	*Lili de Larratea, Spain/UK*	307
⚙	*Rethink Games, UK*	320
▮	*Recycled and FSC paper, vegetable-based inks, low-impact laminates*	
♻	*• Renewable, recycled and recyclable materials* *• Encouraging citizen participation and co-design*	325 324

Felt Rocks

These unusually shaped forms originate from a process of part discovery, part invention. When optical lenses are polished, wool fluff aggregates like snowballs in the tumbler drum with the polishing wheels. This waste ingredient is subject to a felting process, with steam and pressure, to ensure the entangled wool fibres bond to each other forming a solid core. Individual pieces are selected for further processing, some hand dyed, others left in the natural white/grey tones. The resulting 'rocks', varying in size from 10–15 cm (4–6 in), enable tactile exploration and encourage experimental play.

Designer	*Todd MacAllen and Stephanie Forsythe, Molo, Canada*	308
Manufacturer	*Molo, Canada*	308
Material	*100% pure wool felt*	
Eco	*• Recycled, renewable material* *• Waste reduction*	325 326

Solar helicopter

This could be the future of air travel were it not for the fact that this is a hand-held toy! Powered by a miniature photovoltaic panel, the blades of this little wooden chopper whirl around powered by the sun.

Designer	*Nigel's Eco Store, UK*	318
Manufacturer	*Nigel's Eco Store, UK*	318
Material	*Wood, photovoltaic panel, motor*	
Eco	*• Sustainable at play*	328

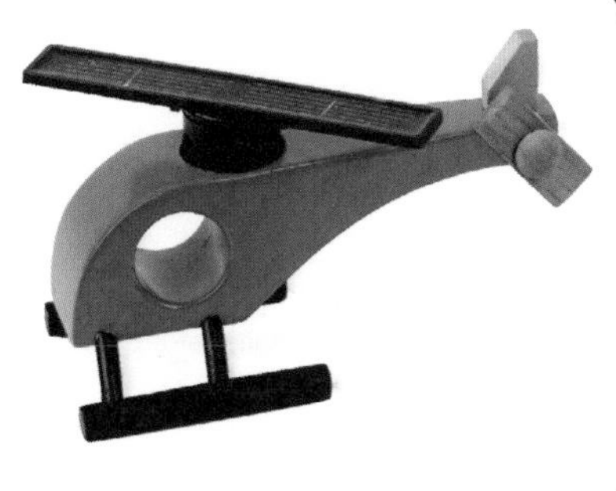

Astrolab

This concept vehicle produces zero emissions and maximizes the capacity of its 3.6 m^2 (38.7 ft^2) of photovoltaic cells to capture large quantities of energy from the sun and power the efficient 16 kWc electric engine, giving a range of 110 km (68 miles) and a top speed of 120 kph (74.5 mph). Special liquid-cooled NiMH Venturi NIV-7 batteries, a high 21% capture ratio of the PV cells, and the ability to accept a charge from the mains, make this a flexible electric hybrid. Technological developments from the Astrolab have contributed to further refinements in Venturi's expanding range of vehicles.

Sacha Lakic for Venturi Automobiles, Monaco	307
Venturi Automobiles, Monaco	322
Various synthetic materials including solar panels	
• *Renewable solar power or electricity from renewable resources*	329
• *Zero emissions if on-board solar power*	328

Eclectic Concept

The Eclectic lives up to its name with an eyebrow-raising design. Labelled the first 'energy autonomous urban vehicle' at its launch in 2006 at the Paris Motor Show, this is definitely a case of form follows function, as the large curvilinear 0.8 m² (8.6 ft²) solar roof panel with 14% yield testifies. It generates on-board power of 7 kWh stored in the 48 V 250 Ah lithium batteries, although mains charging and an optional wind turbine addition can help with powering up. Driven from a centre-positioned seat there are 1-, 3- or 5-seater options depending on whether you want a delivery, utility or passenger vehicle ideally suited to intra-urban transport.

A fully charged battery gives a range of up to 50 km (31 miles) with a top speed of 45 kph (28 mph). The concept has evolved considerably and the more compact Venturi Eclectic is set to be manufactured later in 2009.

✎	*Sacha Lakic for Venturi Automobiles, Monaco*	307
⚙	*Venturi Automobiles, Monaco*	322
	Various synthetic materials including solar panels	
	• *Renewable solar power or electricity from renewable resources*	329
	• *Zero emissions if on-board solar and wind-turbine power*	328

Tesla Roadster Sport

Developed in 2006, this luxury electric zero-emissions sports car, based on a modified Lotus Elise chassis, has blistering acceleration of 0–60 mph in just 3.7 seconds. This is achieved by a 375 volt AC induction air-cooled electric motor delivering 248 hp (185 kW) at 5,000–8,000 rpm. A powerful lithium ion battery pack is fully recharged in 3.5 hours using the High Power Connector Unit and ensures a fantastic range of 220 miles (354 km) giving a fuel equivalent consumption of over 120 mpg (193 kpg). Specification variants ensure you can order a car to suit any personal technological idiosyncrasy as you cruise the California highways in the US state that is leading zero emissions legislation. There's a price to pay for such performance starting from US $128,500, and a waiting list of over 1,000 for 2009 production. Tesla partners with a solar panel installation company, enabling genuine zero emissions by plugging into your very own powerstation.

	Tesla Motors, Inc., USA	322
	Lotus Cars, USA and Tesla Motors, Inc., USA	318
	Various synthetic materials, motor, lithium ion battery	
	• Zero emissions if powered up from renewable sources	328
	• High performance with high efficiency	329

	Sacha Lakic for Venturi Automobiles, Monaco	307
	Venturi Automobiles, Monaco	322
	Various synthetic materials including carbon fibre, aluminium, lithium batteries, electric wheel motors	
	• *Renewable solar power or electricity from renewable resources*	329
	• *Zero emissions if on-board solar power*	328

Volage

Another innovation from Venturi – a French company with a history of racing-car design – the Volage electric sports car was launched at the Paris Motor Show 2008. This carbon-fibre bodied car, with a carbon/honeycomb aluminium chassis, has a compact wheelbase of just 2.7 m (8.9 ft) and weighs in at 1,075 kg (2,370 lb). Michelin Active Drive wheels, each equipped with an individual motor, generator, suspension and braking, provide 55 kW each, a combined power of 220 kW generating a healthy torque of 232 Nm. The 350 kg (772 lb) of polymere lithium batteries have a capacity of 45 kWh enabling 0–100 kph (0–60 mph) in just under five seconds, so this car packs a performance. An 80% recharge of the batteries can be achieved in just four hours with an external recharge booster of eight hours with the on-board charger. This two-seater sports car looks set to challenge other models currently on the market. Expect production models in the near future as the new Venturi factory gets into full production by late 2009.

City Car concept

The dual challenge of rising fuel and energy prices combined with the negative effects of climate change means a radical re-appraisal of our day-to-day mobility and infrastructure requirements is necessary. MIT's Media Lab created the City Car concept as part of a larger collaborative project called Smart Cities. At the core of this concept design is an exoskeleton protecting the passenger cabin and connecting to three robotic electric motor wheels which give 360 degrees rotation of the vehicle. The passenger area includes wafer-thin programmable displays (interior and exterior) and, instead of conventional steering, manoeuvring is achieved by fly-by-wire as in the aeronautics industry. Yet the real coup de grâce is that this cute two-person electric vehicle would be available at transport nodes such as subway, railway or bus stations, patiently waiting to be released from its position in the 'stack' by the next user. Is this the next generation of cars that can chart new success for America's car giants?

✎	*Collaboration between the Smart Cities group at MIT Media Lab, General Motors and Frank O. Gehry*	309
⚙	*Concept design, patent-pending*	
	Various	
	• *Zero emissions (if electricity from renewable sources)* • *Lightweight* • *Use 'on demand' by consumers*	328 326 328

Toyota iQ2

The hybrid Prius has long been the green trailblazer for Toyota, but as the demand for smaller more fuel efficient vehicles increases a new contender has emerged: the iQ2 which uses Toyota Optimal Drive. The attractions are obvious – zero Vehicle Tax Duty (at least in the UK), a carbon dioxide output of 99 g CO_2/km, a frugal consumption of 4.3 l/100 km (67.5 mpg) from the 1.0 l VVT-I petrol engine, and a very sharp turning circle for the 2,000 mm (78¾ in) wheelbase, making for easy parking. Don't expect racing-car performance but as an urban runabout, and for occasional long distance work, the iQ2 makes economic and environmental sense. Those that want a little more performance can look to the slightly larger Aygo, with 4.5 l/100 km consumption and carbon emissions of 106 g CO_2/km.

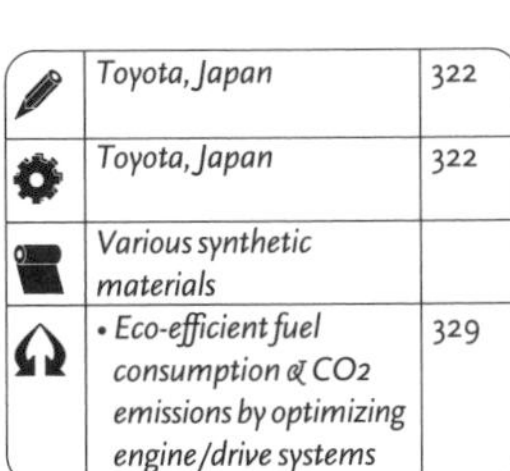

	Toyota, Japan	322
	Toyota, Japan	322
	Various synthetic materials	
	• Eco-efficient fuel consumption & CO_2 emissions by optimizing engine/drive systems	329

Tata Nano

As confirmation of India's status as a global economic power, the country's leading automotive manufacturer, Tata, stunned everyone in 2008 with the launch of the world's most affordable production car. Retailing at just one Lakh Rupees, (Rs. 100,000) or just under US $2,500, this diminutive four-door, rear-engine car can accommodate four adults yet still deliver good ground clearance essential for some of India's roads. The all-aluminium two-cylinder 623 cc, 33 hp fuel-injected petrol engine is controlled by an electronic management system, ensuring that tailpipe emissions are, reputedly, a vast improvement on the ubiquitous two or three wheelers on which many commuters and long-distance travellers rely. This lightweight vehicle with a 4-speed gearbox promises basic affordable motoring for the masses and has been labelled as the eco-friendly 'people's car'. Electric and compressed-air versions are also being developed as a means to reduce air pollution in some of India's leading cities. Expect the demand to be huge. What this means for global car emissions is not quite clear but it harks back to other experiments to give mass automotive mobility to the masses (remember the well-loved Fiat 500 in 1950s Italy). Small, it seems, remains a beautiful concept.

	Engineering Research Centre at Pune, Tata Motors, India	321
	Tata Motors, India	321
	Various synthetic materials	
	• Affordability *• Lightweight construction* *• Moderate eco-efficiency levels*	324 326 329

	Volkswagen, Germany	322
	Volkswagen, Germany	322
	Various synthetic materials	
	• Eco-efficient engine with reduced carbon and particulate emissions *• Efficient fuel consumption*	328 329

VW BlueMotion 2 Polo

Equipped with a 1.4 litre TDI 80PS engine, with a Diesel Particulate Filter (DPF), this version of the Volkswagen Polo offers a combined fuel consumption of 4.5 l per 100 km (62.8 mpg), complies with EURO4 emissions class with just 99 g CO_2/km, and qualifies for zero Vehicle Excise Duty (VED). Compare this with the standard 1.4 TDI diesel version with emissions of 119 g CO_2/km and it is easy to see that the BlueMotion 2 offers a cleaner, leaner car. Available in three or five-door versions, fitted with 14 inch Jerez alloy wheels and capable of 109 mph, this is a very competent performer.

SEAT Ibiza Ecomotive

SEAT is one of the first manufacturers in Europe to exploit the 1.4 TDI 80PS diesel engine in its Ibiza range to just sneak under the psychological threshold of less than 100 g CO_2/km. This low carbon footprint 80 hp engine, coupled with recent vehicle weight savings achieved by subtle re-designs, now permits the latest version the 'new SEAT Ibiza Ecomotive' to achieve less than 98 g CO_2/km and a frugal consumption of just 3.7 litre per 100 km (51.8 mpg) – a world leader in its class. A new Diesel Particulate Filter (DPF) fitted to the engine also means a reduction in fine particulate matter in the emissions. Ecomotive options are now also available for other cars in the SEAT fleet, including the Leon (1.9 TDI engine with 119 g CO_2/km) and family people-carrier Alhambra (2.0 TDI with 159 g CO_2/km). According to the UK government's Act on CO_2 campaign, fuel cost savings of approximately £830 per year are possible on an annual average mileage of 19,310 km (12,000 miles). The SEAT Ibiza Ecomotive becomes a most attractive option for reducing the economic and environmental costs of running a car when the further benefits of zero road tax for this band of CO_2 emissions are included.

	SEAT, Spain	321
	SEAT, Spain	321
	Various virgin and recycled synthetic materials	
	• Eco-efficient engines *• Increase in recycle fraction in new cars* *• Increase in recycling percentages to meet the European Directive 2000/53/CE on End-of-Vehicle Life*	329 325 330

CO_2 emissions of 205 g/km and 173 g/km respectively. Combined fuel consumption figures are 32.8 mpg and 43.5 mpg, but the secret is to ensure that the MPV is full of people. Figures released in 2009 show that Fiat has the lowest average CO_2 emissions for cars sold by a European manufacturer in 2008. These improvements in technology are the key to reducing future emissions, as not everyone will settle for reduced performance.

Fiat Multipla Dynamic

Not all families can squeeze into the smaller eco-efficient car models. The Multipla Dynamic, the second generation of this award-winning MPV, is capable of carrying six passengers in an upright high-strength steel frame. Power unit options include a dual-fuel, 1.6-litre, 16-valve, 103 bhp four-cylinder petrol engine or 120 bhp 1.9-litre MultiJet direct injection turbo-diesel with

✎	*Fiat, Italy*	315
⚙	*Fiat, Italy*	315
▬	*Steel, composite panels*	
♻	• *Reduction in emissions* • *Flexible people and load carrier*	328 328

Honda Insight SE

For many years the Honda Insight, a hybrid petrol/electric car featuring Honda's Integrated Motor Assist (IMA) and backed with 20 years of knowledge, was the main rival to Toyota's Prius. Today there is significant competition for both companies. Nonetheless, the Insight SE (and more luxurious ES and ES-T) four-door family hybrid car maintains a mean performance. The IMA combines a high-efficiency 1.3 litre petrol engine with a brushless DC electric motor and a continuous variable transmission to achieve about 23.2 km per litre (64.2 mpg) and 101 g/km (5.7 oz/mile) CO_2 emissions. During deceleration the batteries are recharged so the car is free from external electricity sources. Acceleration from 0–62 mph is 12.5 seconds and the petrol engine delivers a top speed of 113 mph. New features such as the Eco Assist System, a central digital display that monitors the eco-efficiency of your driving, and an 'Econ' button enable frugal consumption.

✎	*Honda, Japan*	316
⚙	*Honda, Japan*	316
▬	*Various*	
♻	• *Fuel economy* • *Eco-driving information*	329 328

Second generation Prius

Since 1997 the Prius has sold over a million units (first and second generation models), making it the world's most successful hybrid car. Building on this success, the second generation Prius incorporates the Toyota i-Hybrid Synergy Drive®, a sophisticated combined system of energy management that finds the best fit between the drive power of the 1.5 VVT petrol engine, the power of the synchronous, permanent magnet 67 bhp (400 Nm) electric motor, and the balance of power delivery to the wheels with generation of energy by regenerative brake control. The key to this system is the 'power split device' that delivers a continuously variable ratio of petrol engine and electric motor power to the wheels. At start-up the electric motor works alone but as the accelerator pedal is pressed the engine and motor provide power. Hard acceleration is achieved by drawing additional power from the batteries. Deceleration and braking causes the high-output motor to act as a high-output generator driven by the car's wheels. Kinetic energy recovered as electrical energy by this regenerative breaking system is then stored in the high-performance battery. What does all this mean in terms of the car's performance? Combined fuel consumption is 23.4 km per litre (65.7 mpg), resulting in a respectable 104 g/km CO_2 emissions and 0–62 mph in 10.9 seconds. Particulate matter emissions are almost non-existent and CO_2 and NO_x levels are very low. In fact, the hydrocarbon and nitrogen oxide emissions are 80% and 87.5% lower respectively than required by EURO IV regulations for petrol engines. Thanks to this low-pollution footprint the Prius is subject to lower rates of Vehicle Excise Duty in the UK (band B) and is exempt from the Congestion Charge in London, saving up to £2,000 per annum to a daily commuter. This is a proven performer with a credible history.

✎	*Toyota, Japan*	322
⚙	*Toyota, Japan*	322
▮	*Various materials*	
♺	*• Reduced emissions* *• Hybrid power*	328 329

Ford Fiesta ECOnetic

The one to look out for in the range of New Ford Fiestas is the ECOnetic, featuring a 1.6-litre Duratorq TDCi diesel engine complete with Diesel Particulate Filter (DPF). While all the petrol-engined Fiestas exceed 120g CO_2/km, the upper limit for zero Vehicle Excise Tax in the UK, the 1.6 l Duratorq with DPF is rated at 90 hp and yet produces just 98 g CO_2/km with a very efficient fuel consumption (combined cycle) of 3.7 l per 100 km (78.3 mpg). This makes the ECOnetic best in its class for this engine size. Efficiencies are achieved by a specially calibrated engine, lowered sports suspension and aerodynamic improvements, including low rolling resistance tyres. This seems a step in the right direction for this iconic manufacturer. The Duratorq engine is manufactured in the UK using energy from two wind turbines. The Fiestas are constructed in Germany with electricity sourced from renewables, where the aim is to reduce the carbon footprint by up to 190,000 tonnes per annum.

✎	*Ford, Europe*	315
⚙	*Ford, Europe*	315
▬	*Various synthetic materials*	
♻	• *Reduced weight* • *More efficient engines* • *Less particulate matter emissions*	327 329 328

Smart Fortwo diesel

In October 1998 the Smart Car, jointly developed by the German manufacturer Daimler Chrysler/ Mercedes-Benz and Swatch, the renowned manufacturer of colourful watches, passed prolonged safety tests and was launched in the European market. This super-compact car soon achieved cult status and continues to evolve and meet higher environmental standards. Measuring a mere 2.7 m (8 ft 9 in), it is half the length of a standard car so parking is a breeze.

Although the car is designed as an urban runabout, the petrol or diesel engines provide adequate acceleration and top speeds of 133–143 km/h (84–90 mph) to meet the high expectations of today's motorists. The occupants are protected by a strong monocoque steel frame and there is a range of modular panels and interior elements allowing personal customization, catering for changing fashions and facilitating repair. The new Smart Fortwo diesel, with its 799 cc, 45 bhp engine achieves a frugal 85.6 mpg combined cycle fuel consumption, with CO_2 emissions of just 88 g/km.

✎	*Daimler AG, Germany*	314
⚙	*Daimler AG, Germany*	314
▬	*Various*	
♻	• *Fuel economy* • *Improvements in upgradeability and repairability*	329 328

Segway® i2 and i2 Commuter

The Segway Human Transporter (HT) was launched in the US in 2001 and received massive media attention. It's easy to see why; this self-balancing, personal transportation device combines practicality, versatility, mobility and carrying capacity while being a visual curiosity too. Solid-state gyroscopics, tilt sensors, high-speed microprocessors and powerful electric motors link up to create a machine that senses minute shifts in the operator's balance or in the terrain, and immediately adjusts to restore balance. These adjustments happen one hundred times a second when travelling in a straight line or manoeuvring in confined spaces. This intuitive balance is complemented by an equally intuitive stop/start/move system. There is no conventional accelerator, brake or steering: lean forward and the machine moves forward, straighten up and it stops, lean backwards and it begins to reverse. The steering grip on the fixed handlebar is rotated using your wrist. It is possible to turn in situ within the maximum dimension of its footprint, 64 cm (25 in). This sophisticated machine combines innovations from a number of companies: Delphi Electronics provide the controller boards on the main platform with a Texas Instruments digital signal processor; the balance sensor assembly with five gyroscopes is supplied by Silicon Sensing Systems; the 1.5 kilowatt (2 hp) wheel motor, which operates at up to 8,000 rpm, is produced by Pacific Scientific (Danaher); and the two-stage 24:1 reduction gearbox is a joint venture between Segway and Axicon Technologies. There are currently a variety of models, from the i2, i2 Commuter and the i2 Cargo, to off-road models x2 Adventure, x2 Turf and x2 Golf. The Segway® i2 has a range of 38 km (24 miles) – the equivalent of 480 city blocks – on single charge of the rechargeable lithium ion batteries, has a top speed of 20 kph (12.5 mph), and weighs in at 47.7 kg (105 lbs). The i2 Commuter includes a red LED tail-light for safety in traffic, a handlebar bag with room for storage, a lock kit, and an InfoKey controller for security.

✎	*Segway, USA*	321
⚙	*Segway, USA*	321
	Various materials, electronics, motors, battery	
	• Energy reduction *• Improved choice of mobility mode*	329 328

Vectrix VX-1

Vectrix has over a decade of experience in manufacturing high performance electric scooters. The VX-1 is the latest update of the original Vectrix Personal Electric Vehicle with a 28.2 HP (21KW) DC brushless motor delivering 0–50 mph in 6.8 seconds. Requiring just 2½ hours for a full charge, the battery delivers a useful range of between 35–55 miles according to the driving environment. Weighing in at 234 kg (515 lb) this is an all-electric, robust machine that can carry a driver and passenger (with the two-seat option) in comfort.

✎	*Vectrix, USA*	322
⚙	*Vectrix, USA*	322
▬	*Various synthetics and metals, aluminium space frame, brushless motor, NiMH battery*	
♻	• *Zero emissions, zero CO_2 (if electricity from renewable supply)*	328
	• *Eco-efficient power-to-weight ratio*	329
	• *Zero exhaust noise*	328

Brammo Enertia

There are times when the image of an electric scooter just isn't tough enough for the macho culture of the biker. In steps Brammo Motorsports with its daring urban machine, the electric Brammo Enertia motorcycle. Sporting a unique carbon-fibre monocoque chassis, which aerodynamically houses the 3.1 kWh lithium phosphate batteries, the high-output DC 'pancake' motor directly drives the rear wheel through a sealed drive shaft. Weighing in at just 125 kg (275 lb) the Enertia is very nippy, has a top speed of 80 kph (50 mph), and a useful range of 72 km (45 miles).

✎	*Brammo Motorsports, USA*	313
⚙	*Brammo Motorsports, USA*	313
▬	*Carbon-fibre, steel, lithium phosphate batteries, various synthetics and metals*	
♻	• *Zero emissions (when charged by renewable energy sources)*	328
	• *Light weighting*	327

Chameleon range

Despite the well-proven rigidity and ride quality of small-wheeled folding bikes, there are many who prefer the more conventional ride of 24-inch wheels. Airnimal's Chameleon range is a re-examination of folding options to accommodate larger wheels to deliver higher levels of ride stability and comfort without sacrificing portability. These are lightweight high-spec machines, with alloy frames, carbon fibre front forks and 20-speed or 27-speed Shimano gears with Shimano Tiagra or Ultegra STI shifts. The monotube doesn't have any folding components and provides a rigid backbone to the rear triangulated suspension sub-frame and seat post. Weighing about 10 kg (22 lb) without pedals these are 'packable' rather than quick-folding bikes. Modified racer versions, the Ultra and Ultima, offer even lighter options. There are also three possible levels of packing. Level one requires removal of the seat post and front wheel using quick releases, and ensures that two bikes can fit in a car boot. Level two requires removal of both wheels and the handlebar in order to pack it into a 610 x 610 x 280 mm (24 x 24 x 11 in) suitcase. Level three, suitable for air travellers, allows folding the frame and accessories neatly into a carry-on bag of 560 x 360 x 200 mm (22 x 14 x 19 in), although wheels and any saddle bags will have to join the rest of your luggage in the cargo hold. Probably more suited to the long-distance or weekend recreational cyclist they still offer high-spec options for the daily commuter.

✎	*Airnimal Europe Ltd, UK*	312
⚙	*Airnimal Europe Ltd, UK*	312
▮	*Various alloys, carbon fibre, rubber, steel*	
♺	*• Lightweight design* *• Portable*	327 328

Waldmeister bike

This eye-catching engineered frame of laminated wood challenges the convention of steel or aluminium frames. Comprised of over 100 layers of copper beech wood, sourced from German harvested PEFC certified timber, each layer is bonded at right angles with water-based glue, and then the whole frame is pressure treated and protected by three layers of waterproof varnish. In this unique design the upper arc of the frame effectively acts like a leaf spring, while the bottom arc provides extremely good stiffness. All the remaining components are sourced from some of the leading names in bicycle design, known for their performance and durability, including Xentis Cappa carbon fibre wheels from Austria, a Brooks Swift Titanium leather saddle from the UK, Avid Juicy Carbon disc brakes from the US, and a Chris King headset which has been a benchmark of quality for over 30 years. This is a thoroughbred designed to last a lifetime, or two!

✎	*Waldmeister, Germany*	322
⚙	*Waldmeister, Germany*	322
▬	*Recycled and recyclable cardboard. Beech wood*	
♻	• *PEFC and FSC sustainably certified timber*	330
	• *Extremely durable, reliable components*	328
	• *Designed for ease of maintenance and repair*	329
	• *Long-life design*	324
✪	*Red Dot Design award, honourable mention 2008*	

Biomega bicycles

Freedom and vitality are synonymous with that downhill rush on a bicycle. Most people don't need the 18 to 21 gears of a mountain bike as most usage is urban or suburban. What they need is a well-made, reliable, appropriately geared and beautifully designed machine that will provide excellent service and a long life. Enter Biomega, a Danish company that harnessed the skills of top designers and transformed expectations for the cyclist. Marc Newson's MN01 Extravaganza, MN03 Relampago and MN04 Guernica with its non-suspension aluminium fork, are based on a strong, rigid super-formed aluminium monocoque chassis, which is simpler to produce than the normal method of welding metal tubes. The Boston, a streetwise hybrid combining BMX with urban utility features, was designed by Jens Martin Skibsted, and Beatrice Santiccioli's Amsterdam is a shaft-drive bicycle for women.

	Various designers including Marc Newson, Jens Martin Skibsted and Beatrice Santiccioli	308
	Biomega, Denmark	313
	Super-formed, aluminium, metal alloys, rubber	
	• Human-powered *• Durable, tough design*	329 328

Brompton

Folding bicycles are not a new invention but Brompton has manufactured durable products over the last three decades and is probably one of the most popular brands in the UK. The robust 'full-size' steel frame can be folded within 20 seconds to make a compact package that weighs between 9.5–12 kg (20–28 lbs) and measures only slightly bigger than the 40 cm (16 in) wheels. Currently the range includes straight handlebars, traditional bars or sports/touring bars, with options of 1-, 2-, 3- or 6-speed gears. There are options for the normal steel frame of lightweight titanium, depending upon needs. Optional extras allow customization but the design remains fundamentally little changed since its inception, making it less prone to the whims of fashion. Owning a robust, well-tested and versatile folding bicycle allows you to cut overall journey times by combining cycling with public or other private transport. Folding bicycles make a viable contribution towards a more integrated and sustainable transport system.

	Brompton Bicycle Ltd, UK	31
	Brompton Bicycle Ltd, UK	31
	Rubber, steel or titanium, plastic	
	• Multifunctional *• Durable*	32 32

Feetz®

While individual folding bicycles have been around since the 1970s, Feetz developed the concept further to create a range of versatile three- and four-wheeled cycles that double as strollers and shopping carts. The original Feetz model has a tough steel frame that supports the '3 dynamic front wheel suspension®' giving handling akin to a two-wheeled bike and providing support for children or goods riding in the front basket. When folded, the length of the bike is reduced from 187 cm to 110 cm (73½ to 43⅓ in), allowing it to be used as a stroller or shopping cart. Each wheel is braked, fitted with low rolling resistance tyres and there's a handbrake for parking. A four-wheeled BakFeetz has additional carrying capacity at the back of the cycle, and another four-wheel version, the SupportFeetz, is slightly narrower, but provides even greater stability and is suitable for an older generation. There's also a five-year warranty on the frame as the manufacturer is confident in this durable, robust design.

✎	*Feetz, the Netherlands*	315
⚙	*Feetz, the Netherlands*	315
	Steel, wood, aluminium or wicker basket	
	• Multifunctional bicycle	328
	• Robust, durable	328
	• Easily maintained and repaired	329

Strida 5.0

Unfolding the Strida 5.0 takes ten seconds and immediately reveals its radical triangular frame, a departure from the typical arrangements in other folding bicycles. It weighs just 10 kg (22 lb), with tubes of 7000 Series aluminium and wheels and other components made of glass-reinforced polyamide, a durable, lightweight polymer. Folded size is a convenient 114 x 51 x 23 cm (45 x 20 x 9 in), and the latest version is fitted with 16-inch alloy wheels and disc brakes. A conventional chain is replaced by a Kevlar Greaseless Belt Drive over low-friction polymer cogs, making for an oil-free and low-maintenance bicycle. Tyre and belt repairs are facilitated by the offset frame-wheel arrangement. Eye-catching and unique, the Strida has found a cult audience.

	Mark Sanders, UK	309
	Ming Cycle, Taiwan	318
	Glass-reinforced polyamide, aluminium, rubber, stainless steel	
	• *Reduction in materials used* • *Ease of maintenance*	325 328

Windcheetah

A cruciform frame enables the rider to adopt a low centre of gravity, which, when coupled with carbon-fibre fairing, provides very efficient aerodynamics. Pinpoint accuracy of steering is achieved by means of a unique joystick system that gives good stability in cornering. Lightweight materials and precision engineering make this the Rolls Royce of recumbents. The efficiency of the design has attracted interest from courier and local delivery companies who wish to develop zero-emissions transport policies for urban areas.

	Advanced Vehicle Design, UK	31
	Advanced Vehicle Design, UK	31
	Metal alloys, rubber, carbon fibre, Kevlar	
	• *Human-powered*	32

Delite

A standard frame fitted with a range of modular components can suit touring or racing bicycles or a hybrid. The Delite Urban has an 8-speed hub gear, and the Delite Hybrid HS is fitted with a 500W motor enabling speeds of up to 45 kph. Suspension is provided by front fork dampers and a damped rear sub-frame assembly.

	riese und müller GmbH, Germany	320
	riese und müller GmbH, Germany	320
	Various	
	• *Modular design*	328

Avenue

Ironically 'Mountain' or 'All-terrain' bikes are the dominant type of bicycle in many cities and urban areas. Yet these bicycles are rarely suited to the demands of cycling in work clothing or transporting young children and heavy loads. Avenue has been designed by riese und müller to take the best features of classic Dutch bicycles but to conceal the latest suspension and transmission features to improve ease of use. This means a swinging rear arm and fully enclosed chain to maximize comfort and cleanliness. Standard specifications include Shimano Nexus 8-speed gears and Aluminium 7005 T6 tubing. The Avenue City weighs 18.8 kg (41.5 lb) and the Avenue Light is 17.4 kg (38.3 lb). This is an intelligent re-appraisal of our cycling mobility needs.

✎	*riese und müller GmbH, Germany*	320
⚙	*riese and müller GmbH, Germany*	320
▮	*Various metals, rubbers, plastics*	
♺	*• Multifunctional urban cycle and load carrier*	328

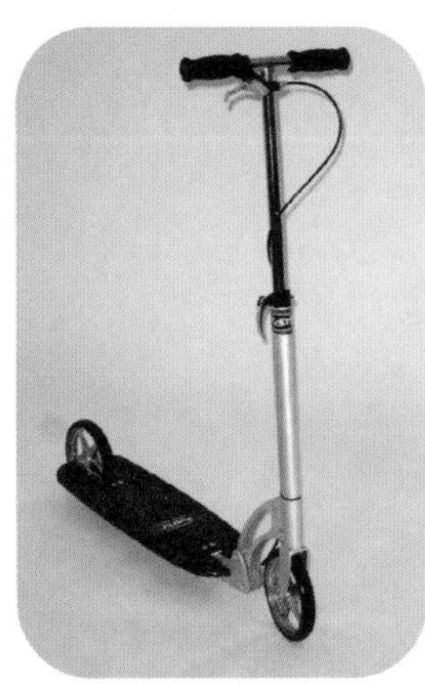

Xootr Cruz

Skateboard culture meets the bicycle in this resurrection of the old push scooter. Lightweight aluminium frame, cast wheels and a low-slung laminated birchwood deck ensure manoeuvrability and stability. This vehicle is very portable, weighing just 4.5 kg (10 lb) and folding to a package less than 800 mm (31 in) long.

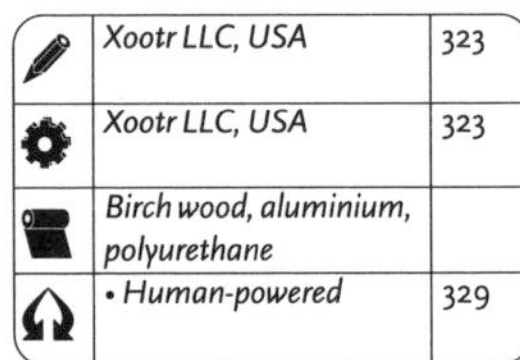

✎	*Xootr LLC, USA*	323
⚙	*Xootr LLC, USA*	323
▮	*Birch wood, aluminium, polyurethane*	
♺	*• Human-powered*	329

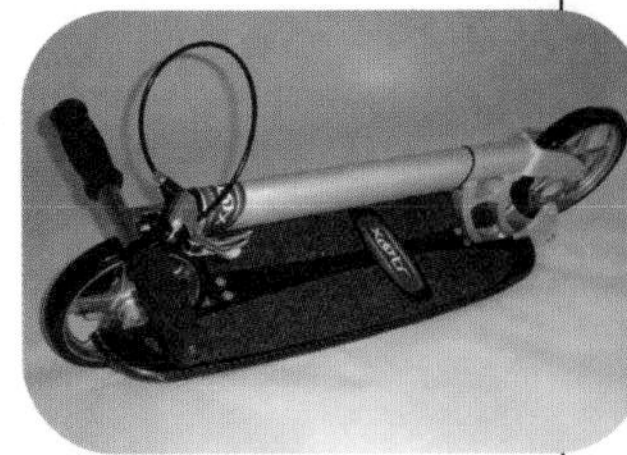

X-bike, Eurobike, City bike, Powatryke Cruiser

Powabyke is a range of robust, dependable electric-assist bicycles, with 5- to 24-speed gear options to suit all ages, commuters and recreational users. The Eurobike has been in production since 1999 and now offers a 6-speed Shimano gearing with a lightweight 36-volt lithium ion 15 amp battery pack. The distance range without pedal assist varies according to the model from 16–36 km (10–20 miles).

✎	*Powabyke Ltd, UK*	320
⚙	*Powabyke Ltd, UK*	320
▭	*Various*	
♲	*• Hybrid human-/ electric-powered transport*	329

Pedalite

Three-quarters of all bicycle accidents happen at a road junction when cyclists are stationery or less visible to other road traffic. Pedalite battery-free pedal lights alert other road users by continuous flashing of a clear (forward), amber (side) and red (rear) LED encased in the pedal providing warnings throughout 360 degrees. Critically, these lights flash when the cyclist is stationery or freewheeling so vehicle drivers are alerted at all times. Fitting couldn't be easier – simply screw the pedal on to each cog shank.

✎	*Pedalite International, UK*	319
⚙	*Pedalite International, UK*	319
▬	*Steel, plastic, LEDs, dynamo*	
♻	*• Increased safety, health and well-being for cyclists* *• Renewable, human-energy powered*	328 329

SRAM 9.0 sl

SRAM manufactures brakes, gears and gear shifts to high standards of aesthetics and functionality, using between 30 and 50% recycled content for many of the sub-components, which can be disassembled for pure-grade recycling in the future.

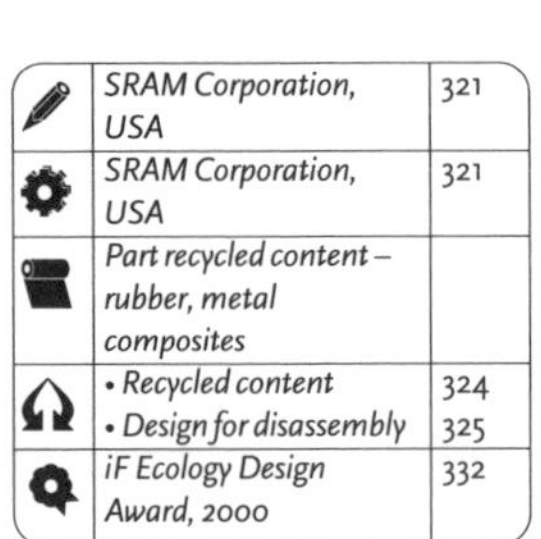

✎	*SRAM Corporation, USA*	321
⚙	*SRAM Corporation, USA*	321
▬	*Part recycled content – rubber, metal composites*	
♻	*• Recycled content* *• Design for disassembly*	324 325
✿	*iF Ecology Design Award, 2000*	332

Baglite and Ankelite

Improving visibility of the cyclist is essential to encourage more people to take to their bikes and leave their cars at home. Baglite is a harness that can fit over a backpack and everyday cycling clothing giving you a superbright red LED module at the rear and two white LED modules to your front. Charging of the LED modules, via the integral photovoltaic cell, takes one hour in direct sunlight, up to three hours in a cloudy sky and can even be slowly charged under artificial light. Three lighting modes can be selected – constant, flashing or off – with three hours output in flashing mode. The Ankelite, with a single amber LED module, fits around the ankles or arms by a Velcro fastening. Combining the two systems ensures that the cyclist is visible through 360 degrees, ensuring a higher level of safety than mono-directional lighting.

✎	*Pedalite International, UK*	319
⚙	*Pedalite International, UK*	319
▬	*Plastic, LEDs, photovoltaic cell, integrated circuit*	
♻	*• Increased safety, health and well-being for cyclists* *• Renewable, solar-energy powered*	328 329

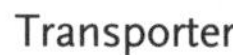

Transporter

As the costs of transport generally spiral ever upwards, many of us could be (re)turning to human power for the daily commute or shop. Dominic Hargreaves' concept combines both options with this backpack/ scooter. As with any multifunctional product the ease of converting from one function to another is critical. This is achieved by unlocking the skateboard located in the rear of the backpack, then pulling the steering handle that neatly telescopes upwards. Reversing the motions is just as easy. Ideal for speed shopping or leaving the more pedestrian commuters wondering what flashed past them.

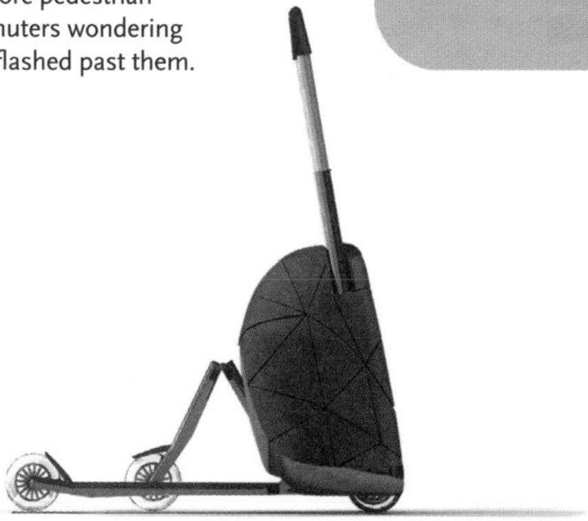

	Dominic Hargreaves, UK	306
	Prototype	
	Various synthetics	
	• Multifunctional *• Encouraging local shopping*	328 327

Curraghs

Imagine making a lightweight, versatile rowing or sailing boat using traditional methods for less than £150. Look no further than the Irish curragh 'skin-boats', whose pedigree stretches back to the early Neolithic period. All curraghs are built upside-down, with the willow withies or hazel rods stuck into holes in a wooden gunwale. The sides are shaped by weaving in a pattern called French randing, then the withies are bent over and tied with tarred hemp twine to form the keel-less base of the boat. Today, the frame structures are covered with heavy calico (canvas) tacked to the gunwale and covered with tar for waterproofing. Until the 19th century the traditional covering was cowhide. At least 12 different types of curragh are found on the Atlantic seaboard of Ireland, varying from Bunbeg and Dunfanaghy curraghs of 3–6 m (10–20 ft) length, to the great 8 m- (26 ft-) long, four-man, Kerry naomhóg capable of taking loads of nearly two tonnes. Usually the curraghs are rowed: each pair of oars are thin-bladed with a wooden counterweight inserted over a wooden pin. Occasionally a simple lugsail is hoisted on a detachable mast. Curraghs are reliable sea-going boats but are also at home in estuarine waters or inland lakes. Designs evolve as makers innovate with local materials and respond to local conditions. Such boats are a good example of economic, environmental and social sustainability.

✎	*Traditional and hobby designer–makers, Ireland and UK*	
⚙	*Serial one-offs*	
▮	*Willow, hazel, pine, deal, canvas, tar, hemp twine*	
♻	*• Renewable materials* *• Biodegradable* *• Local sustainable design* *• Retention of traditional craft skills*	325 325 325 325

Greenlight Eco-surfboard

✎	*Greenlight Surfboard Supply, USA*	316
⚙	*Greenlight Surfboard Supply, USA*	316
▮	*Bamboo, bamboo fibreglass, biopolymers, recyclable EPS*	
♻	*• Renewable materials* *• Reduction in use of VOCs and other toxins*	325 326

Surfing appears to be a fairly innocuous activity, but the truth is all that jet-setting about and the continual trashing of synthetic polymer boards and wet suits tells a different story. But there's a way forward, especially for American surfers... Greenlight Surfboard Supply can hand-craft you a surfboard from laminated bamboo stringers and veneers using special 'knitted' bamboo fibreglass around a recyclable EPS core. Even the leash plug is made of a corn-starch biopolymer. All the raw materials can be bought by aspirant board makers and there's a new range of bamboo fins. It's not a completely toxin-free process as the Resin Research Epoxy does exhibit low levels of VOCs, but the Ecohesive used to laminate the bamboo stringers has less than 18 times lower VOC levels than EPA and doesn't contain any ozone depleting chemicals.

Dopie sandals

Cheap flip-flops bought at a discount seaside shop rarely last more than a season before being destined for landfill, so why not consider a lightweight sandal with good protection, flexibility and durability. Dopies are made to a high ergonomic standard by the ethically driven Terra Plana shoe manufacturers, and are made of a mixture of EVA and rubber. Production facilities try to source materials as locally as possible.

Designer	*Terra Plana, UK*	322
Manufacturer	*Terra Plana, UK*	322
Materials	*EVA, rubber*	
Eco	*• Eco-design strategies* *• Lightweight but durable* *• Simple construction* *• Improved ergonomics*	324 325 327 328

Veloland

Part of a project called Switzerland Mobility, Veloland is an information provider and service company responsible for a network of over 6,300 km (4,000 miles) of national and regional cycling trails in Switzerland. Maps, guides and a website provide wide access to information. Trails have been linked with public-transport networks and bicycle rental at railway stations, and with 60,000 km of hiking paths.

Designer	*Veloland Schweiz, Switzerland*	310
Manufacturer	*Various*	
Materials	*Various*	
Eco	*• Encourages alternative modes of transport (integrating cycling with public transport)*	327

NIGHTEYE®

This lightweight headlamp is fitted with the patented Ultralight using a low-energy Xenon bulb with a special reflector to provide a quality light source.
A rear red LED personal safety light fits to the back of the headband.
All components, including the polycarbonate casings, clip together and so can be separated for recycling.

Designer	*PROFORM Design, Germany*	309
Manufacturer	*Nighteye GmbH, Austria*	318
Materials	*Polycarbonate, elastic, LEDs, Xenon bulb*	
Eco	*• Low energy consumption*	329
Award	*iF Design Award, 2000*	332

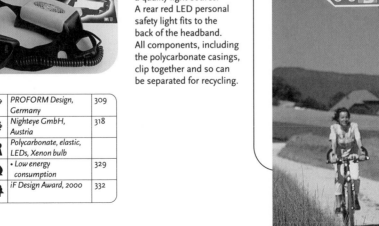

Kelly Kettle®

The ingenious design of the Kelly Kettle® holds any natural fuel (twigs, heather, dry grass, pine cones) in the bottom fire basket on which the kettle sits, making it a safer option and easier to light in extreme weather. Flames funnel up the inside

of the kettle providing rapid heating of the water. Once you've heated your water, you can cook on the embers. Available in three capacities, 1.42, 1.0 and 0.57 litres, a cork stopper on a chain contains the water and doubles as a guide to steady the wooden handle when pouring. This kettle will last a lifetime.

✎	*Kelly Kettle Company, Ireland*	317
⚙	*Kelly Kettle Company, Ireland*	317
▬	*Aluminium, brass, cork, wood*	
♺	*• Lightweight yet durable* *• Eco-efficient consumption using local natural renewable fuels*	326 329

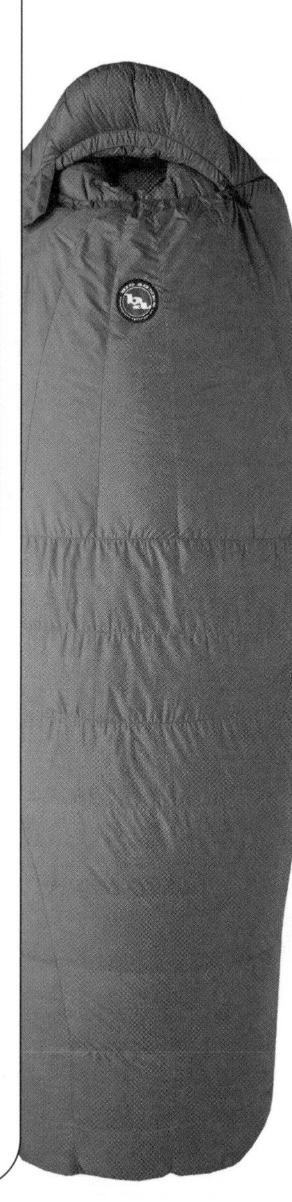

Big Agnes Battle Mountain

The Battle Mountain is a unique sleeping bag with an integral pad that slides into a sleeve at the bottom of the bag so you'll never slide off the pad when you turn in the night. Rated down to -15° F (-26° C), the 600-fill goose down provides insulation on the top of the bag while the pad is insulated by a synthetic (man-made) fill derived from bamboo called Bamboo Charcoal Synthetic Insulation, formed by spinning bamboo charcoal into a fibre. A protective shell fabric of water-repellent treated nylon microfiber keeps everything snug and dry. Weighing in at 4 lb 10 oz to 5 lb 1 oz (2.1 to 2.3 kg), it compresses into a package 30.5 x 23 cm (12 x 9 in). Not the smallest sleeping bag on the market but then it really is a bag and integral mat. This is a real attempt to explore how more renewable materials can be integrated into bag design.

✎	*Big Agnes, USA*	313
⚙	*Big Agnes, USA*	313
▬	*Goose down, synthetic bamboo fill, nylon*	
♺	*• Increased proportion of renewable materials* *• Improved ergonomics*	325 328

Patagonia Synchilla® range

Patagonia produced the original Synchilla products in 1993 using post-consumer recycled (PCR) fleece, with each garment saving 22 two-litre PET soda bottles from landfill. Patagonia has consistently increased its use of recycled materials and today the new Synchilla range includes between 85–87% recycled fraction in the 100% polyester fleece. Snapping or zipping together, these garments are well made, lightweight, and durable. When they do wear out, consumers can return them to Patagonia under their Common Threads Recycling Program and ensure that the old garment is reborn again.

✎	*Patagonia, USA*	319
⚙	*Patagonia, USA*	319
▮	*Recycled polyester fleece, zippers, trim*	
♻	• *Recycled materials* • *Encouraging closed loop production* • *Take-back programme*	325 326 329

Patagonia Eco Rain Shell Jacket

Patagonia was one of the first global manufacturers of outdoor clothing to use significant quantities of recycled fibre in their range. This trend continues today with these lightweight mens and womens jackets which are made from 100% recycled polyester, originating from post-consumer soda bottles, second-quality fabrics and worn-out garments. The result is a two-layer waterproof yet breathable 3.5 oz (99 g) fabric with a mesh lining and microfleece details. Suitable for everyday use in the rain and wind, in the city, on the trails or canoeing down a river, you'd never guess the Eco Rain Shell Jacket is 'second time around'. Return a jacket to Patagonia at the end of its life, and it will be recycled yet again.

✎	*Patagonia, USA*	319
⚙	*Patagonia, USA*	319
▮	*Recycled polyester, zippers, water-repellent finish*	
♻	• *Recycled materials* • *Encouraging closed loop production* • *Take-back programme*	325 326 329

Icebreaker base layers and clothing range

Wool fibre from Merino sheep raised on selected farms in the Southern Alps in the South Island of New Zealand provides the material for Icebreaker's base layers. Merino wool fibre is soft, has excellent breathability, wicks away moisture, insulates well, and also contains natural anti-bacterial agents to reduce odours. Icebreaker creates a range of base and over-layers for insulation, or for wind-proofing or protection against rain/snow. Each user can build up a personal specification by mixing different layers. The company's ethos embeds sustainability and transparency. Garments have a 'Baacode' that enables you to trace your purchase right through the supply chain so you can 'virtually' meet the farmer who tended the sheep. As Icebreaker is now a global company, manufacturing now takes place in New Zealand and offshore, but all manufacturers have to comply with a strict ethical code, including no child labour, and most are already engaged with ISO standards or are actively working towards them. Most of the clothing range is certified to comply with Oeko-Tex European environmental standards for textiles.

	Icebreaker, New Zealand	316
	Icebreaker, New Zealand	316
	Merino wool fibre	
	• *Certified clean production systems* • *ISO 14001 certified* • *ISO 9001 quality assurance certified* • *Ethical code of production & supply chain* • *Sourcing bio-regional materials* • *Durable*	326 330 330 326 325 324

Eco-Blue Ski Jacket

Technological developments in sportswear and equipment continue at a meteoric pace, but there's been a consistent difficulty in recycling composite, laminate fabrics because intimate contact between the layers prevents appropriate waste stream separation. SympaTex Technology has overcome this challenge by creating Ecocycle-SL, a synthetic membrane using recycled polyester from PET bottles (one of the easiest synthetic materials to recycle) that is bonded to an outer and inner fabric of polyether copolymer without using solvents.

This triple membrane is very breathable, anti-allergenic and non-toxic, ensuring that it is certified to the Oeko-Tex Standard. The tailored and practical ski jackets are made in Turkey by Barco Texstil, an environmentally aware producer.

	SympaTex Technologies, Germany	321
	Laminate: SympaTex Technologies, Germany; jacket: Barco Texstil, Turkey	321
	SympaTex® recyclable laminate Ecocycle-SL	
	• Use of recycled materials *• Recyclable at end-of-life* *• Elimination of toxic solvents or additives*	325 329 326
	Winner of ispo European Ski Award for winter 2007/2008	

Helix-20 and ReSource Series bags

Osprey Packs, USA & Europe	319
Osprey Packs, USA & Europe	319
Recycled PET fabric, zippers, trim	
• Increased recycled content	325

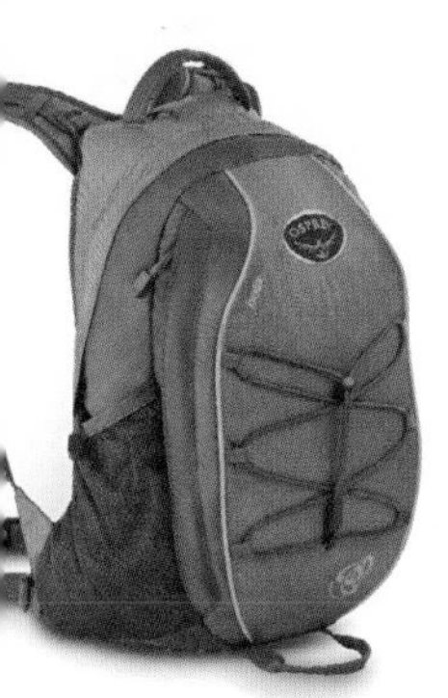

The ReSource Series of backpacks, shoulder bags and courier bags utilizes recycled fabrics. The Helix-20 – a simple daypack – is made of 450D recycled PET fabric, binding tape, webbing and mesh plus a 60% recycled Spacer mesh giving an overall recycled content of 78%. That's an impressive achievement and signals the direction that bag manufacturers in general should pursue. Weighing just 0.62 kg (22 oz) this is a commodious 17-litre capacity bag that is suitable for urban use or weekend trail hiking. Aside from the numerous internal compartments there are external mesh pockets, a bungee retainer and even an ice-pick loop.

Howies Scramble Ripstop Trousers

Howies is a UK company servicing the snowboarding, surfing and globetrotting generation with well-made, tough, organic fibre clothing. The Scramble Ripstop Trousers, made from organic cotton ripstop, typifies Howies' philosophy and approach with a design that embraces a loose slouch fit, multiple pockets, button fly and strong stitching. Most garments are made overseas but the company has a strong ethical approach and pledges 'to give 1% of its turnover or 10% of pre-tax profits (whichever is the greater) to grass-root environmental and social projects'.

	Howies, UK	316
	Howies, UK, with garments made in Portugal, Turkey and China	316
	Organic cotton	
	• Renewable organic cotton *• Durable garments* *• Supports grass-roots initiatives*	325 324 327

Terra Grass Armchair

A subtle merging of man and nature is embodied in this witty outdoor seat reminiscent of some mini Bronze Age burial mound. The structural framework is provided by corrugated cardboard to which locally sourced soil is added and grass seed applied. Just a few weeks later succulent grass covers your very own green throne for the garden.

✎	N Fornitore, Italy	308
⚙	Purves & Purves, UK	320
▮	Corrugated cardboard, grass seed, soil	
♻	• Renewable and compostable • Locally sourced materials	325 325

Grow Your Own Greenhouse

There is an on-going debate about issues of food miles and food security. Coupled with the realization that over 50% of the human population now lives in an urban environment, it seems that localizing food production is an important priority. Making it easy to grow vegetables, even in the tiniest of space, is the key aim of this project by Jochem Faudet. Water is harvested via the timber-frame supported Perspex funnel at the top of the greenhouse, then stored in the 300 litre base tank, and ultimately pumped to the plants at specified times. There's an inbuilt temperature and ventilation control, a shading system, a range of pot sizes to accommodate different kinds of plants and even a set of tools to assist with cultivation. Time to start sowing those seeds today.

✎	Jochem Faudet, UK/ the Netherlands	305
⚙	Prototype	
▮	ABS rotation moulded water tank, Canadian red cedar, Perspex, water pump, timer	
♻	• Encouraging self-sufficiency • Easy to assemble, disassemble and repair	327 329

Eglu

The innovative company Omlet is intent on encouraging householders to pop a couple of chickens in their back gardens so there are fresh eggs every day. The daunting task of becoming a chicken farmer, albeit small scale, is facilitated by the well-designed Eglu that can house two to five bantam chickens. With easy-to-remove components for cleaning, and accessories that include a shade-protected chicken run, the Eglu makes chicken husbandry easy. The chicken chauffer who delivers your Eglu, complete with reliable organically reared chickens, will show you the basics.

✎	Omlet, UK	319
⚙	Omlet, UK	319
▮	Various polymers	
♻	• User-friendly design to encourage organic home food production • Polymers are 100% recyclable • Web forum for detailed advice	330 325 325

Terracotta incense burner

Technology from the Middle Ages is not often revered these days, but many design solutions from then have yet to be bettered. This terracotta incense burner originating from France traditionally burns cade wood incense from the small juniper-like shrub found in southern France. Cade is a natural air purifier and insect repellent that also actively absorbs bad aromas. Pure cade wood is easily mixed with other fragrant powders or flower parts such as lavender, rosemary or orange. These mixtures are completely hazard-free, unlike deadly aerosol insecticides. Two versions of the burner are produced, an indoor one just 110 mm (4⅓ in) diameter, or a larger outdoor burner 255 mm (10 in) diameter. Either version is an ideal way of 'transporting' oneself to the verdant aromatic slopes of Les Alpes Maritimes or Provence.

	Indigenous makers, France	
	Natural Collection, UK	318
	Terracotta, herbs	
	• *Abundant geosphere and renewable materials* • *Safe, non-toxic*	325 328

E-TECH II

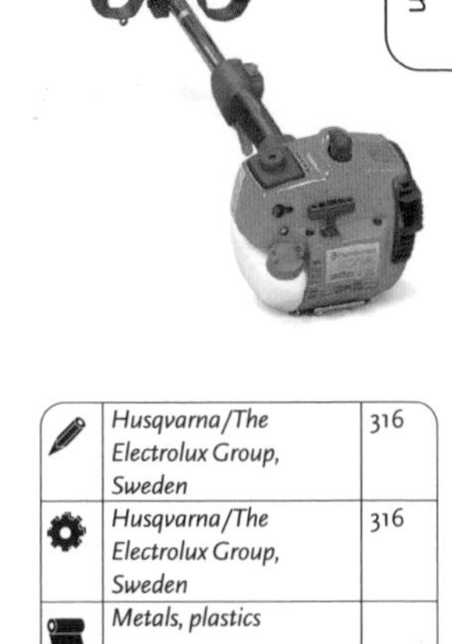

The E-TECH II two-stroke motor for chainsaws and trimmers combines efficient power production with a new catalytic converter, ensuring that these motors meet the world's strictest standard for emissions for motorized hand-held garden and forestry equipment, the 1995 California Air Resources Board (CARB) standard. Electrolux, the world's largest manufacturer of chainsaws, has also forged partnerships with petroleum companies to develop fuels that reduce emissions. For example, the Finnish company Raision offers a vegetable chain oil for chainsaws through Husqvarna.

	Husqvarna/The Electrolux Group, Sweden	316
	Husqvarna/The Electrolux Group, Sweden	316
	Metals, plastics	
	• *Reduction in emissions* • *Reduction in consumables*	328 329

FSC Fencing

A decade ago most consumers had not heard of the Forest Stewardship Council, the organization dedicated to encouraging sustainable management practices in forests, timber producers and manufacturers. Today the FSC endorse a wide range of products. These individually designed fencing panels are manufactured from planed, pressure-treated European timber from FSC certified wood. Pressure-treating wood by injecting preservatives is the industry's standard way of extending durability, since these products are made from fast-growing softwoods, mainly conifers. In the longterm, the pressure-treatment cycle could be omitted if larger quantities of hardwood, such as oak (*Quercus spp.*) and sweet chestnut (*Castanea sativa*), are grown in small roundwood plantations that are regularly harvested or coppiced. Sweet chestnut, often used as fencing posts, is renowned for its durability.

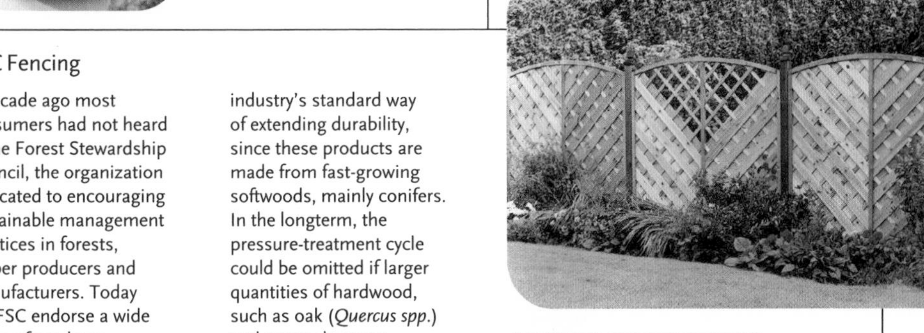

	Natural Collection, UK	318
	Natural Collection, UK	318
	Timber	
	• *Certified timber from sustainably managed forests*	325

Can-O-Worms

A compact self-assembly series of nested circular trays, made from 100% post-consumer recycled plastic, is supplied with a coir fibre block. This block is moistened and broken up, then placed in the bottom tray to provide 'bedding' for a colony of native composting worms. As a tray is filled with household or garden waste, another is added to build up the stack. The worms migrate up and down the stack through the mesh in the bottom of each tray, digesting waste and turning it into compost.

✎	*Reln, Australia*	309
⚙	*Reln, Australia, with Wiggly Wigglers, UK*	323
▬	*Plastics, live worms*	
♻	*• Encourages local biodegradation of waste* *• Recycled and renewable materials*	329 325

Compost converter

Manufactured from 95% recycled plastic, the 220-litre (48-gal) capacity Compost converter is a strong, rigid bin with a wide top aperture, which facilitates disposal of biodegradable domestic and garden waste and its conversion into compost, and a wide hatch at the bottom for removing mature compost. Blackwall make a range of compost bins from 200 litres to 708 litres (44–156 gals) capacity for domestic use, from injection-moulded, flat-pack, wood-grain-effect, recycled thermoplastic bins to blow-moulded, cylindrical bins mounted on a tubular galvanized-steel frame, permitting aeration of the compost by regular inversion or tumbling. Bins are guaranteed for at least ten years.

✎	*Blackwall Ltd, UK*	313
⚙	*Blackwall Ltd, UK*	313
▬	*Plastics, tubular galvanized steel*	
♻	*• Encourages local composting* *• Recycled materials*	329 325

Adjustable spades and fork

An adjustable telescopic shaft provides a variable length of 105 cm to 125 cm (41–49 in) to accommodate people of different heights. Anyone who has done any serious gardening will know that inappropriate tools can cause back and posture problems. These well-made garden tools ensure efficient and healthy work practices.

✎	*Fiskars, Finland*	315
⚙	*Fiskars, Finland*	315
▬	*Metal, plastics*	
♻	*• Universal design* *• Improves health*	328 328
✪	*iF Design Award 2003*	332

MicroBore

Porous piping placed in or on the soil surface provides a means for the precise delivery of water direct to the plant root zone. The flexible, rubber-based hose made from shredded, recycled tyres is perforated with holes, allowing water to trickle into the soil. Pipe diameters vary from 4 mm or 7 mm (⅙ or ¼ in) for the MicroBore, which is ideal for watering window boxes and office plant displays, to 13 mm, 16 mm or 22 mm (½, ⅓ or 9/10 in) for the HortiBore and ProBore, which are for commercial horticulture and landscaping. System accessories allow you to customize the irrigation system and permit the use of stored rainwater with in-line filters, taps, tee-junctions and end-stops.

	Porous Pipe, UK	320
	Porous Pipe, UK	320
	Recycled tyres	
	• *Water conservation* • *Recycled materials*	329 325

Versodiverso

Nicolas Le Moigne was winner of a design competition entitled 'RE-think + RE-cycle' organized by the online webzine Designboom in 2005. His well-executed watering can design was quickly snapped up by an Italian manufacturer of products for the home, Viceversa, who continue to supply it. No wonder: this is a crisp piece of design thinking that transforms a redundant product into something useful and beautiful. It even has the potential to encourage people to love their plants more too.

	Nicolas Le Moigne, Switzerland	308
	Viceversa, Italy	322
	Plastic	
	• *Extends lifespan of used water bottle* • *Mono-material* • *Simplicity with functionality*	329 325 328
	2005, Designboom competition winner	

Automower™ Solar Hybrid

This is an extraordinary device that offers the lazy gardener the ultimate mowing machine. Weighing just 10 kg (22 lb), this hybrid solar/electric machine is powered from the mains and/or recharged by the solar panels. It can chomp its way through up to 90 m^2 of lawn an hour working on uneven grass and inclines up to 35%. The 2.2Ah NiMH battery is recharged from the mains in just 45 minutes, giving up to 60 minutes of mowing. Amazingly, if the battery needs more power it makes its own way back to the charging station. The Automower™ is capable of working in the rain and around pets, is very quiet in comparison to regular lawn mowers, and is even fitted with an anti-theft alarm and a pin code lock.

	Husqvarna, Sweden	316
	Husqvarna, Sweden	316
	Photovoltaic cells in solar panel, brushless DC motor, ASA plastic, NiMH battery	
	• *Hybrid solar/electric power (zero emissions if electricity source is renewable)* • *Lightweight*	329 325

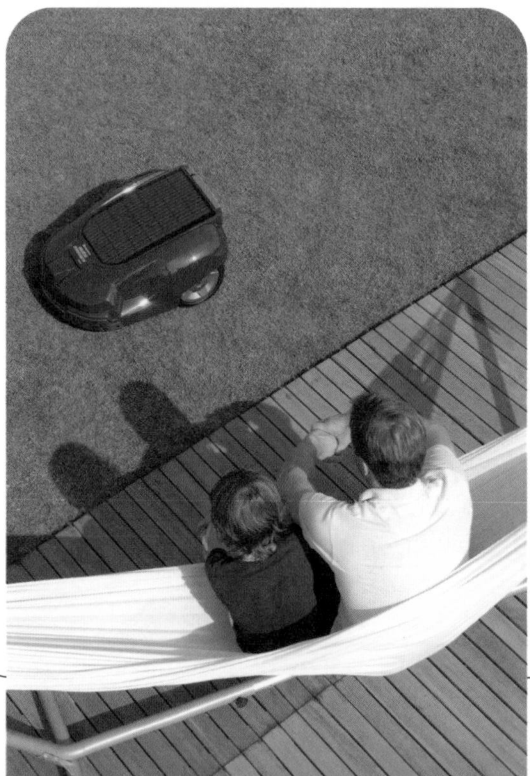

MUJI CD player

Consistent with MUJI's philosophy of reductionism and minimalism, this CD player has a pared down form while simultaneously creating an almost surreal piece of wall sculpture. This intriguing object does some semiotic gymnastics. Pull the power cord and watch the CD spin like a fan of an air-conditioning unit. But it is more than a whimsical design. Speakers are integral to the lightweight, 550 g (19 oz), player so every effort has been made to minimize the consumption of materials.

✎	*IDEO Japan, Naoto Fukasawa and MUJI, Masaaki Kanai, Japan*	318
⚙	*MUJI (Ryohin Keikatu Co. Ltd), Japan*	318
	Various polymers, metals and electronics	
	• Lightweight design *• Reduction in materials used*	327 325
	iF Design Award, 2002	332

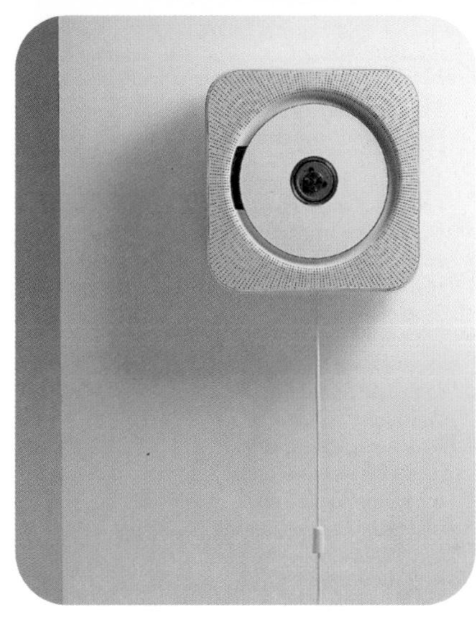

iPod Shuffle

This third generation iPod Shuffle is remarkably small, at just 45.2 mm (1.8 in) high, 17.5 mm (0.7 in) wide and 7.8 mm (0.3 in) deep, and weighs 10.7 g (0.38 oz) – even less than an AA battery. Compared to the classic iPod, at 140 g (5 oz) this represents over a Factor 10 saving in weight. Such realizations are important as sales of iPods continue to increase: more than 100 million were sold between November 2001 and April 2007. The rechargeable lithium polymer battery enables 10 hours of play, ensuring you can enjoy the 1,000 song capacity. Six music formats are supported including the protected AAC iTunes and MP3/MP3 VBR. The new VoiceOver feature gives the iPod a voice (in 14 different languages) announcing what song is playing, the performer, and the name of the playlist. Controls are conveniently located on the earphone cord enabling full control of this truly miniature device. If Apple continues to improve their product-take back record, enabling further recycling, then this is a true story of 'small is beautiful'.

✎	*Apple, USA*	312
⚙	*Apple, USA*	312
	Various synthetic materials, electronics	
	• Light weighting *• Efficient battery for long play time*	327 329

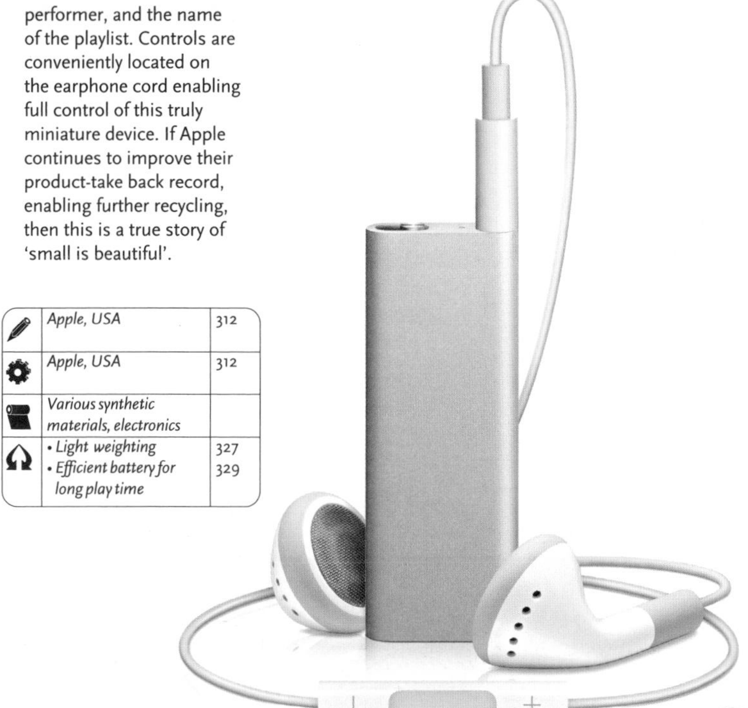

Freeplay Ranger/ Coleman Outrider

This dual-band AM/FM Freeplay Ranger radio is compact but rugged. The wind-up handle on the back of the radio provides about 35 minutes playing time at normal volume for each 30-second wind and a small LED indicator lets you know the best wind speed. In direct sunlight a small photovoltaic panel on the top allows continuous playing without recharging but if the NiMH battery ever runs down you can wind up or recharge it from the mains using the AC/DC adapter. A fully charged battery can give up to 25 hours playing time. The radio is available in the USA as the Coleman Outrider.

✎	*Freeplay Energy, South Africa/Europe*	315
⚙	*Freeplay Energy, South Africa/Europe*	315
▮	*Various materials*	
♲	*• Renewable energy options*	329

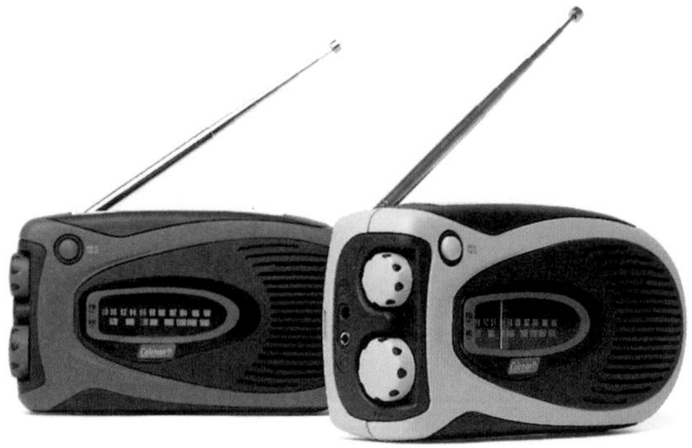

Glass Sound

Chunky speaker cabinets are redundant in this ultimate minimalist sound system. Suspended from stainless-steel wires, which carry the signal, a thin glass diaphragm emits sound waves. This system uses NXT technology to deliver the sound.

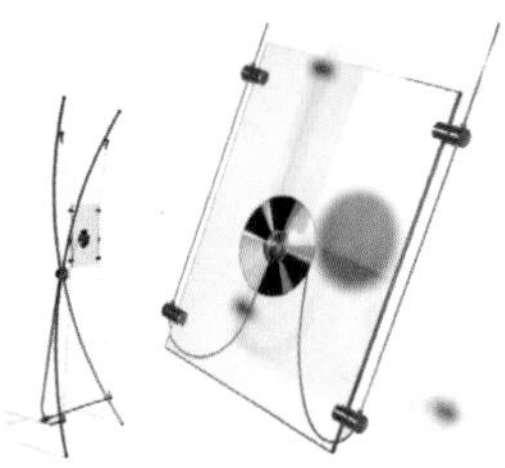

✎	*Christopher Höser, Designteam, Glas Platz, Germany*	316
⚙	*Glas Platz, Germany*	316
▮	*Glass, stainless steel*	
♲	*• Reduction in materials used*	325
✪	*iF Ecology Design Award, 2000*	332

SmartWood guitars

Gibson Guitars are renowned for the quality and sound of their acoustic and electric guitars. This model uses hard maple, Honduras mahogany and Chechen woods certified under the SmartWood and FSC schemes and supplied by EcoTimber, Inc. There's good evidence that Gibson guitars are cherished by their owners and accordingly have an in-built longevity. Use of certified woods for high-value products doubly reinforces the message about designing for longevity using materials from sustainable sources.

Gibson Guitars, USA	316
Gibson Guitars, USA	316
Various woods, metal, electronics	
• *Renewable materials* • *Certified SmartWood, FSC timber*	325 325

…o

…h thermoplastics …ominate in casings for …onic goods but Marc …ier demonstrates …ucking convention …ces a new sexy look …s AM/FM radio. …er also confers …its over plastics by …ng some shock …ance and weather- …ing.

Marc Berthier, France	304
Lexon Design Concepts, France	317
Rubber, electric components	
• *Renewable and synthetic material*	325

Companion Radio

This multifunctional device is an all-in-one radio, flashlight and mobile phone charger weighing a modest 230 g (8 oz) and 125 mm (4.9 in) long. There are three charging options – self-charge by winding the alternator, solar charging by the photovoltaic panel, or emergency charging via an external USB AC adaptor connected to mains power. There's an LED charge indicator, but you'll soon discover that just one minute of winding gives 30 minutes of torchlight through the three LEDs, or 20 minutes of radio listening. Alternatively, just sit in the sun and the photovoltaic cell will keep the NiMH battery topped up and the music flowing.

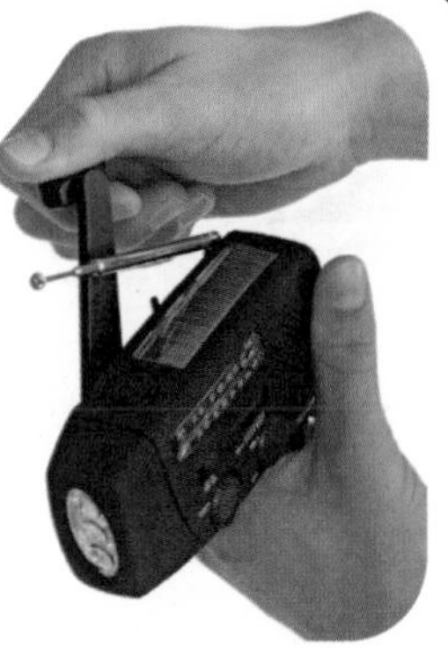

Freeplay Energy, South Africa/Europe	315
Freeplay Energy, South Africa/Europe	315
Impact-resistant plastics, rubberized mouldings, alternator, rechargeable NiMH battery, photovoltaic panel, LEDs	
• *Renewable human power* • *Multifunctional*	329 328

Freecharge 12V

This hand-held battery-free charger relies on a wind-up alternator to generate power to charge mobile phones, PDAs, iPods and GPS receivers, and can charge them through any device that can be powered through a generic cigarette lighter. You need the correct converter for your specific device. There's an LED charge indicator so just wind until you've enough electrical juice.

✎	*Freeplay Energy, South Africa/Europe*	315
⚙	*Freeplay Energy, South Africa/Europe*	315
▬	*Impact resistant plastics, alternator*	
♻	*• Renewable human power*	329

Nokia 3110 Evolve

✎	*Nokia, Sweden*	318
⚙	*Nokia, Sweden*	318
▬	*Various synthetic polymers, biopolymers, electronic components, recycled packaging*	
♻	*• Lifecycle thinking* *• Improved energy efficiency* *• Use of biopolymers in casing* *• Good ergonomics and functionality*	324 329 325 328

Nokia has adopted a design strategy of lifecycle thinking to improve their sourcing of materials and reducing energy consumption that has kept them in the number one position in Greenpeace's Guide to Green Electronics. Nokia products comply with EU and China Restriction of Hazardous Substances (RoHS) legislation, the company has a strict substance management programme, and is moving towards recovery of 65–80% of materials at the end of the phones' life. The 3110 Evolve sets new benchmarks by making the cover from 50% biopolymers, using the very efficient AC-8 charger (94% less energy requirements than specified under Energy Star compliance), and achieving 15–20% lower energy consumption than its competitors over the whole lifecycle. With large keys, a good screen, a 1.3 megapixel camera with x8 digital zoom, an MP3 player, Bluetooth and Infrared connectivity, and FM radio, this is an amazing package weighing just 87 g (3 oz).

›lio™

odern global nomads ınt to be entertained, ·tworked and capable uploading to a weblog ıvel diary, so powering › all those devices such phones, iPods, cameras ıd GPS receivers seems sential. The original ·lio™ charger was created ' just that purpose, ıether on the trail, in the ·tel or at home, one hour sunshine can give new ce for your devices.

The photovoltaic cells in each 'leaf' are unfurled in a sunny spot then conveniently stowed away when the internal rechargeable battery is fully charged. Weighing in at just 156 g (5.5 oz) you'll hardly notice it's in your backpack. There's a tough Magnesium edition, which generates 9W, rather than the standard 6W, for those travellers likely to encounter challenging conditions.

	Better Energy Systems, Ltd, UK	313
	Better Energy Systems, Ltd, UK	313
	Rechargeable lithium ion battery, synthetic polymer or magnesium, photovoltaics, electronic circuit	
	• Renewable solar power *• Lightweight*	329 327

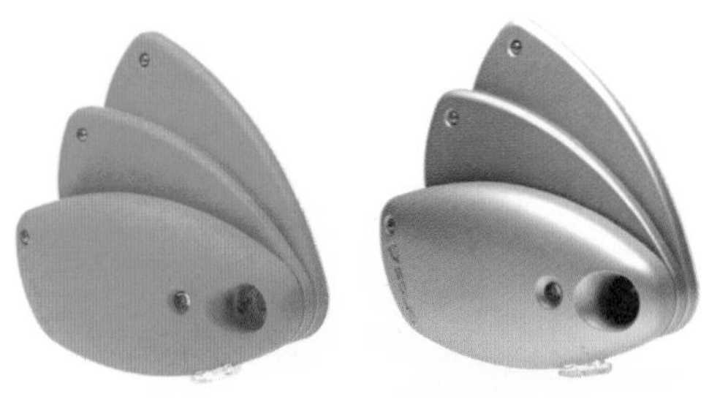

REVOLUTION™ Eco Media Player

Trevor Baylis, who invented the first wind-up radio for use in developing countries back in 1993, launched the versatile Eco Media Player in 2007. It is part of the Next Generation range that extends the functionality, while reducing the environmental impact of the wind-up multimedia device. Nestling neatly in the palm of your hand, the 170 g (6 oz) device offers 40–45 minutes playing for one minute of cranking, and a fully charged battery means continuous playing for 48 hours. With 4Gb or 8Gb flash memory, thousands of multimedia files can be stored, including the latest file formats from MP4 video to audio books. Other features include FM radio, phone charger, memo recorder, LED torch, photo viewer and USB charger.

	Tony Davies, Trevor Baylis Brands plc, UK	322
	Trevor Baylis Brands plc, UK	322
	Various synthetic polymers, electronic circuits, 1,000 mAh lithium ion rechargeable battery, USB connector, recyclable cardboard packaging	
	• Renewable human powered energy *• Multifunctional media device*	329 328

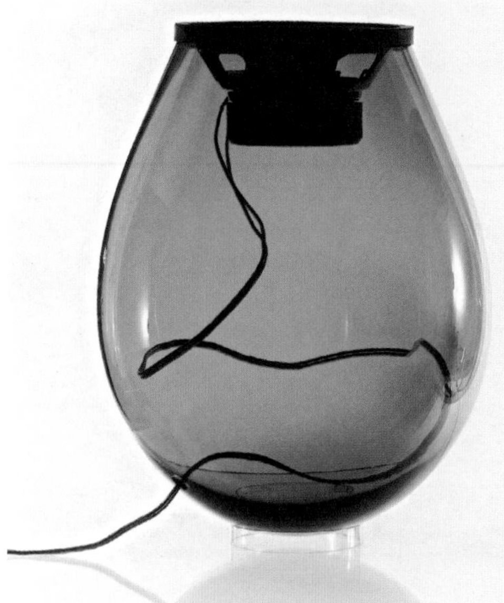

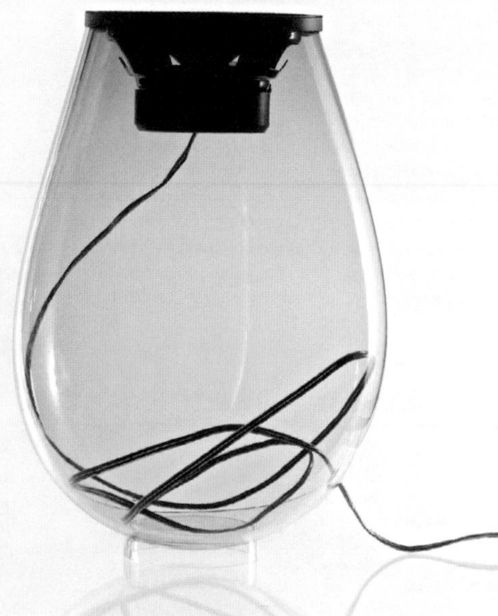

Front MP3 player and speakers

Front, an all-female collective of Swedish designers, contests the unimaginative technological seductions of the marketplace with this project. By embedding an MP3 player into a beautifully crafted case of wood, lead and antler bone they query the endless shifts of form, fashion and new technologies. The device is imbued with the possibility of a long life in which the materials develop patina, and one user hands it on to the next generation. By placing speakers into high-density glass jars the normal aesthetic and acoustic rules of hi-fi speakers are challenged, and a beautiful new sound emerges. Front's MP3 player and speakers encourage us to re-examine our relationship with the services that these devices provide.

	Front, Sweden	306
	One-offs	
	MP3 player, wood, lead, antler	
	• *Casing from natural or recyclable materials* • *Creation of longer lasting emotional bonds*	325 328

Wattson 01

Described as a 'personal energy monitor' that reveals the amount of electricity you are using in your home or business, the original handmade wooden Wattson was launched in 2006 by DIY Kyoto. Since then they have developed the prototype and it is now available as a sleek manufactured unit, about the third the size of a laptop PC. The portable monitor changes colour to display the watts or money being consumed – an incentive that encourages users to quickly realize up to 25% electricity savings by changing habits and, if necessary, appliances too. Data transmitted wirelessly from an electricity meter is received by the monitor, so it can be placed anywhere in your building. Four weeks worth of consumption data can be stored by Wattson's memory then downloaded via USB, viewed by Holmes' software, and even uploaded to the DIY Kyoto website where you can compare your energy savings and habits with other 'Wattsoners'. It is also possible with Wattson to record the amount of renewable energy you generate and track it against energy received from external suppliers.

	DIY Kyoto, UK	305
	DIY Kyoto, UK	305
	Synthetic polymers, electronic circuits and memory, wireless receiver	
	• *Encouraging awareness and reduction of energy consumption*	328

B. M. vase, fruitbowl and spoon

B. M. refers to the Bell Metal Project, 1998–2000, for which Pakhalé applied and refined the *cire perdue* or lost-wax casting technique popular in central India for hundreds of years. Wax harvested from jungle trees is purified and extruded to produce round profile strings, which are wound around a sand/clay/dung core. Once the desired shape is attained the whole wax structure is encased in more sand/clay/dung mixture, to make a mould, which is then fired. Molten scrap brass, bronze and copper are poured into the mould and allowed to cool. Once extracted from the mould the cast piece is hand-cleaned and brushed. Pakhalé breathes new life into these time-honoured techniques by sympathe and skilfully creating contemporary forms. T B. M. objects elicit a renewed joy in design, reclaiming its purpose satisfy physical, mental emotional and spiritual needs rather than just deliver profits and cultu status.

✎	*Satyendra Pakhalé, the Netherlands and India*	3
⚙	*Atelier Satyendra Pakhalé, the Netherlands*	3
▰	*Brass, bronze, copper scrap*	
♻	• *Recycled metals* • *Low-energy manufacturing* • *Innovation of traditional technology*	3 3 3

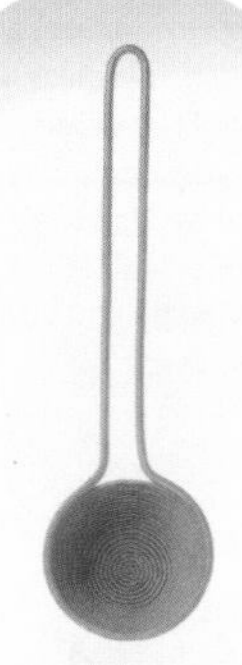

Pinch

In homage to the beauty and functionality of the clothes peg, the artist Jos van der Meulen brings our thoughts full circle – from the twigs in the forest to an artefact in the service of man. Like little statues these unique branches stand silently waiting for something to happen, for something to appear in their little sprung jaws – a keepsake, photo or other object to become momentarily a point of personal reference, a focus of conversation, or more.

✎	*Jos van der Meulen, the Netherlands*	310
⚙	*Goods, the Netherlands*	316
▰	*Wood, steel*	
♻	• *Renewable and recylable materials*	325

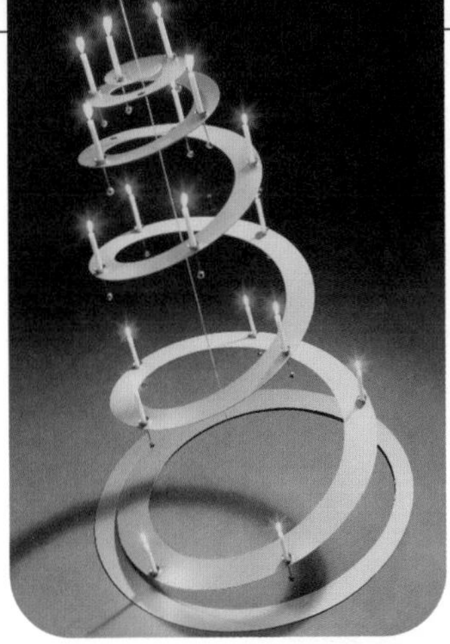

Spiralbaum

This flat-pack, laser-cut, plywood square unfolds as a helix when suspended. It is the ultimate in minimalist Christmas trees, is easily stored away for the next festive season and saves another Sitka spruce from being consigned to the landfill site each New Year.

✎	*Feldmann + Schultchen, Germany*	306
⚙	*Prototype for Werth Forsttechnik, Germany*	323
▮	*Plywood, steel*	
♺	*• Reusable product*	329

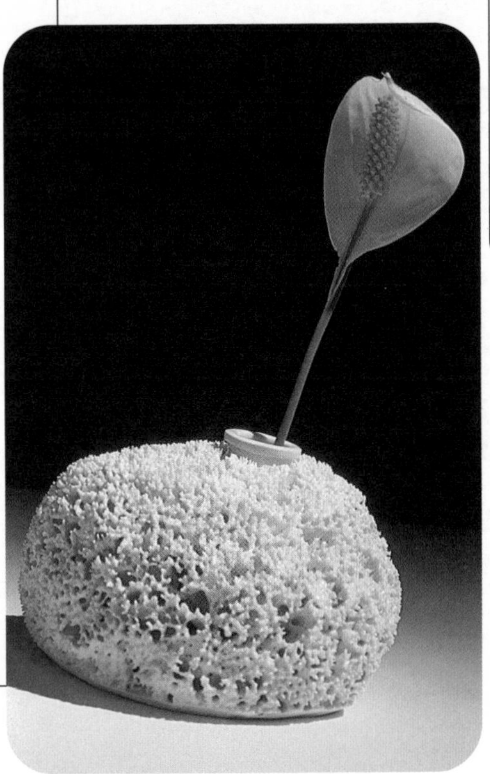

Christmas tree poster

The dilemma of whether to have a cut Christmas tree or an artificial one has been resolved by this minimalist design. Simply hang your 120 x 160 cm (47 x 63 in) 'Christmas tree' on the wall and temporarily attach all the Christmas cards you've received. Roll it up after the festive season, recycle your cards, and bring the poster out again next year.

✎	*Ivan Duval and Jean Sebastian Ides, Atypyk, France*	304
⚙	*Atypyk, France*	304
▮	*Paper, printing inks*	
♺	*• Encourage reduction in timber consumption* *• Reusable and recyclable*	325 329

Sponge vase

Marcel Wanders commandeers nature's own manufacturing, adds his own porcelain tube and sets vase design on a new course. Designers should actively seek opportunities for 'harvesting' nature's products, which, with minimal energy input or modification, can be re-manufactured into new objects. Can we look forward to specialist 'product farms' where sponges are 'bio-manufactured' in neat rows, bamboos grow to EU regulation size in specially built moulds and bio-plastics spontaneously grow to predetermined forms?

✎	*Marcel Wanders, Droog, the Netherlands*	310
⚙	*Droog / Moooi, the Netherlands*	314/ 318
▮	*Natural sponge, porcelain*	
♺	*• Renewable material*	325

Aladdin

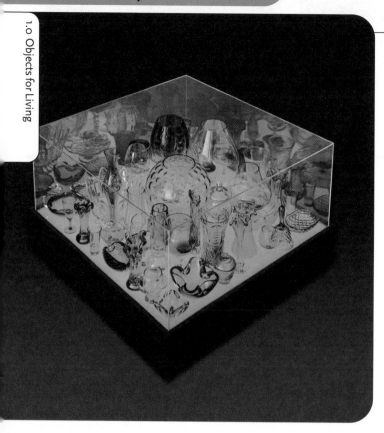

Designer Stuart Haygarth collected diverse glassware forms and objects from flea markets, car boot sales and junk shops, and then grouped them into colour-themed collections. The collections are housed in lightbox vitrines of different sizes – part museum case, part shrine – each of which functions as a dramatic light source and a table. This lovely and unexpected combination creates a powerful presence for aesthetic and sensory pleasure.

Stuart Haygarth, UK	306
Limited edition, 2006	
Reclaimed glassware, glass vitrine, lighting unit, MDF lightbox	
• Reuse and rejuvenation *• Multifunctional artefact*	327 328

Droplet vessel

Ash and sycamore wood from sustainably-managed forests is hand-turned into seductive shapes and offset with colourful bases made of salvaged acrylic plastic. The juxtaposition of these two elements creates appreciation of the inherent qualities and beauty of natural and man-made materials, and the value of sourcing sustainably and recycling.

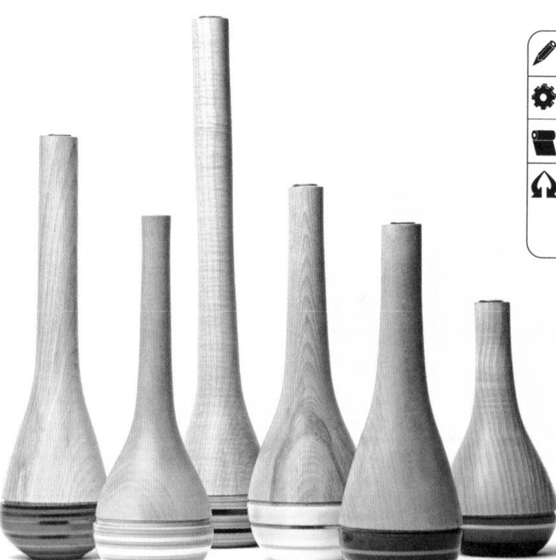

Sarah Thirlwell, UK	310
Small batch production	
Ash and sycamore wood, reclaimed acrylic plastic	
• Renewable and recycled materials *• Durable hand-crafted artefacts*	325 328

Interactive Wallpaper™

Interactive Wallpaper™ is silk-screen printed by hand on a recycled flax or white base paper. The designer's role in these 'halfway' products is to create a design interface for the user so that they might complete and customize the design process by attaching a series of stickers from a broad range of designs.

Recent commissions by Rachel Kelly include work for the neo-natal unit at the Royal London Hospital, leading to the idea that perhaps there is significant potential for products of this genre to help in healthcare environments, encouraging patients to explore their creativity as part of the healing process.

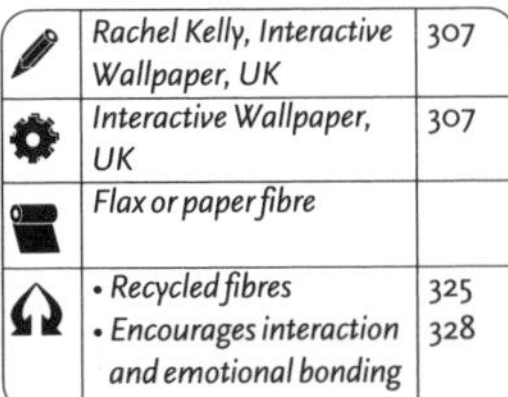

✎	*Rachel Kelly, Interactive Wallpaper, UK*	307
⚙	*Interactive Wallpaper, UK*	307
▬	*Flax or paper fibre*	
♻	*• Recycled fibres* *• Encourages interaction and emotional bonding*	325 328

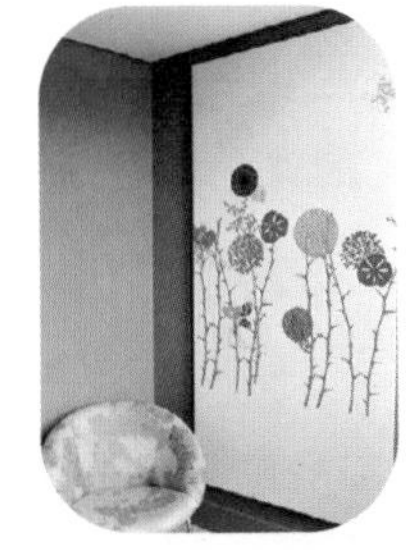

Tangent™ 3D wallpaper

A core philosophical premise at MIO is the idea of 'responsible desire', that is creating attractive, desirable products using readily available urban resources. Tangent™ is manufactured entirely from waste paper in a series of tiles that can be configured in different patterns and tinted with water-based paints. MIO transform a humble material into a visual, tactile wall sculpture adding a new dimension, literally, to the concept of wallpaper. Bored of last year's wallpaper? Strip off the Tangent™ and dispose it at your local recycling point.

✎	*Jaime Salm, MIO, USA*	309
⚙	*MIO, USA*	308
▬	*100% waste paper*	
♻	*• Renewable and recyclable materials*	325

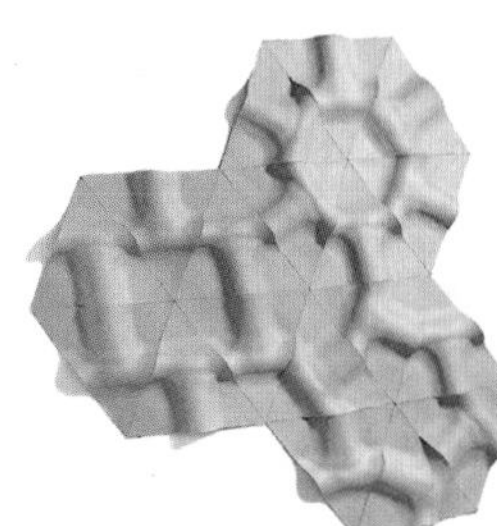

Bicicleta

Even in India a bicycle's inner tubes can only soldier on for so long before they become worn out. At that point in steps Nani Marquina to experiment with possibilities of using this material for new creations. The handmade rugs of the Bicicleta collection, measuring 170 x 240 cm (5½ x 8 ft), are hand-loomed using 100% recycled rubber from 130–140 inner tubes. Looped or cut pile attached to three warps enables simple but sculpturally appealing designs. The company has consistently sought to utilize natural, renewable and recyclable materials, and use renewable energy and encourage craft production, especially in India and other developing countries.

	Nani Marquina and Adriana Miquel, Spain	318
	Nani Marquina, Spain and local manufacturing in India	318
	Recycled rubber	
	• Recycled materials *• Hand production to maintain traditional skills*	325 325

John Deere and Silence rugs

If raising a smile is a design service then Tore Vinje Brustad achieves it with his witty rugs, handmade with 100% New Zealand wool by a family-run business in India. The tractor and footprint patterns are generated by controlling pile depth. All rugs are labelled with the Kaleen label, a hallmark of commitment towards the eradication of child labour, and all workers in the business are adults and receive wages above the minimum.

	Tore Vinje Brustad, Permafrost, Norway	30
	Handmade in India for Permafrost, Norway	30
	Wool	
	• Renewable material *• Ethical production system*	32 32

Roses

This delightful, playful rug appears alive, an effect achieved by hand weaving the wool felt shapes with a manually actuated loom where rods can be inserted to control the height of the pile. The Roses rug is a typical example of innovation from the portfolio of Nani Marquina, and it also supports a vibrant craft production system in India, ensuring new skills are combined with old. The rug is available in two sizes, 170 x 240 cm (5½ x 8 ft) and 200 x 300 cm (6½ x 10 ft), and three warm colours (red, brown, orange) or two neutrals (ivory and grey).

✎	*Nani Marquina at Nani Marquina, Spain*	318
⚙	*Nani Marquina, Spain, and local manufacturing in India*	318
▮	*Wool felt*	
♻	*• Renewable materials* *• Hand production to maintain traditional skills*	325 325

Kala

This rug design emerges from a collaborative project between Nani Marquina and Care&Fair, an association of over 450 organizations that works to improve working conditions and ensure against child labour abuse in the Indian, Pakistani and Nepalese rug-making industries. Original drawings by children at Care&Fair schools were used by the Nani Marquina design team to create Kala, which in Hindi means both 'tomorrow' and 'art'. Kala is a hand-tufted rug made of 100% New Zealand wool with a density of 38,000 knots per m² and a pile height of 12 mm (½ in). The sales from the rug will be invested in a new Care&Fair school.

✎	*Nani Marquina + Care&Fair, Nani Marquina, Spain*	308
⚙	*Nani Marquina, Spain, and local manufacturing in India*	318
▮	*Wool*	
♻	*• Renewable materials* *• Hand production to maintain traditional skills* *• Sales go to a social enterprise*	325 324 325

Constantin and Laurene Boym, USA, with Rebecca Wijsbeek, the Netherlands	304
Moooi, the Netherlands	318
Second-hand ceramics, adhesive	
• Reused objects *• Low-energy manufacturing*	327 326

Salvation Ceramics

Celebrating everyday mass-produced things, the sum of the parts creates a greater whole as each 'new' Salvation Ceramic comes to life. The Boyms design a range of basic shapes that are assembled by stylist Rebecca Wijsbeek. Inviting us to examine the original components, these objects transform from everyday to special, from industrial production to unique one-offs, from the ordinary to the beautiful. Moooi challenge the concept of mass production.

The Soft Vase

Challenging our perceptions about polymers and the way we use them, Hella Jongerius experimented with flexible rubberized polyurethane to create a traditionally styled vase. It provokes us to question how we value plastics. Objects made from plastics can be highly valued in the case of 'designer' objects or regarded as a throwaway item in the case of the ubiquitous plastic bag. Instead of using hard durable ceramics or tough shiny ABS, the traditional materials for a vase, Jongerius has chosen soft flexible polyurethane, thus encouraging the user to experiment with altering the shape of the vase.

Hella Jongerius, Jongeriuslab, the Netherlands	307
Droog Design, the Netherlands	314
Polyurethane, elastomers	
• Improved user-friendliness *• Durable* *• Recyclable*	328 328 329

PS Jonsberg

Although these four vase designs are produced in large numbers for the global IKEA brand, each one is in fact an individually hand-finished piece bearing the marks of the artisan who skilfully decorated each artefact. IKEA commissioned Hella Jongerius of Jongeriuslab – who has an impressive history of eclectic independent work and collaborations with the likes of Droog Design – to create mass-produced decorative objects. Each of the patterns used on the vases represents and celebrates a different part of the world and techniques that reflect specific geographies, and help revitalize the varied traditions for a global audience. These vases are set to become a classic 'collectable' and will undoubtedly be cherished by future generations.

Hella Jongerius, the Netherlands, for IKEA, Sweden	307
IKEA, Sweden, using manufacturing facilities in China	317
Terracotta, stoneware, feldspar porcelain, slips and glazes	
• Abundant lithosphere materials *• Revitalization of traditional techniques*	325 324

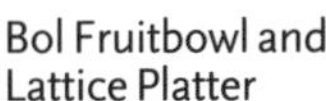

Bol Fruitbowl and Lattice Platter

In its vertical configuration the Bol is a sculptural, sentient object made of delicate strips of laser-cut hoop pine plywood – sustainably forested in Australia – held together with steel and rubber rings. When a piece of fruit is dropped into the closed Bol it gracefully opens out and metamorphoses into a dynamic bowl form that permits good air circulation to keep up to ten pieces of fruit fresh. Collapse it back to its sculptural form when not in use. The Lattice Platter is a flat-pack self-assembly shallow bowl made of the same laser-cut pine. Like Airfix scale model kits, each piece is pushed out from the pine sheet and assembled using the instructions provided. The result is a lightweight but very strong bowl.

✎	*Christopher Metcalfe, New Zealand*	308
⚙	*Christopher Metcalfe, New Zealand*	308
▮	*Hoop pine plywood, steel, EPR O-rings*	
♺	• *Lightweight* • *Self-assembly and cold-construction (Lattice)* • *Interactive object (Bol)*	327 326/ 327 328

SoftBowls

Going under the delightful names of Beehive, Swoop and Wobowl, these tactile bowls invite play, encouraging the user to shape and mould them. Made from locally sourced wool, these bowls are handcrafted by local crafts people in Philadelphia at one of the last millineries in the USA. The skills and techniques of hat production are re-positioned towards new markets through the use of a renewable, recyclable and compostable material in vibrant colours. Manufacturing SoftBowls requires less than one tenth of the energy used for an equivalent ceramic bowl.

✎	*Jaime Salm and Roger Allen, MIO, USA*	309
⚙	*MIO, USA*	308
▮	*Wool felt*	
♺	• *Renewable material* • *Utilization of existing manufacturing facilities for new purpose* • *Low-energy manufacturing* • *Maintaining a local skill base*	325 325 326 327

Eco Cooler

This three-part kitchen-top larder, with a 280 mm (11 in) diameter footprint, is made of slip-cast terracotta.
It is made up of a vegetable cooler, contained in a water reservoir, and an upper fruit bowl, which doubles as a lid. Cooled by the water, the main chamber is kept at 3–4° C (37–39° F), ideal for storing root vegetables. Designed for DeWeNe, the Eco Cooler is made in the UK by a social enterprise engaged in employing and training people marginalized because of physical and mental disabilities.

Designer	*David Weatherhead, UK*	311
Manufacturer	*DeWeNe, UK*	314
Materials	*Terracotta, slip glaze*	
Sustainability	*• Abundant lithosphere materials* *• Passive cooling system* *• Social enterprise manufacturing*	325 324 327

Tap Tap Tongs

This beautiful and minima two-piece cherrywood set can be used as a single utensil for stirring, or can be fixed together as salad tongs. Gary Allson combines traditional and digital technologies to create balanced elementa forms for a range of kitchenware products in locally sourced woods finished with natural ingredients.

Designer	*Gary Allson, UK*	3
Manufacturer	*Gary Allson, UK*	3
Materials	*Cherry wood*	
Sustainability	*• Locally sourced FSC-certified wood* *• Natural finishes*	3 3

New Heirlooms

Cj O'Neill sources vintage ceramic tableware from German manufacturer Scherzer in the markets and charity shops of the UK and Denmark, and then gives them a new lease of life in a collaborative project with Scherzer. Each object is rejuvenated by applying new hand-cut and printed transfer patterns to create a collection of one-off pieces, and serve as a repository for memories. As the price of energy for industrial production steadily rises, the concept of re-manufacturing might become more popular.

Designer	*Cj O'Neill, UK*	308
Manufacturer	*Scherzer, Germany*	320
Materials	*Porcelain, transfer prints*	
Sustainability	*• Reuse and rejuvenation* *• Low-energy manufacturing*	327 326

Lavabowl

Buying tableware and kitchen ceramics is always fraught with a catalogue of dilemmas. Do you choose this season's fashionable colour, go for impractical but funky styles or settle for a 'classic' design? Other dilemmas arise around issues of price and quality. Lavabowl avoids all these difficult decisions by being solid and dependable, modern yet traditional, neutral but not boring, and it also manages to produce a porcelain-like 'sound'. De Leede developed a special clay with Royal Tichelaar Makkum by mixing crushed gravel into the clay. This prevents the thick walls of the bowls from cracking

but also imparts its own texture and colour flecks to the high-temperature fired bowls.

✎	*Annelies de Leede, the Netherlands*	308
⚙	*Goods, the Netherlands*	315
▮	*Clay, gravel*	
♻	*• Abundant geosphere materials*	325

Bone China Line

Just a generation ago the best china tableware would sit in the sideboard waiting for special visitors and guests before it made its appearance. Today the idea of 'best china', and the sense of occasion and pleasure that this activity brought, have faded. Everyday ceramic tableware often doesn't do justice to the food or drink being served but Bodum has sought to change that in their Bone China Line by combining the concepts of 'everyday' and 'best' with modern eating habits. This seven-part dinner service comprises dinner and dessert plates, soup/pasta/rice bowls, a mug, coffee cup and saucer, and espresso cup and saucer. Subtle design detail emerges on inspection, the surfaces of each piece forming an inner and an outer parabola. Tough and durable, bone china is also fired at relatively low temperatures compared to other ceramics, such as porcelain.

✎	*Carsten Jørgensen*	316
⚙	*Bodum, Switzerland*	313
▮	*Bone china*	
♻	*• Durable* *• Abundant geosphere materials*	328 325
✿	*iF Design Award, 2003*	332

The Body Shop range

Since its formation in the 1970s The Body Shop has aimed to provide a holistic, natural approach to body care and hygiene, with due consideration to the environmental, ethical and social responsibilities of the business. That approach still drives what has become the role model for an ethical international business and encouraged L'Oreal, the French global cosmetics group, to acquire The Body Shop in 2006. The Body Shop supports community trade, addresses climate change and promotes issues such as their Ethical Trading Initiative, supply chain management with respect to Human Rights, HIV and health campaigns. Product information is generally printed directly on to the bottles to eliminate the need for stick-on labels and to facilitate recycling.

✎	*The Body Shop International, UK*	313
⚙	*The Body Shop International, UK*	313
▬	*Natural oils, conditioners, soaps and recyclable HDPE*	
♻	• *Reusable and recyclable containers* • *High natural-content ingredients*	329 325

Preserve®

Everything about the Preserve® toothbrush and its packaging suggests that Recycline is a company that takes its environmental responsibilities seriously. A clear, rectilinear box is made from a cellulose polymer derived from trees; the removable cap is HDPE and is intended as a carrying case rather than disposable packaging. A paper insert in the box carries the brand information. It is printed with soy inks and is made from 100% recycled paper fibre including 50% post-consumer waste.

The toothbrush handle and head are ergonomically designed with soft bristles on the outside to protect the gums and stiffer bristles to penetrate between the teeth.

✎	*Preserve, USA*	320
⚙	*Preserve, USA*	320
▬	*Recycled plastics and paper, cellulose-based polymers, nylon, HDPE, soy bean inks*	
♻	• *Recycled, recyclable and renewable-source plastics* • *Product take-back* • *Lifecycle analysis*	325 329 324

Although the bristles are virgin nylon, the shaft is made from 100% recycled plastics (minimum 65% from recycled yogurt pots). When the bristles are worn out the used brush can be returned to Recycline in a postage-paid envelope available at the retailer or direct from Recycline. This is responsible closed-loop manufacturing where the manufacturer and consumer form a partnership in the lifecycle of the product.

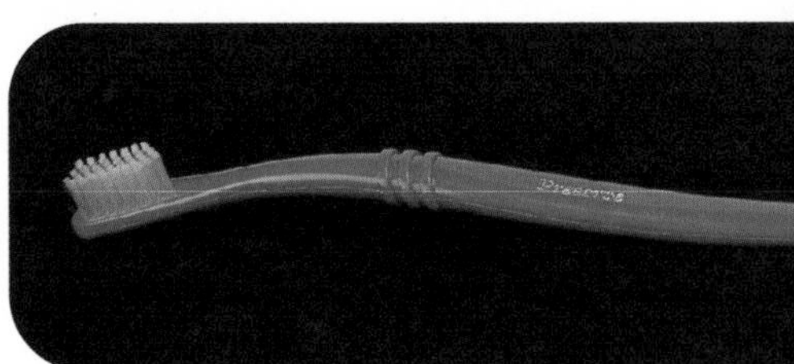

Durex Avanti

Natural or synthetic latex has long been the preferred material for manufacturing condoms but the material still suffers from an image problem. Latex produces its own distinctive odour and, owing to the thickness of material required to ensure full protection during intercourse, can result in a lack of sensitivity to the wearer. It also produces an allergic reaction in some people. After considerable research a version of polyurethane proved itself in tests. It is as strong as latex but 40% thinner, is odourless and almost transparent. Add a little flavouring – do you fancy tangerine, strawberry, spearmint? – and here is a little self-help device guaranteed to assist in population control and the fight against the spread of sexually transmitted diseases including AIDS.

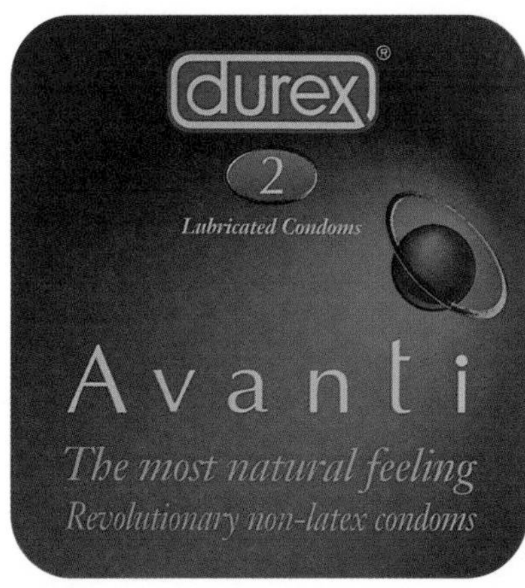

✎	*Durex, UK*	314
⚙	*Durex/ LRC Products, UK*	314
▰	*Polyurethane*	
♻	*• An aid to reduce population growth and improve human health*	327

Aquaball®

Place two pellets inside the spiky plastic ball, the Aquaball, and put it with your dirty washing into your machine. The pellet ingredients include fatty alcohol polyoxyethylene ether, higher alky sulphate, CMC, sodium carbonate, EDTA, fragrance (derived from essential oils) and FWA (fluorescent whitening agent), which 21st Century Health claim are free from harsh chemicals and therefore safer for people prone to allergies, eczema and sensitive skin. The pellets dissolve faster than detergents, increasing the level of ionized oxygen and raising the pH, so the ingredients get to work early by lifting dirt from fabrics. Rinse cycles can be shortened by up to 30 minutes for an average wash, so saving lots of water. There are fewer residues left over than with conventional washing powders and a very low phosphorus content, so there is a comparative reduction in water-borne pollutants per wash.

✎	*Aquaball®*	312
⚙	*21st Century Health, UK*	312
▰	*HDPE casing, ionic crystals and salts*	
♻	*• Reduced consumables • Reduced water consumption and pollution*	329 328

Dr Bronner's soaps

2009 marks the 60th anniversary of Dr Bronner's Magic Soaps with news that all its liquid and bar soaps are now certified under the United States Department of Agriculture Organic label. The company is certified Fair Trade with the 'Fair for Life' programme of the independent Swiss Institute for Marketecology. The body pump range of liquid soaps includes the Organic Shikakai mixed with Dr Bronner's classic castile-base (olive and other vegetable oils) and fragranced with zingy scents from tea tree to lime, lavender and peppermint.

✎	*Dr Bronner's Magic Soaps, USA*	314
⚙	*Dr Bronner's Magic Soaps, USA*	314
▮	*Plastic containers, soap*	
♻	*• Organic and Fair Trade certification • Plant-based, renewable ingredients*	325 325 325

eco-ball™

Those sensitive to today's chemically based washing powders have an alternative method available in the form of the eco-ball™. This is a plastic ball that contains ionic powder that releases ionized oxygen into the water and so facilitates penetration of water molecules into fabrics to release the dirt. A little washing soda helps deal with very dirty washing but it is claimed that a set of three balls will help clean the equivalent of 750 washes before losing their activity.

✎	*Ecozone, UK*	314
⚙	*Ecozone, UK*	314
▮	*Plastic, ionic powder*	
♻	*• Reduction of consumables*	329

DURAT® design collection

Those with minimalist and modernist aspirations should look no further than the DURAT® design collection of tens of different washbasins, bathtubs, shower trays and kitchen worktops with integral sinks. Individual designs are available in 46 standard colours but bespoke patterns, edge designs and other finishes are possible. DURAT® is a smooth, hard polyester-based material that contains 50% recycled plastics and is 100% recyclable. It is warm to the touch and suitable for a wide variety of applications in any 'wet zone' in the home or work environments. The manufacturing technology permits jointless customized panels, worktops and one-off pieces.

✎	*Tonester, Finland*	322
⚙	*Tonester, Finland*	322
▮	*Recycled and virgin polyester-based plastics*	
♻	*• Recycled content • Recyclable*	325 329

Ecover®

The name Ecover, like The Body Shop, needs little introduction to those who became green consumers in the 1980s. Established in 1979, Ecover has always espoused a business policy that recognizes that economics must be in harmony with ecology. This policy extends to product development, the green architecture of its 6,000 m² main factory in Belgium (see p. 240), and the international distribution network through thousands of small health food shops, as well as the supermarket giants. Company policy dictates that products must originate from a natural source with a low level of toxicity to minimize their burden on the environment and they must be equally efficient as conventional, more polluting products. Ecover products are not permitted to include petrochemical detergents/ perfumes/solvents/acids, polycarboxylates, phosphonates, animal soaps, perborates, sulphates, colourings, phosphates, EDTA/NTA, optical brighteners or chlorine-based bleaches. Animal testing is also banned. As the market share for green consumption grows Ecover's product range has expanded and now includes products for washing (clothes, dishes and even cars), household cleaning and personal care. Sponsors of the *Observer* newspaper's Ethical Awards, Ecover also supports other ethical, social and environmental projects, such as a collaboration with international agency WaterAid. Ecover's environmental policy gives specific attention to sourcing vegetable or mineral ingredients that are totally degradable, to minimizing packaging and ensuring that it is all recyclable.

✎	*Ecover, Belgium*	314
⚙	*Ecover, Belgium*	314
▬	*Various cleaning agents, plastic containers*	
♻	• *Reduction of water-borne toxins and pollutants*	328
	• *Recycling of containers*	329
	• *Retailing system geared to small and large outlets*	327
	• *Degradable products*	325

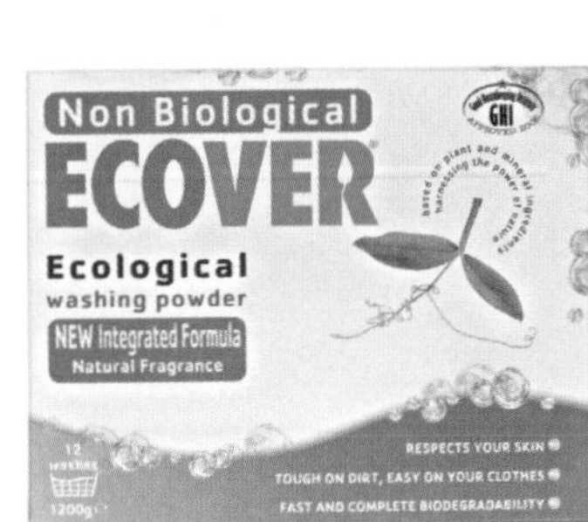

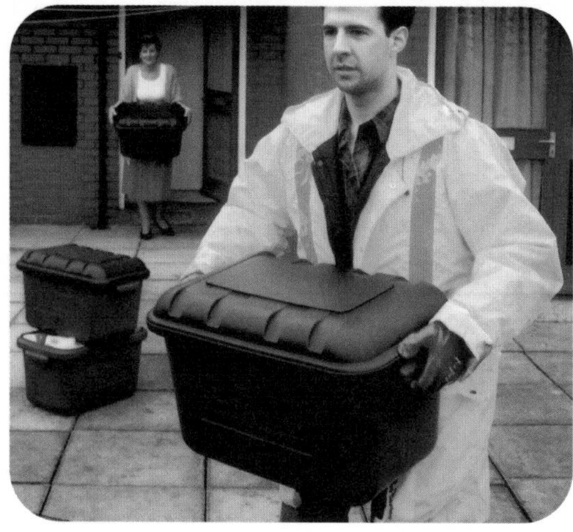

LINPAC Environmental kerbside collection box

Since the introduction of the LINPAC Environmental kerbside collection box in 1996 to the city of Sheffield, UK, over 20 million plastic bottles have been diverted from landfill sites to recycling plants where the plastic is reused to create yet more boxes. This robust box, with high-impact and -deformation haracteristics, encourages greater recycling by local authorities and private contractors.

✎	*LINPAC Environmental, UK*	317
⚙	*LINPAC Environmental, UK*	317
▮	*Used HDPE bottles*	
♻	*• Recycled materials* *• Encourages recycling*	325 327

Zago™

Recycling domestic waste has an image problem, so anything that can elevate this activity into fun is welcome. Three Zago™ triangular rubbish bins made from flat-pack, recycled cardboard neatly sit together to form a functional separator for different waste streams. The photographic exteriors clearly indicate each particular waste stream and reinforce the message that waste is a valuable resource.

✎	*Benza, Inc., USA*	312
⚙	*Benza, Inc., USA*	312
▮	*Recycled cardboard*	
♻	*• Recycled materials* *• Encourages recycling*	325 327

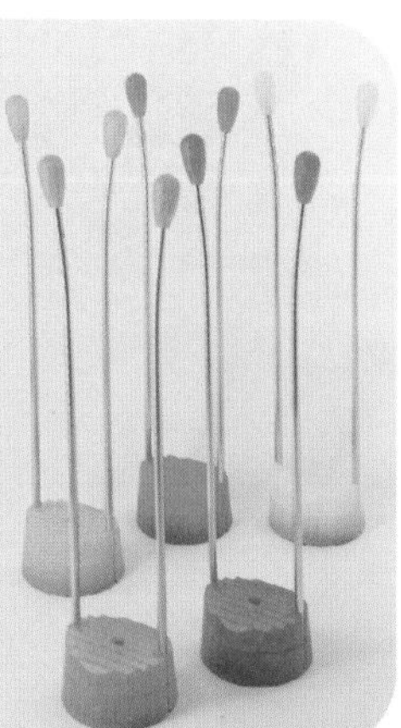

Cricket

Consumption of bottled water and soft drinks contained in PET plastic bottles has risen dramatically in the last decade, so any device tha facilitates recycling is to b welcomed. This witty bot crusher makes recycling fun and improves storag capacity of containers for collecting waste bottles.

✎	*Julian Brown, Studio Brown, UK*	304
⚙	*Rexite SpA, Italy*	320
▮	*Steel, plastic*	
♻	*• Encourages recycling*	327

Attila

Consumers' voracious appetite for convenience drinks will ensure that the humble steel or aluminium drinks can will be a feature of the 21st-century landscape. While recycling of these cans improved significantly during the 1980s, any device that actively encourages people to recycle more is a good thing. Attila is a durable crusher that is a pleasure to use: simply place your can in the bottom of the translucent column and enjoy that satisfying crumpling noise as the 'anvil' crushes the can with the downward push of the arms.

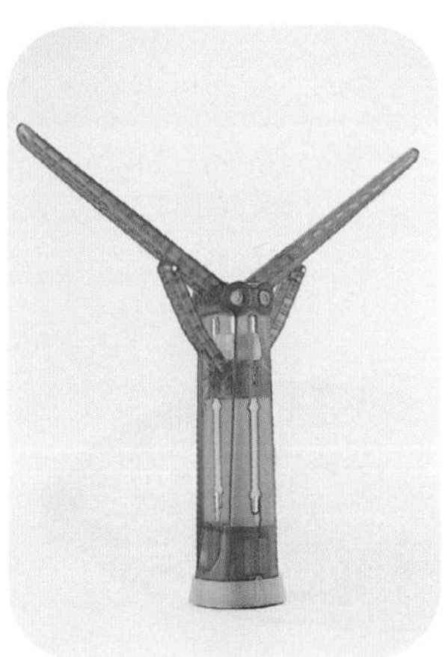

✎	*Julian Brown, Studio Brown, UK*	304
⚙	*Rexite SpA, Italy*	320
▬	*Injection-moulded ABS, polycarbonate, Santoprene*	
♻	*• Encourages recycling*	327

Polyrap

This sculptural origami-style wastepaper bin is manufactured from one sheet of polypropylene and is available in a range of colours. The ingenious design means you don't need a bin liner, as the bin can be unwrapped simply, rinsed and wiped instead. Furthermore, the company has a zero-landfill policy and will take back your old bin when you no longer want it and recycle the materials. The Polyrap is also shipped to the user in 100% recycled packaging.

✎	*Blue Marmalade, UK*	304
⚙	*Blue Marmalade, UK*	304
▬	*Polypropylene, 100% recycled packaging*	
♻	*• Single material • Lightweight • Product take-back for recycling*	325 325 329

Designer	*Nigel's Eco Store, UK*	318
Manufacturer	*Nigel's Eco Store, UK*	318
Material	*100% recycled paper*	
Eco	*• Recycled and biodegradable materials*	325

Ecopod

Treading lightly in life, and death, is something to aim for, and this lightweight biodegradable coffin can help you achieve that ambition. Made of 100% recycled paper that has been naturally hardened then decorated, the Ecopod is ideal for a non-toxic burial and for use in green field sites. The Ecopod weighs only 14 kg (30.8 lb) and comes with a natural calico mattress.

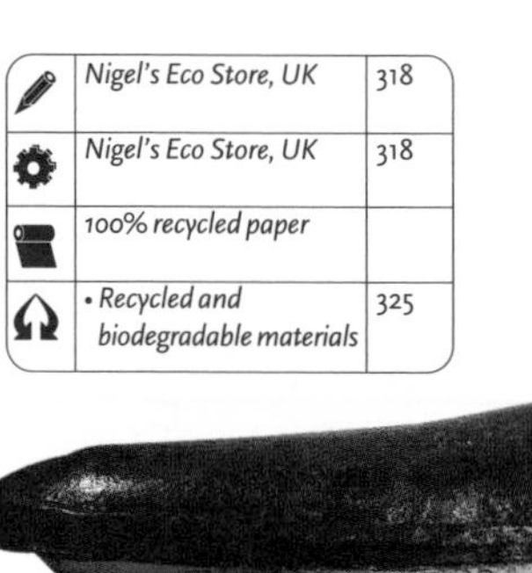

Earthsleeper™

Made entirely of Sundeala board from recycled newsprint with wood corner joints and wood nuts and bolts, these coffins are highly biodegradable and make less environmental impact than conventional wooden coffins. Coffins are available ready-assembled or as flat-pack, self-assembly units.

Designer	*Vaccari Ltd, UK*	322
Manufacturer	*Vaccari Ltd, UK*	322
Material	*Sundeala board*	
Eco	*• Recycled materials* *• Compostable*	325 325

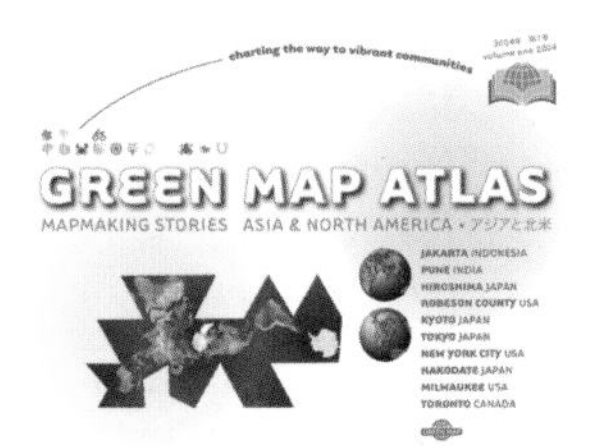

Green Maps and Open Green Map

Not only does the Green Map System encourage greener forms of consumption but it fosters a new view of the totality of the local environment – the people, the shops, the diverse businesses and culture – that nurtures the long slow journey towards more sustainable ways of living. Originated by Wendy Brewer in New York, the Green Map System has rapidly expanded around the globe and now includes hundreds of maps in cities and towns. Its success is not only in the core concept and graphical icons but in its willingness to let local design and local solutions have their voice. In 2008 Green Map added the Open Green Map system which now features 2,971 sites for 'green living' using a Google-based interactive map. This makes it easy to locate sites near you where you can recycle and get proactive.

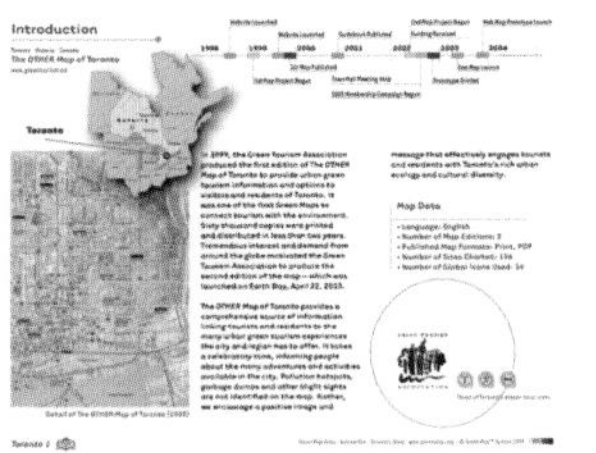

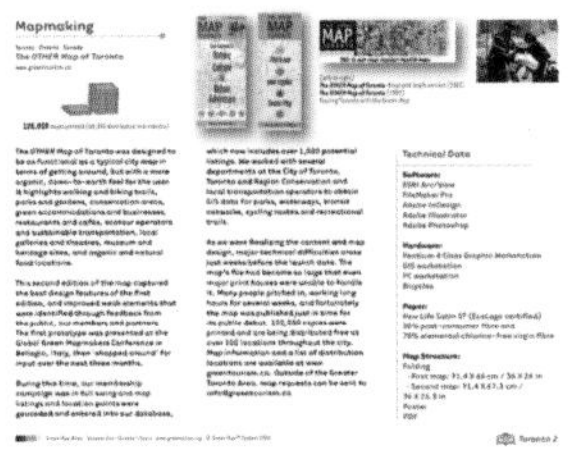

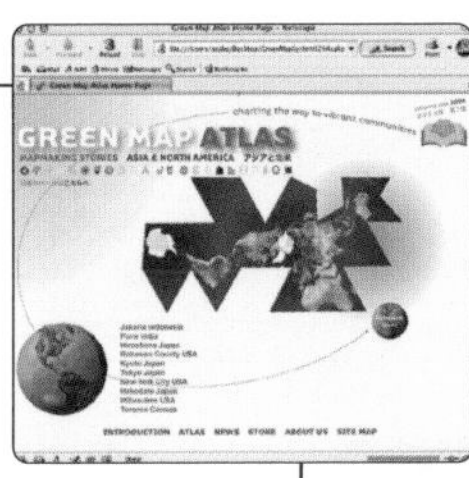

	Green Map System, USA	306
	Green Map System, USA	306
	Virtual or real versions, CD or printed paper	
	• *Dematerialized product/product-service-system* • *Educational service* • *Local economy-focused* • *Open source*	324 328 324 324

Ananda

These totemic bookmarks represent the iconographic language of Satyendra Pakhalé , full of energy, life and vitality. While most bookmarks are reduced to rectilinear nonentities, the Ananda almost wills the user to pick up a book, if only to have the tactile experience of inserting them inbetween the pages. Here's one bookmark that won't be consigned to the bin but becomes a personal token, a reminder of one of the true values of objects in our lives.

	Satyendra Pakhalé, the Netherlands and India	309
	De Vecchi, Italy	314
	Silver	
	• *Precious material* • *Durable*	325 328

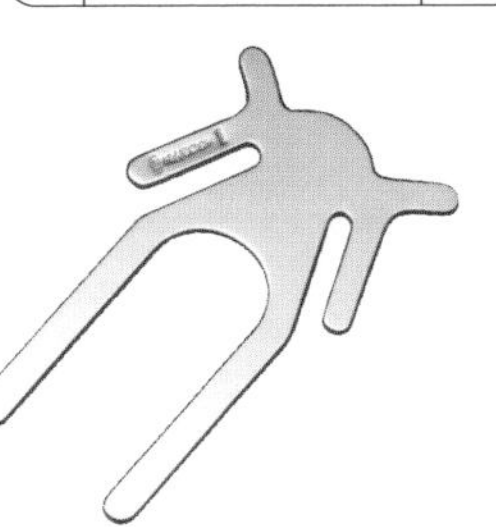

Power Grip

Unlike 'pistol grip' cordless drills and screwdrivers the Power Grip has a very compact but ergonomic handle, ensuring it is easily used with one hand and is much more versatile in confined spaces. A reduction in weight prevents user fatigue without loss of power. This design demonstrates significant reductions in materials used compared with the established leading designs for this market.

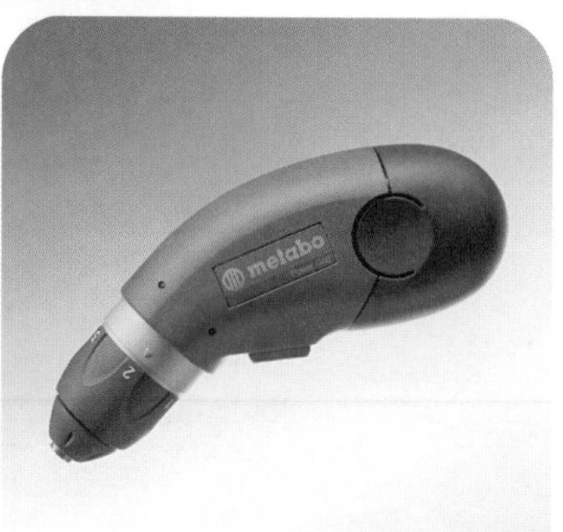

	Design Tech, Germany	305
	Metabo, Germany	318
	Various materials	
	• Reduction in materials used	326
	• Improved ergonomics	328
	iF Design Award, 2003	332

Fingermax

These finger brushes offe creative opportunities to those who find holding a conventional paintbrush difficult. Universal fitting is achieved by moulding a thermoplastic resin polymer in a spiral shape with an elliptical cross-section.

	Büro für Form, Germany	30
	Fingermax, Germany	315
	Polymer	
	• Universal design	328
	• Design for need	327
	iF Design Award, 2000	332

Pet Pod™

This quirky design makes a comfortable shelter and living space for a cat or small dog. The papier mâché gives insulation, so the Pet Pod is the ideal solution for pets housed in unheated buildings.

	Vaccari Ltd, UK	322
	Vaccari Ltd, UK	322
	Papier mâché	
	• Recycled materials	325
	• Compostable	325

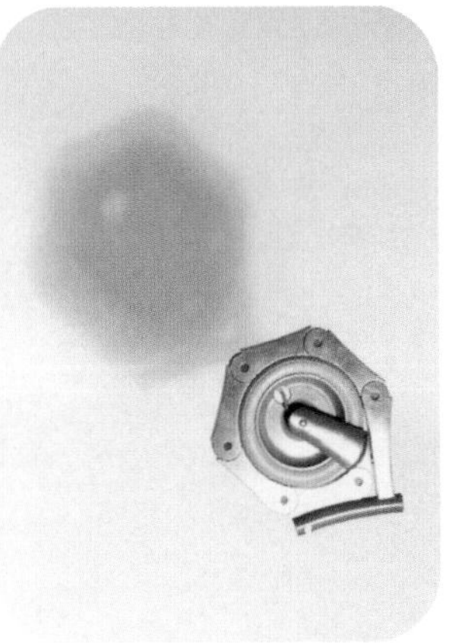

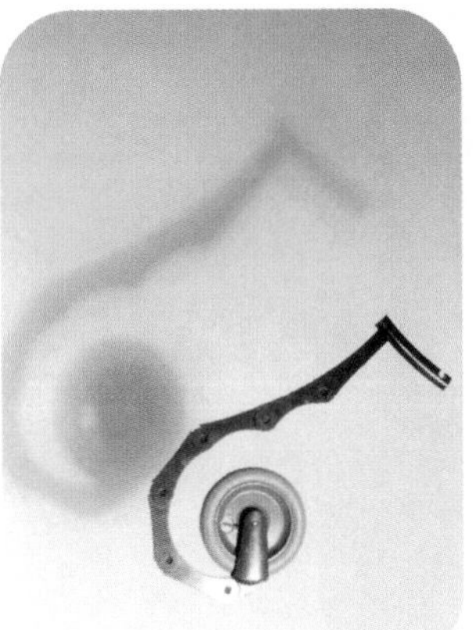

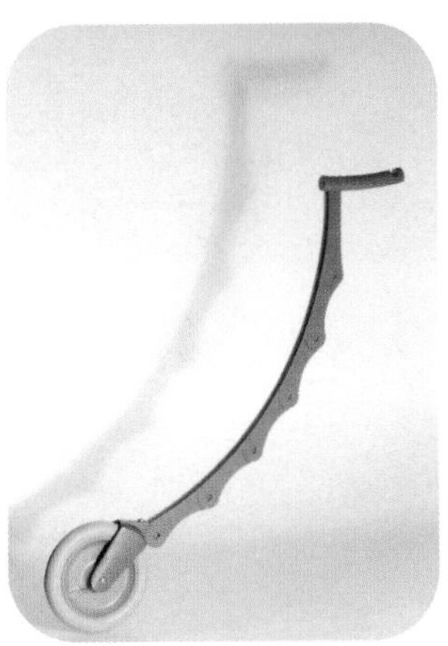

Kango

Here's a means of cutting down on the number of car journeys to local supermarkets. This mono-wheel trolley is capable of carrying a week's worth of groceries, is easily manoeuvred and, after use, is folded up into a handy package. Ideal for regular or casual use, for leisure or travel purposes, the Kango leaves all those ugly, two-wheel, tartan leatherette trollies in the shade.

✎	*Feldmann + Schultchen, Germany*	306
⚙	*Patented prototype – Feldmann + Schultchen, Germany*	306
▬	*Cordura fabric, rubber, fibre-reinforced plastic*	
♲	*• Encourages energy-efficient shopping*	327

2.0 Objects for Working

Work: An Evolving Concept

Although there is a tendency to think of the world as one huge post-industrial society, the reality is that there are myriad societies, some still firmly rooted in feudal agrarian systems, others heavily industrialized and still others dominated by service industries. It is therefore untrue, and possibly dangerously misleading, to think that everyone perceives problems of sustainability and work in the same way.

In our globalized economy, 'information' is just as much a raw material as timber, iron, steel or chemicals are in an industrialized society. The main difference is in their environmental impacts. In the information society the worker needs access to a workstation, which may be in the office, at home or somewhere in between. Mobile phone and wireless technology now means that the worker may not need to travel to a physical workplace, but on the other hand may be travelling more from the central office in order to actually do his/her job. Calculating whether there is a net energy reduction as a result of these behavioural shifts rapidly becomes complicated. In contrast, in places where manufacturing is still a significant industry, such as in India, China and other developing nations, the worker still has to travel to the factory for the purpose of fabricating a product. Information and industrial workers all consume finite resources and energy and produce waste, toxins and hazardous chemicals. Much of the industrial production in developing countries is to service the needs of consumer societies in developed nations, so the only shifts have been where the energy is consumed and where the environmental destruction occurs. All societies must therefore consider the design of products, materials, consumables and services that reduce their 'holistic' environmental impacts, and that include the embodied energy of transporting them. Retrofitting old systems with improved eco-efficient technology is also an important strategy to help re-utilize existing capital, infrastructure and resources.

Transporting people and distributing goods

Work involves transporting people, distributing goods or both. While electronic networks can reduce the need to move people physically, most work involves some travel. More efficient transport systems are therefore critical. This has been recognized by companies supplying buses, trains and other systems for public transport with significant advances in hybrid and diesel engine technologies accompanied by more experimentation with renewable technologies. Yet these technologies cannot deliver the kind of efficiency levels that we need to deal with dwindling energy and mineral resources. Above all, these public transport systems have to be coordinated to provide people with real flexibility and freedom.

The average supermarket, furniture store or trade outlet, especially in the developed North, will have products from all over the world. Transported over great distances, expending vast quantities of energy, a product's transport energy can sometimes exceed the energy used to make it. Reduction in packaging weight and volume is a perennial challenge to distributors. Even the smallest savings in packaging can represent huge savings in transport energy and waste production for the retailer or middle-man in the distribution chain. One-way-trip packaging can often be replaced by lightweight, reusable packaging systems, and an emphasis on local products sold in local markets could also result in large savings.

Working lightly: a sustainable day

More efficient working practices in the office and factory are aided by well-designed, durable, easily maintained products. It is now possible to have one machine to serve an office network with facilities to fax, photocopy, print and scan. Digital files can be shared on local and international networks making the paperless office a partial reality. Offices can be equipped with durable, modular furniture systems and carpets can be replaced under a lease-maintenance contract. Office consumables can use recycled content and reused components.

In industrial production facilities designers, in coordination with environmental managers, can reduce inputs of energy and materials and increase efficiencies in production and distribution. Waste streams provide another source of raw material and closed-loop recycling of process chemicals and materials ensures improved eco-efficiency, better profits and improved worker health. Design can help deliver reduced impacts on the environment, improved social benefits as well as profitability.

As a post-peak oil world is now challenging our imaginations, we might look to a shift in work practices that encourage massive energy savings, new office buildings and homes that become net energy generators rather than consumers and closed-loop systems to ensure materials are completely recycled. A sustainable working day in 2025 might involve some of the products that follow...

EcoClassic 30 and EcoClassic 50

As a phasing out of traditional tungsten lightbulbs begins in some members of the European Union, consumers are looking for viable replacement bulbs that give high-quality lighting, meet contemporary aesthetic needs, and save energy. The EcoClassic range by Philips, who have been producing energy saving lamps since 1981, aims to meet all these requirements. EcoClassic 30 is a bayonet fixture bulb that reduces energy consumption by 30% compared to traditional bulbs. It is available in a variety of watt ratings and surface finishes. EcoClassic 50 is a cap type, candle-shaped range of bulbs using dimmable halogen technology that gives 50% energy reductions compared to normal halogens, available in a range of surface finishes, that gives good choice for ambient, task or room lighting.

✎	*Philips Lighting, the Netherlands*	319
⚙	*Philips Lighting, the Netherlands*	319
▬	*Glass, metal, Electronics, inert gases*	
♻	*• Reduction of energy consumption*	326
✪	*iF GOLD award 2008*	332

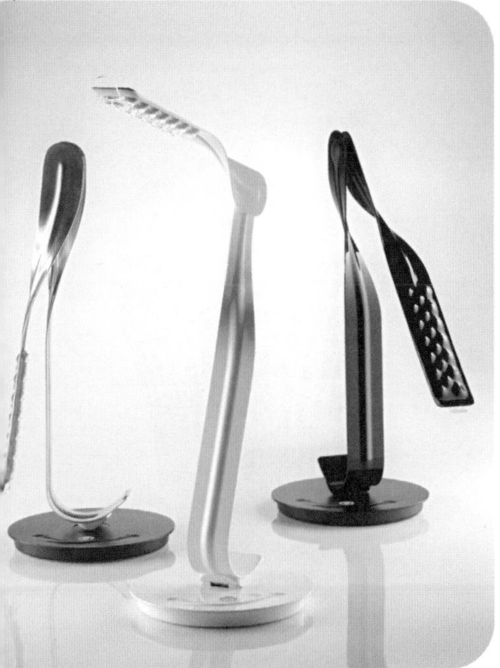

LEAF™ lamp

Combining eco-efficient technologies with ergonomic excellence that works with human intuition is an ambition for anyone striving to design more sustainable products. This sculptural low-energy LED task/ambient desk light offers users a choice between warm or cool light. Twenty LEDs use 8–9 Watts, 40% less than a compact fluorescent bulb, and will last for over 60,000 hours. Just touch the light to turn it on, slide a finger down the grove in the base to control intensity and adjust the upper and lower blades by 180 and 210 degrees respectively to achieve the desired angle of lighting. With its sleek, elegant design, this light attracts attention and will convert anyone to the benefits of eco-design.

✎	*Yves Behar, fuseproject, for Herman Miller Inc, USA*	304
⚙	*Herman Miller Inc., USA*	316
▬	*Steel, plastic, aluminium*	
♻	*• Minimization of energy consumption and materials*	326
	• Some recycled material content	325
	• Design for assembly, disassembly and long life, with a 5-year guarantee	326
	• 95% recyclable at end of life	325

LED lighting

LED technology is driving rapid developments in domestic and industrial illumination products. Maintaining backwards compatibility in the lighting industry is important for competitive, economic and environmental reasons. Bayonet and other industry-standard fittings to clusters of LEDs fitted to specialist bulbs produce a low-voltage, low-current, high-energy-efficiency option for everything from task lighting to advertising panel illumination. Since 1983, LEDtronics's research has developed LED technology. LEDs offer power savings of up to 80–90% compared with incandescent lamps, a wide range of lux and lighting spectra, and little or no heat output, so they are suitable where a cool environment needs to be maintained.

✎	*LEDtronics, USA*	317
⚙	*LEDtronics, USA*	317
▮	*Solid-state technology, optical-grade epoxy, electronic circuitry*	
♻	*• Low-voltage lighting* *• Reduction in materials used*	329 325

Master LED

Light Emitting Diodes (LEDs) have revolutionized office and commercial space lighting. The low energy requirements of successive generations of LEDs is constantly being improved, but what is most important is that it is easy to retrofit LED lamps to existing lighting systems. The Master LED series from Philips aims to make transition to LED lighting easy with this range of GU10, E27/NR63 and E27 bulb types that fit the majority of current luminaries. Based on the ultra-compact Philips Leuxeon Rebel power LEDs, the 7W Master LED has a light output equivalent to 40W GLS, 35W halogen or 9W CFLi lamp, offering energy savings of up to 80%. For a 24/7 lighting situation payback time is around one year and the bulbs have an ultra-long life at 45,000 hours at 50% failure.

✎	*Philips Lighting, the Netherlands*	319
⚙	*Philips Lighting, the Netherlands*	319
▮	*Various synthetic materials, LEDs*	
♻	*• Long-life product* *• Reduction in energy consumption*	329 326

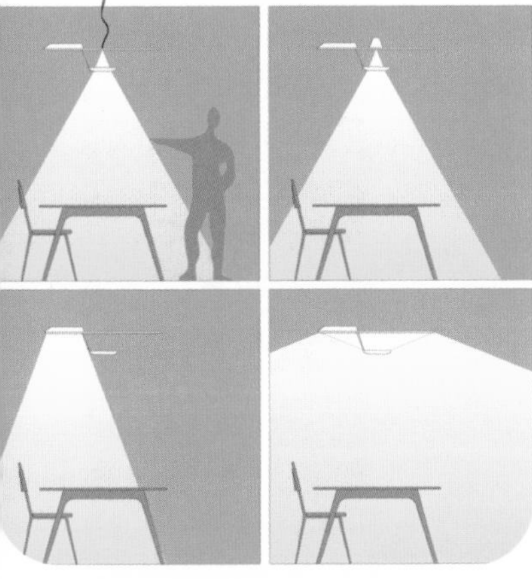

Zeno

Seasonal Affective Disorder, otherwise known by the appropriate acronym SAD, is a condition resulting from hormonal imbalance caused by inadequate exposure to sunlight. It is thought that many people suffer from SAD in temperate climates during the winter months. Office workers, who predominantly work in artificially lit environments, are a particularly susceptible group. Zeno addresses the issue of light quality by creating luminance using natural sunlight and a variety of artificial lighting spectra. Sunlight is captured using an external array and fed down optic fibres to a complex reflecting disc whose surface has a high reflection co-efficient. Natural sunlight can be blended with light from compact fluorescent bulbs (100W halogen or 70W iodine) to vary the luminosity and quality, and hence the atmosphere created by the lighting.

✎	*Diego Rossi and Raffaele Tedesco, Italy*	309
⚙	*Luceplan SpA, Italy*	318
	Various technosphere materials	
	• *Reduced energy consumption* • *Improved well-being*	326 328

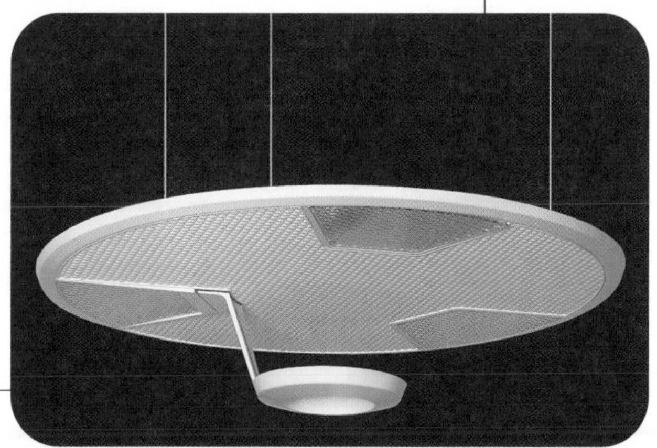

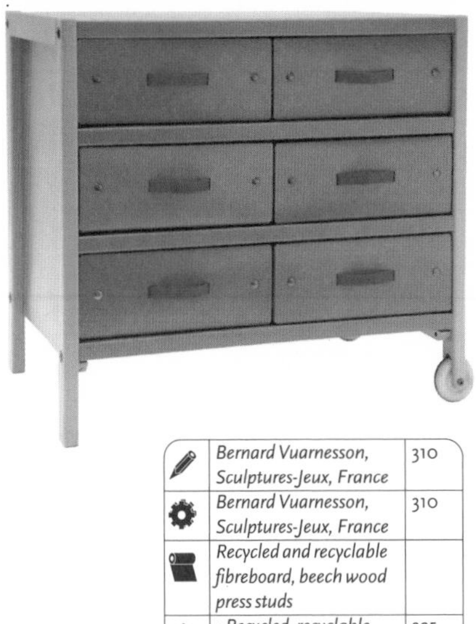

CLARK collection

It takes 10 to 15 minutes to assemble this flat-pack storage unit for home and office, made of 100% recycled and recyclable fibreboard and varnished solid beech wood. Drawer dimensions vary, so a number of variations are possible. Wheels facilitate moving the chest of drawers for cleaning or moving furniture.

✎	*Bernard Vuarnesson, Sculptures-Jeux, France*	310
⚙	*Bernard Vuarnesson, Sculptures-Jeux, France*	310
▮	*Recycled and recyclable fibreboard, beech wood press studs*	
♻	*• Recycled, recyclable and reusable materials* *• Self-assembly* *• Low-embodied energy*	325 327 325

Viper

Elliptical cross-section cardboard tubes made from recycled paper are connected to each other at top and bottom by a specially moulded plastic capping. Extensive articulation between adjacent tubes permits the screen to be rolled up when not in use.

✎	*Hans Sandgren Jakobsen, Denmark*	307
⚙	*Fritz Hansen A/S, Denmark*	315
▮	*Cardboard, plastic*	
♻	*• Recycled material*	325

Cartoons

Cartoons is a flexible, free-standing screen suitable for partitioning in domestic and office spaces. Corrugated paper board extracted from pure cellulose is stiffened at the edges with a closure of cold-processed, CFC-free polyurethane and at the ends with die-cast aluminium. This configuration allows the screen to be positioned in a sinuous style to suit the user and to be rolled up when not in use.

	Luigi Baroli, Italy	304
	Baleri Italia SpA, Italy	312
	Corrugated paper board, aluminium, CFC-free technopolymer	
	• *Renewable and recyclable materials* • *Clean production*	325 326

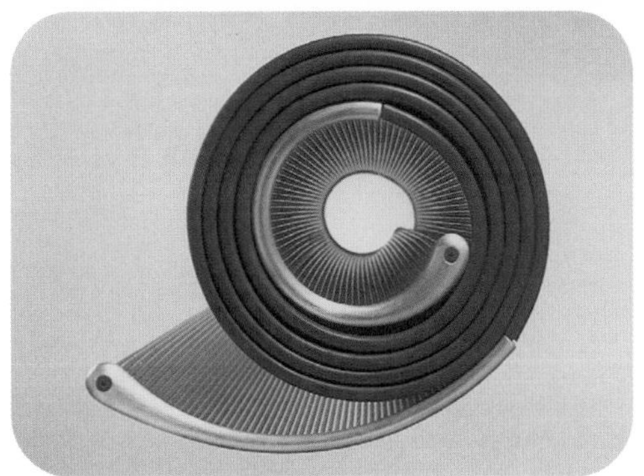

Cloud

Cloud was created as part of an ongoing project called Creative Meeting Places by OFFECCT. It is a portable, inflatable meeting room with an integral floor that is contained in a transport bag and comes complete with an integral fan to inflate the space. Made of white ripstop nylon, it is lined with an acoustically absorbent textile surface. Cloud aims to forge a relaxed, unpretentious atmosphere for fostering a different environment for communication. Inside this ethereal form, normal barriers and traditional 'roles' evaporate as people are cocooned in the whiteness of the interior. Although aimed at the corporate market, this sociable space has considerable potential to serve a wide range of communities.

	Monica Förster, Sweden	306
	OFFECCT, Sweden	319
	Ripstop nylon, acoustic textile, fan	
	• *Lightweight, portable, multifunctional building* • *New social space*	325/ 328 325

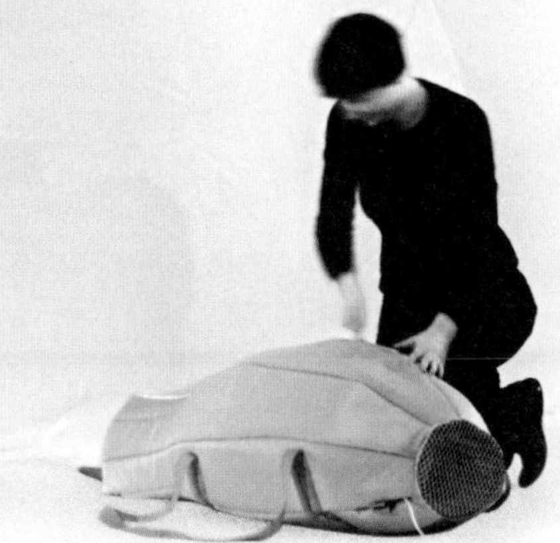

Jump Stuff, Jump Stuff II

Everybody customizes their domestic space, so why not the work space too? Through an extended series of projects and development of conceptual prototypes in the mid- to late 1990s, such as the Flo & Eddy workstation, Haworth examined the cognitive ergonomics of the desk area. The outcome is the Jump Stuff system, which allows individuals to select the components they require to maximize the functionality and comfort of their own desks. The spine of the system is a free-standing or panel-/wall-mounted rail to which the modular components can be attached. Whatever your regular tasks, you can attach and orient the appropriate accessory to the mounting rail. Different types of task lights can be attached to the rail and all the accessories can be easily adjusted for a 'hot desking' role. Although there are four basic variations to the system it is also possible to purchase each module independently so you can 'grow' the system to suit your needs.

✎	*Haworth, Inc., USA*	31
⚙	*Haworth, Inc., USA*	31
▮	*Various metals and polymers*	
♻	• *Multifunctional, modular system* • *Design for need*	32 32

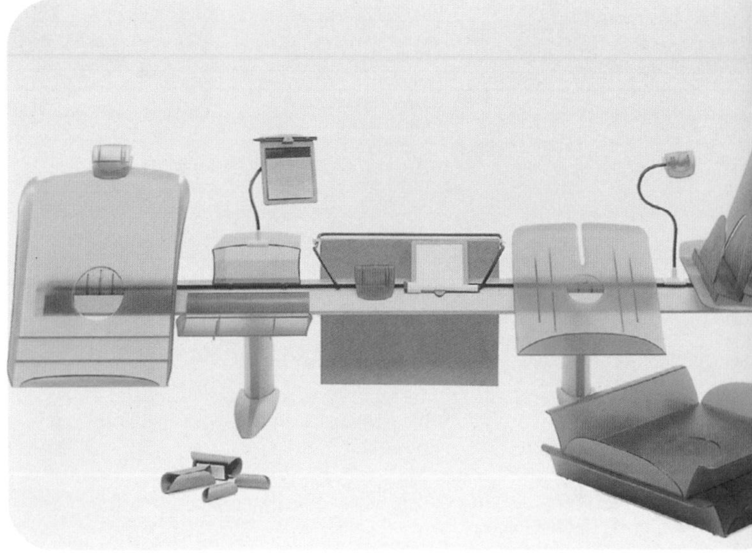

Unitable

Those who work from home or have occasional need for an office at home will appreciate the versatility of the Unitable. Height, length and angle of inclination of the work surface are all adjustable. Table sections are modular and hollow or solid depending on the user's requirements. Natural beech and birch, birch linoleum and laminates are available together with a birch ply stained black. All tops fix to a metal frame that accommodates table tops of 1.5–4 m (5–13 ft) length and 750–900 mm (30–35 in) width. For the tilting version of the table, the Unidrawing table, a lightweight multiplex webbed panel is formed of corrugated aluminium sandwiched between ply or laminate. And yet another variant, the Unifoldingtable, can be folded for use as a display board or conference table.

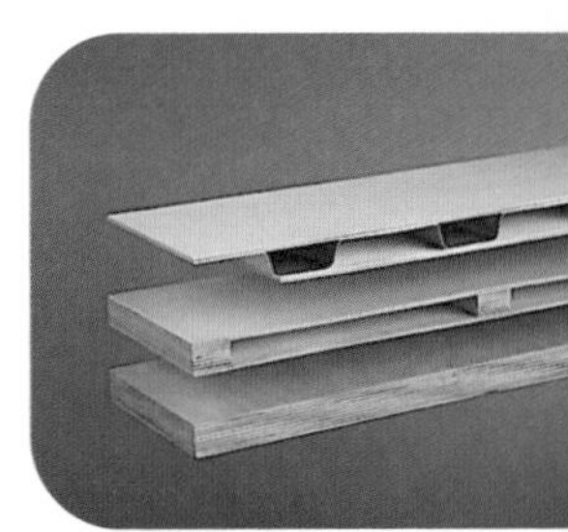

✎	*Atelier Alinea AG, Switzerland*	312
⚙	*Atelier Alinea AG, Switzerland*	312
▮	*Various woods, plywoods and wood laminates, corrugated aluminium, steel*	
♻	• *Multifunctional* • *Reduction in materials used*	328 325

Kant

This solid maple frame desk has a distinctive top with zigzag rear storage area where personal and professional office paraphernalia can be tucked away. This feature recognizes the instinctive behaviour of clearing a space for working while conveniently providing somewhere to store what's not needed. Further personalization is possible through the addition of accessories such as under the top drawers or a computer station shelf. The birch plywood top is covered with durable and easily cleaned laminate, FU, or linoleum finishes, available in a range of colours. It is available as a desk with a height of 74 or 111 cm (29 or 43⅔ in), or as a low table, 45 cm (17⅔ in) high.

✎	*Patrick Frey and Markus Boge, Nils Holger Moormann, Germany*	306
⚙	*Nils Holger Moormann, Germany*	308
	Solid maple, birch plywood, linoleum or FU laminate surface	
	• *Simplicity and modular accessories for personalization*	328
	• *Mostly made from renewable materials*	325
	• *Durable*	325
	iF Design Award, 2005	332

Celle chair

Herman Miller continue to innovate in ergonomic sustainable chair design for working environments with their latest model, the Celle chair, designed for 99% recyclability at end of life. Comfort is improved with Jerome Caruso's new Cellular Suspension™, pliable moulded polymer 'cells' that respond to the movement of the body. It is produced in Herman Miller's GreenHouse factory in Michigan that is ISO 14001 certified, uses renewable energy, and is also certified to the US LEED standard. Celle is getting close to being the world's leading sustainable office chair that can be part of a genuine closed loop business operation encouraging responsible resource use in manufacturing.

✎	*Jerome Caruso, for Herman Miller Inc, USA*	304
⚙	*Herman Miller Inc, USA*	316
	Steel, plastic, polyester textile	
	• *GREENGUARD™ certified for low indoor emissions*	330
	• *Cradle2Cradle Gold (some models Silver) award certification*	329
	• *Production facility uses 100% renewable energy*	325
	• *Returnable/recyclable packaging*	325
	• *Design for ease of assembly, disassembly & repair, and long life with a 12-year guarantee*	326

GUBI Chair II

The simple, modern, Scandinavian aesthetic forever finds new expression, in this case in the GUBI Chair II which is a delightful marriage of tubular steel tubing and recycled PET fibre. Two 'felt' mats of the PET fibre are moulded around the steel frame, creating a tactile, easy to clean, seamless seating pad that offers good comfort. There are several versions, including a stackable option, armchair or lounge chair and plain or two-colour variants. Elsewhere in the GUBI collection there is a stackable table and range of free-standing room dividers.

✎	*Komplot Design, Denmark Wilkhahn, Germany*	307
⚙	*GUBI A/S, Denmark*	316
	PET fibres, steel tubing	
	• *Recycled and recyclable materials*	325

Leap™ seating

Building on the lessons learnt from the design of the Protégé Chair in 1991, Leap™ Seating is one of Steelcase's leading products with respect to recycling, waste reduction and low-impact manufacturing. The basic design is very durable but parts can be easily removed for repair or upgrading if required. At least 92% of the chair's parts are recyclable and the cushioning used in the upholstery is made of 50% recycled PET. During manufacture adhesives and paints with no or limited volatile organic compounds (VOCs) and water-based metal-plating processes considerably reduce aquatic and aerial emissions. Leap is Greenguard® Indoor Air Quality certified. Employee working conditions at the Grand Rapids, Michigan, factory have also been re-engineered to provide a more healthy environment. What will happen to the chair at the end of its life has not yet been defined, but leasing and take-back are options all responsible manufacturers will have to consider in the near future.

✎	*Steelcase, Inc., USA*	321
⚙	*Steelcase, Inc., USA*	321
▬	*Recycled PET, plated steel, polyester*	
♻	*• Design for disassembly and ease of repair* *• Recycled and recyclable materials* *• Low-impact manufacturing*	326 325 326

X & Y chairs

Available in X or Y configurations, these versatile stacking office or conference chairs are robustly constructed. Improved comfort is afforded by the moveable backrest that is on spring bearings. The range comes with or without upholstery depending on user requirements. Both X and Y versions can be assembled into benches of two, three or four seats by the inclusion of an additional beam.

✎	*Emilio Ambasz & Assoc., USA*	305
⚙	*Interstuhl Büromöbel GmbH, Germany*	317
▬	*Metal, plastics, textiles*	
♻	*• Durable* *• Multifunctional*	325 325
✪	*iF Design Award, 2003*	332

Aeron

The Aeron chair represents a steep change in the way office chairs are designed. It is manufactured in three sizes to accommodate diversity of the human form and weight, making it suitable for users up to 136 kg (200 lb) in weight and from the first percentile female to the ninety-ninth percentile male. It has very advanced ergonomics. Pneumatic height adjustment, a sophisticated Kinemat tilt system and the Pellicle, a synthetic, breathable, membrane, are components of the seat pan, which adjusts to individual body shapes. The manufacturing process uses less energy than conventional foam construction and the use of discrete components of synthetic and recycled materials facilitates disassembly and ease of repair for worn components (which are subsequently recycled). Such design improves the longevity of the product. Components are made of one material rather than a mixture of materials to facilitate future reuse and recycling.

✎	*Bill Stumpf and Don Chadwick, Herman Miller Inc, USA*	310
⚙	*Herman Miller Inc., USA*	316
▬	*Plastic (PET, ABS, nylon and glass-filled nylons), steel, aluminium and foam/fabric*	
♻	*• Improved ergonomics* *• Design for disassembly, recycling and remanufacturing* *• Single-material components*	328 326 327

Picto, FS line

Wilkhahn initiated a project in 1992 entitled 'Environmental Control' with the support of the Ministry for Environmental and Economic Affairs for the state of Lower Saxony. Following an audit of their corporate eco-balance of inputs and outputs, teams were set up to reduce environmental impacts in production and to select materials within an integrated IT framework. The Wilkhahn range of office seating is designed to minimize polluting processes during production. Chrome plating of metals is avoided and upholstery is made from durable, wear-resistant wool and polyester fabrics without gluing or welding. All furniture is easily assembled, disassembled and maintained, and individual components can be recovered for recycling upon disassembly.

✎	*ProduktEntwicklung Roericht, Germany*	309
⚙	*Wilkhahn, Germany*	323
▬	*Pure-grade metals, thermoplastics*	
♻	*• Clean production* *• Design for disassembly and recycling*	325 326/327

‹irra™ work chair

›llowing on from the ‹ccess of the upmarket ‹ron® office chair in ‹94, the Mirra™ aims provide levels of ‹mfort, performance and ‹cyclability to the mid- ‹ice work chairs. Herman iller's participation in ‹ extensive study of body ‹ape and size conducted ‹ CAESAR (Civilian ‹merican–European ‹rface Anthropometric ‹esource) provided ‹ckground research for ‹e Mirra™. Laser-scanning ‹e seating positions of ‹ndreds of people in ‹fferent postures provided ‹w data to assist with ‹ design. In particular, ‹e differences between the ‹ck shape of men and women in different postures helped develop the one-piece TriFlex™ backrest with built-in flex zones and FlexFront™, a flexible, adjustable front edge that permits individual control of the depth of the seat. Further features that allow the user to control configuration include the Harmonic™ tilt mechanism. Herman Miller claims that Mirra™ is designed to suit 95% of the world's population. In accordance with the manufacturer's stringent Design for the Environment (DfE) protocols the chair's lifecycle has been examined to assure that its material chemistry, recyclability, manufacturing, packaging and ease of disassembly minimize environmental impacts. At end of life 96% of the materials can be recycled.

✎	*Studio 7.5, Germany*	309
⚙	*Herman Miller Inc, USA*	316
▮	*Various recyclable materials*	
♻	• *Recyclable materials* • *Lifecycle analysis*	325 324

Torsio

Comprising just two separately moulded wooden parts, the Torsio range is designed to be stackable and can be linked in rows for conference seating. The twisted back section tapers from 22 mm (1 in) at the feet to 7 mm (¼ in) for the backrest, giving it a subtle flexing for improved comfort. Surface finishes include maple, walnut or black.

✎	*Hanspeter Steiger Designstudio, Switzerland*	306
⚙	*Röthlisberger, Switzerland*	320
▮	*Wood laminates*	
♻	• *Reduction in materials used* • *Renewable materials* • *Durable*	325 325 325
✪	*iF Design Award, 2003* *Swiss Design Prize, 2003*	332

‹undeala medium ‹oard screen

‹he original Sundeala ‹ompany began ‹anufacturing fibreboard ‹om waste cellulose in ‹898, and for the last 70 ‹ears Sundeala boards ‹ave utilized recycled ‹ewsprint as the primary material. 'K' quality unbleached natural board is for interior use, while 'A' quality with natural binders and colouring to reduce moisture penetration is suitable for sheltered exterior use.

✎	*Celotex, UK*	313
⚙	*Celotex, UK*	313
▮	*Recycled newsprint, natural binders*	
♻	• *Waste materials used* • *Recycled and recyclable material*	326 325

softseating & softwall

These sensuous, sculptural designs enliven any office environment, provide flexibility and an intelligent solution to enable existing spaces to be altered as needed. In the words of the molo design team, 'We think the most effective answer to a sustainable architecture is to create interior environments that people can adapt for change in use, occupancy or fashion'. Each flat-pack component of the softseating or softwall is a honeycomb structure made of tissue paper (15% recycled fraction from old office paper), unbleached kraft paper (50% recycled cardboard box fibre) and a non-woven polyethylene textile (5% recycled fraction), which is compressed into a 'book' less than 50 mm (2 in) thick. The 'books' are unfolded like Chinese lanterns, and then arranged into the desired interior landscape. The wall range is available in paper and textile variations, in translucent white and opaque black, and the cellular honeycomb structure also serves to dampen sound. Seating consists of stools, benches and sofas, all of which have magnetic end panels allowing them to be connected to other pieces in the series.

✎	*Molo Design, Canada*	308
⚙	*Molo Design, Canada*	308
	Tissue paper, unbleached kraft paper, polyethylene non-woven textile	
	• *Recycled and recyclable materials*	325
	• *Flexible, multifunctional design*	328
	• *Lightweight, efficient use of materials*	325
	• *Non-toxic elements*	326
	INDEX award 2005	

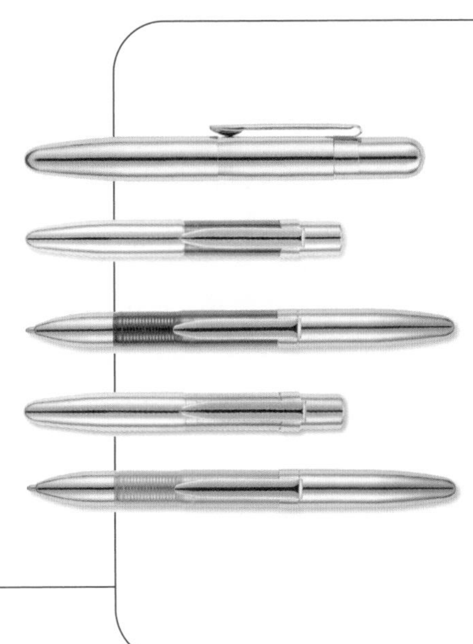

Millennium Pens

Since 1948 the Fisher Space Pen Company has manufactured high-quality pens using their unique pressurized brass-cased ink refills. The company claims that the seals in the refill are so good that the thixotropic ink will not dry out for over a hundred years. Pressurized with nitrogen gas to 50psi, the ink is delivered to an ultra-hard tungsten carbide ball held in a stainless-steel collar and is capable of flowing at temperatures of –45 to +121° Celsius. Released for the millennium 2000 celebrations, the Millennium Pens are made of extremely durable titanium nitride and chromed steel in a black or gold finish. These are luxury pens that should last a lifetime.

Designer	*Fisher Space Pen Co., USA*	315
Manufacturer	*Fisher Space Pen Co., USA*	315
Materials	*Titanium nitride, chrome, brass, tungsten carbide, stainless steel, ink*	
Sustainability	*• Durable materials and product* *• Repairable and refillable*	325 329

[R]arkable Medium [Coll]ection

[Foll]wing on from the [succ]ess of the Remarkable [recy]cled pencils the [com]pany has diversified to [prod]uce a range of rulers, [writi]ng pads, pencil cases, [pens] and colouring pencils [usin]g recycled or certified [mate]rials. Rulers are made [from] seven recycled [vend]ing cups; pads from [recy]cled paper and board; [penc]il cases from recycled [t]yres and colouring [penc]ils from FSC certified [sust]ainable timber. As [Euro]pean legislation comes [into] effect regarding the [disp]osal and recycling of [elect]rical and electronic [equi]pment (the WEEE Directive) Remarkable has produced pens made from recycling plastic components from computer printers. Similarly old fridges are being tested for recycling materials to produce new fridge magnets. Significantly, Remarkable struck a deal with two leading supermarket companies in the UK – Tesco and Sainsbury's – to sell recycled branded products, showing that retailer and consumer perceptions are shifting in favour of certain types of recycled goods.

Designer	*Remarkable (Pencils) Ltd, UK*	320
Manufacturer	*Remarkable (Pencils) Ltd, UK*	320
Materials	*Various recycled materials*	
Sustainability	*• Recycled materials*	325

Paperfile

Unused billboard posters find a new life as abstract graphical folders for storing precious notes, documents and paperwork. These tough everyday files have out-lived the semantics of the advertising hoarding, their corporate messages deconstructed and abandoned.

Designer	*Jan Neggers, the Netherlands*	308
Manufacturer	*Goods, the Netherlands*	316
Materials	*Paper, printing inks*	
Sustainability	*• Reused materials*	327

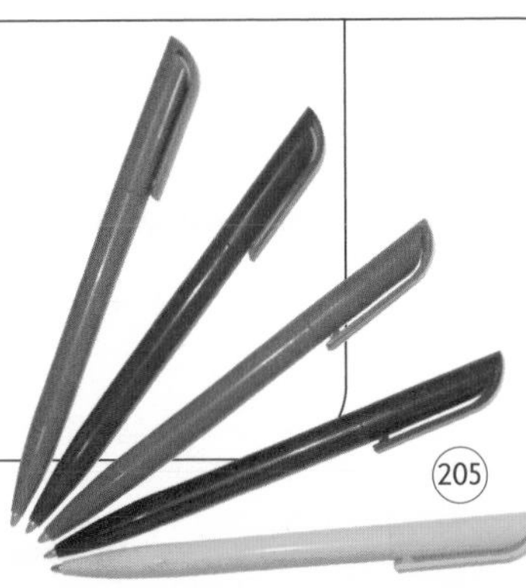

Save A Cup

Drinks vending machines daily consume vast quantities of standard 80 mm (3 in) polystyrene cups to satisfy the thirst of office workers and users of public spaces. All those spent cups – what a waste! Save A Cup has organized direct or third-party collection of used cups in all the major UK cities, using specially designed bins and machines to shred the cups. Companies registered with the UK's Environment Agency can obtain a Packaging Recovery Note (PRN) for the tonnage recycled to comply with the UK Packaging Waste Regulations. The feedstock recyclate is suitable for low-grade use such as pens, rulers and key rings.

✎	*Save A Cup Recycling Company, UK*	320
⚙	*Save A Cup Recycling Company, UK*	320
▮	*Polystyrene*	
♻	• *Recycled materials*	325

EcoStapler™

This nifty little 'stapler' device holds paper together without the need for a staple by cutting and folding the corners of the sheets to join them together. It is a small gesture, but it is estimated that if everyone in offices in the UK alone decided to use the EcoStapler over 72 tonnes of metal would be saved each year. It might even save time in the hunt for the elusive staple box that never seems to be there when you need it...

✎	*Ecozone (UK) Ltd, UK*	314
⚙	*Ecozone (UK) Ltd, UK*	314
▮	*ABS plastic*	
♻	• *Saves resources* • *Facilitates office paper recycling*	325 325

USBCELL™

An estimated 15 billion Alkaline batteries end up in landfills each year, but there really is no need to buy a disposable AA battery ever again. The ingenious USBCELL™ 1.2v 1300 mah NiMH battery can be fully recharged within five hours, wherever you can find a powered USB connection on a PC or other device. Simply hinge back the plastic cap, plug it into a USB port, and the in-built intelligent charger in the battery does the rest. (It can also be recharged in an approved NiMH charger at 250 ma for seven hours.) Even completely depleted batteries can be recharged in a few minutes for a short boost, enabling you to keep using the batteries for cameras, game stations, TV remotes and so on.

✎	*Moixa Energy Ltd, UK*	318
⚙	*Moixa Energy Ltd, UK*	318
▮	*NiMH rechargeable battery, plastic cap, charger, metal*	
♻	• *Material reduction – saves on disposable batteries* • *Saves landfill and non-recycling of precious materials*	325 329
✪	• *iF Product Design Award 2008* • *Observer Ethical Awards 2008*	332

Remarkable recycled pencil

Used polystyrene cups from vending machines are shredded and re-processed into a new 'plastic alloy', in which graphite and other materials are mixed with polystyrene and extruded in a special die to create a new type of pencil. It performs as well as traditional 'lead' pencils and helps reduce consumption of the timber that traditionally encases the lead.

Designer	*Edward Douglas Miller, UK*	308
Manufacturer	*Remarkable (Pencils) Ltd, UK*	320
Materials	*Recycled polystyrene, graphite, additives*	
Eco	*• Recycled materials* *• Reduced resource consumption*	325 326

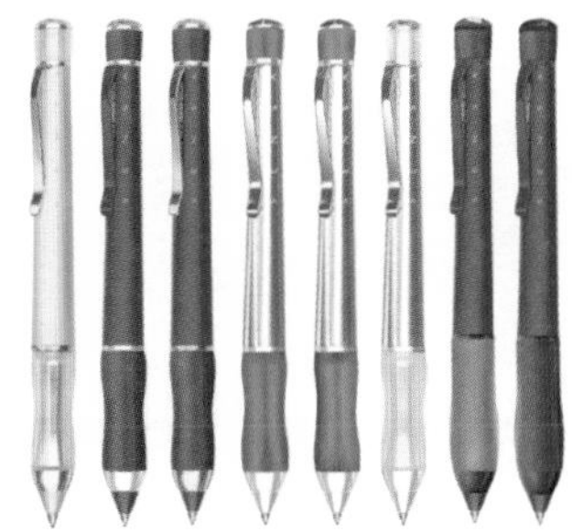

Sensa™ pen

Gripping a pen for an extended period can cause discomfort. A soft, non-toxic gel around the grip area moulds itself to the user's fingers as it warms up and consequently improves comfort. Once the gel cools it returns to its original shape.

Designer	*Boyd Willat, USA*	311
Manufacturer	*Willat Writing Instruments, USA*	323
Materials	*Metal, gel*	
Eco	*• Improved ergonomics*	328

Karisma

Sanford Corporation is the world's largest manufacturer of pencils based upon waste wood products, a mixture of wood flour and polymers. All wood-cased pencils manufactured by Sanford UK use wood from managed forests and, where possible, pencils are protected by water-based varnishes, which are hardened by ultraviolet light, rather than using solvent-based inks. Packaging and plastic waste are recycled at the production plant.

Designer	*Sanford, US*	320
Manufacturer	*Sanford, US*	320
Materials	*Wood, water-based varnishes*	
Eco	*• Recycled materials* *• Clean production* *• Supply-chain management*	325 326 326

Digital iR series

Twenty-nine of Canon Deutschland's copiers are certified to the German Blue Angel eco-label, including 24 copiers in the Digital iR series, all-in-one networked devices capable of fax, scan, email and scan-to-database operations. Even the smallest in the range of black/white copiers, the iR1020, offers a very capable specification including 20 ppm output, network printing, double-sided document creation and a small machine footprint. At the top end of the range the Blue Angel label iR5075N is a comprehensive multi-functional device capable of printing at 75 ppm, 'one pass' duplex scanning and document folding. In standby mode it consumes just one watt of electricity. In addition to providing a range of new copiers to meet every office requirement, Canon has copier remanufacturing plants in the USA and Germany, and re-conditioning plants in Japan and China, so reliable used copiers are available too. Take-back of old copiers is most advanced in Germany where regulations, in the form of the Electronic Scrap Ordinance, ensure appliance manufacturers are responsible for the collection and recycling of electronic products no longer required by consumers. Copier consumables have also been targeted to reduce waste. Canon Bretagne in France is one of three global plants that act as regional collection centres for old cartridges.

(designer)	*Canon Deutschland GmbH, Germany*	313
(manufacturer)	*Canon Deutschland GmbH, Germany*	313
(materials)	*Various materials and electronics*	
(strategies)	*• Improved functionality*	328
	• Reduced materials and consumables	325/ 328
	• Blue Angel Eco-label	330

Xerox WorkCentre™ M118 and 7655

Xerox has long seen itself as a document management business rather than a manufacturer of copiers, printers and other office equipment. This is a strategic position that encourages the design and manufacturing of machines that are efficient, easily serviced and repaired. The company operates a worldwide take-back and reuse system, and the Xerox Green World Alliance is a network focused on encouraging recycling of printer supplies. Xerox's extensive range of WorkCentre™ multi-functional copier, printer, scanner and fax machines enable one machine to replace the work of several. Two models in particular stand out: the WorkCentre™ M118 and the WorkCentre™ 7655. The M118 is a desktop black-and-white letterhead machine with automatic two-sided copying/printing at a speed of 18 pages per minute (ppm) with a toner cartridge capacity of 9,000 pages, and is Energy Star compliant. The 7655 is a stand-alone system, with colour reproduction up to 40 ppm and black and white up to 55 ppm with a print quality of 2,400 dpi, that can be customized to suit the users' needs by adding additional modules to increase functionality. With its optional 250-sheet Duplexing Automatic Document Feeder, multiple binding and folding options, this is a veritable printing press. Xerox produces Product Safety Data Sheets and Materials Safety Data Sheets regarding information on mechanical, electrical, hazardous substances and other environmental attributes of its machines. Under normal operating conditions the toners, which are typically styrene-acrylic, styrene-butadiene or polyester polymers, are entirely stable with good adherence to paper.

(designer)	*Xerox Corporation, USA*	323
(manufacturer)	*Xerox Corporation, USA*	323
(materials)	*Various electronic components, synthetic polymers, metals*	
(strategies)	*• Reducing energy consumption, consumables*	326
	• Multifunctional Product-Service-Systems	328
	• Take-back and recycling programmes	329
	• Safety Data Sheets as standard	325

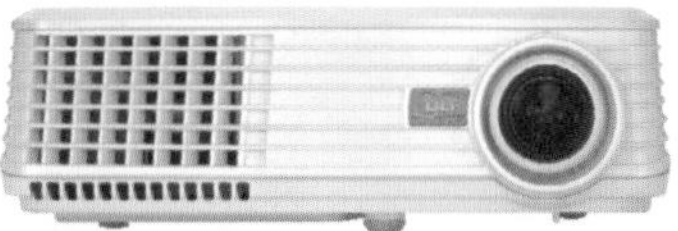

EC NP100

ıis durable digital ojector with SVGA solution (800 x 600 ‹els) for high image ıality and 2,000 ANSI mens brightness, has amp lifetime of up to ooo hours in 'Eco' mode. e removable control anel doubles as a remote ntrol and security control prevent unauthorized e. Perfect for small terprises that need have robust, energy- ficient equipment tailored suit their needs.

	NEC Corporation, Japan	318
	NEC Corporation, Japan	318
	Various	
	• Long life *• Reduced energy consumption*	325 326

ESPRIMO E series

Fujitsu was one of the first manufacturers to implement a large-scale recycling programme in 1988 and to introduce a Blue Angel labelled personal computer to the market in 1993. The company has ranked consistently in the top quartile of global PC manufacturers in Greenpeace's Guide to Greener Electronics, with very good scores for reduced energy consumption. The ESPRIMO E is a compact computer that can be orientated vertically to save desk space, has a halogen-free mainboard and achieves much improved energy efficiency with the Intel® Q43/Q45 chipset and an optional 80% efficiency power pack. The ESPRIMO E 5925 EPA is compliant with the new Energy Star® Version 4.0 rating, currently only met by approximately 20% of the world's computers.

	Fujitsu Technology Solutions, Germany	315
	Fujitsu Technology Solutions, Germany	315
	Various electronic components, synthetic polymers, metals	
	• Reduction of energy consumption *• Product take-back and recycling programme*	326 329

ngle-handed ɛyboard

ccess for all' is the clarion y of the proponents of e Information Age but nventional keyboard sign denies access to dividuals with disabilities. altron's single-handed d head/mouth stick yboards are tools to lp them overcome this ırdle and to enjoy what hers take for granted. e Etype keyboard is also tool for those suffering om repetitive strain injury SI), that most modern of ailments. A curved keyboard and palm resting pads ensure less tiring movements.

	PCD Maltron Ltd, UK	319
	PCD Maltron Ltd, UK	319
	Various plastics, electronics	
	• Improved user health *• Improved access to information for those with disabilities*	328 328

IFCO returnable transit packaging

Eleven standard sizes of flat-pack, reusable plastic containers with ventilated sides are meant for transit packaging for all types of fresh produce. The IFCO system is used in over 30 countries. Compatible with loading on Euro and ISO pallets, the units are of constant tare and weight-to-volume ratios are economical, they are easily cleaned, and when folded reduce storage space requirements by 80%. At the end of their useful working lives the polypropylene is recycled. A study by Ecobalance Applied Research GmbH revealed significantly less environmental impact from the IFCO system than from conventional one-way corrugated cardboard boxes.

✎	*International Food Container Organization (IFCO), Germany*	317
⚙	*International Food Container Organization (IFCO), Germany*	317
▮	*Polypropylene*	
♻	• *Recyclable single material* • *Closed-loop system* • *Reductions in transport energy* • *Blue Angel eco-label*	325 326 327 330

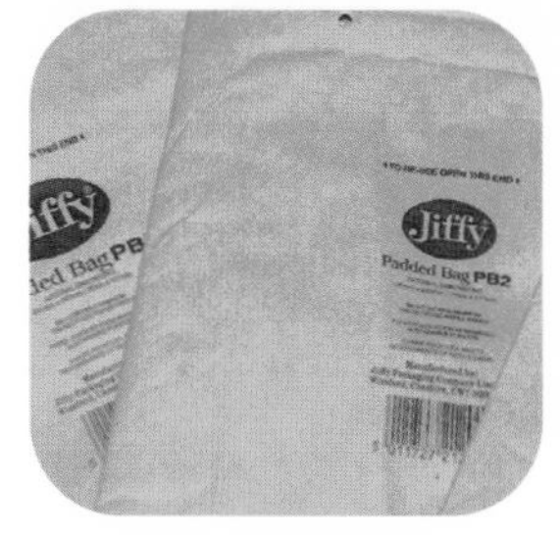

Jiffy® padded bag

Jiffy produces a range of bags to protect goods transported via postal and courier systems. The padding is made of 72–75% recycled, shredded newsprint and the exterior is a tough brown paper, which permits the bags to be reused.

✎	*Pactiv Corporation, USA*	319
⚙	*Pactiv Corporation, USA*	319
▮	*Recycled newsprint*	
♻	• *Recycled materials* • *Reusable product*	325 325

Pallecon 3 Autoflow

Made of sheet and solid steel, this container is suitable for transporting a wide range of industrial liquids from pharmaceutical products to foodstuffs. It is emptied via a sump through valves, which are recyclable, and is easily cleaned between consignments. It can be stored into one-third of its original size by folding the sides, thus saving valuable cargo space. Meeting EU standards and with an expected service life of ten years, the Pallecon 3 Autoflow can be entirely recycled at the end of its useful life.

✎	*LSK Industries Pty Ltd, Australia*	318
⚙	*LSK Industries Pty Ltd, Australia*	318
▮	*Steel*	
♻	• *Single material* • *Reusable product* • *Recyclable* • *Potential reduction of transport energy*	325 325 325 327
✪	*iF Design Award, 2000*	332

esswood pallet

ılike traditional timber llets, the 'Inka' pallets n't need to be fixed :h staples or nails since ey are manufactured m recycled timber ste bonded with water- istant synthetic resins. her advantages over nventional pallets lude more compact cking and lower tare ight. Standard pallet es meet current ropean regulations d are recyclable. of January 2000, wood is included within the EU packaging waste recovery and recycling regulations, so the reduced wood content of these pallets lowers the costs associated with these obligations.

	Inka Paletten GmbH, Germany	317
	Inka Paletten GmbH, Germany	317
	Waste timber, resins	
	• *Recycled materials* • *Reduction in transport energy*	325 327

VarioPac® / Ejector®

The struggle to extract CDs from their protective covers is consigned to the past thanks to this well-conceived and -manufactured product. Simply press the lever in the corner to eject the CD. An assessment by FH Lippe of the VarioPac Rover conventional cases revealed that the tactile, translucent and shatter-proof Metocene (metallocene polypropylene) requires 46% less energy during manufacturing by injection moulding and reduces transport volume by 33% – ample proof that this redesign reduces environmental impacts.

	VarioPac Disc Systems GmbH, Germany	310
	Ejector Systems GmbH, Germany	314
	Metocene X 50081	
	• *Reduction in materials used* • *Reduction of embodied and transport energy*	325 326/ 327
	iF Ecology Design Award, 2000	332

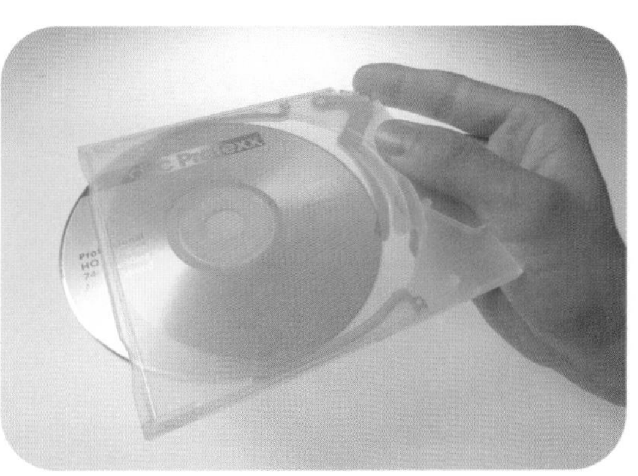

Disk Lev

There are numerous synthetic plastic CD 'jewel' cases on the market but most need secondary packaging in transit, which is often discarded by the user and inherently high in embodied energy. Disk Lev is a lightweight cardboard design that can be mailed but also provides protection for archiving and storage. When folded, a central tab fixes the CD in position and on opening, the CD is automatically raised from the tab, without applying any pressure, for easy removal.

	Philipp Prause, Austria	309
	Ernst Schausberger & Co. GmbH, Austria	315
	Cardboard	
	• *Reduction in materials used* • *Recyclable or compostable*	326 325
	iF Design Award, 2003	332

Recopack, Recozips, Recocubes, Expandos and Geami

Rent-a-Green Box claims that the Recopack lidded storage box, made from 100% recycled plastic trash from local landfills, is the first zero-waste range of packing boxes. Lightweight, strong and nestable, these boxes are delivered to the user then picked up after use, whether for moving products or personal goods. The Recozips, used to secure the box lids, are made from 100% recycled bottle tops. If you have fragile items to transport in the Recopacks, or wish to ensure safe transit for anything you've posted or entrusted to courier services, then there is a range of packaging materials that can assist. Recocubes are compostable packing cubes made of recycled newsprint 'sludge' that is not suitable for another printing cycle, and they are an ideal replacement for Styrofoam. During manufacturing the Recocubes are injected with sugars, starches and B12 vitamin to ensure easy biodegradation at end-of-life. Another packing material option is Expandos, made from 100% recycled paper fibre: these unique triangulated shapes interlock around the objects to be protected. And if you need a replacement for the ubiquitous bubble wrap that is often difficult to recycle, then use the company's Geami textured paper wrap, made from 100% recycled cardboard sludge, that 'self locks' when wrapped around objects so avoiding the use of tape.

✎	*Rent-a-Green Box, USA*	320
⚙	*Rent-a-Green Box, USA*	320
▮	*Recycled plastics, cardboard, paper*	
♲	• *100% recycled materials*	325
	• *Zero-waste packaging products*	327
	• *Reusable, multi-size containers*	325

Schäfer Eco Keg

Die-cast, injection-moulded, thermoplastic base and top clip on to the stainless-steel body of this beverage container, avoiding the need to glue in place rubber or polyurethane sealing rings. All parts can be disassembled for repair, replacement and pure-grade recycling. The container is suitable for all automated KEG plants and can be stacked more easily than conventional containers, making for space savings and improved transport efficiency.

✎	*Schäfer Werke GmbH, Germany*	320
⚙	*Schäfer Werke GmbH, Germany*	320
▮	*Stainless steel, thermoplastic*	
♲	• *Design for durability, disassembly and recyclability*	325/326
	• *Reduction in transport energy*	327
✪	*iF Ecology Design Award, 2000*	332

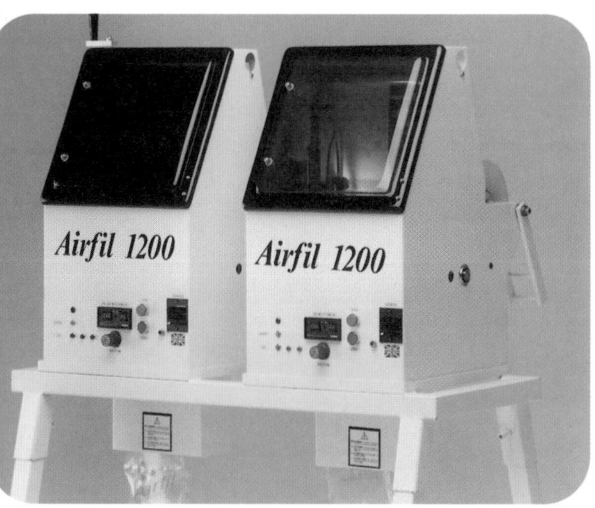

Airfil

This lightweight packaging made of recyclable PE uses air as the shock-absorbing material to protect goods in transit. Airfil air bags are produced in standard and bespoke sizes, providing a viable, less expensive alternative to polystyrene 'chips' and 'bubble wrap'. The Airfil system significantly reduces storage space requirements and allows a reduction in the thickness of the outer packaging material. It is also reusable, clean and free of dust.

✎	*Amasec Airfil Ltd, UK*	312
⚙	*Amasec Airfil Ltd, UK*	312
▬	*Polyethylene*	
♺	*• Reduction in materials used* *• Recyclable materials (plastic)*	326 325

3M Spezialfolie 8000/8000HL

Contamination of recycled plastic feedstock with unknown types of plastic can render recyclate unusable and damage the production plant. It is not always possible to create labels by embossing the information on components or products, so 3M have produced a stick-on label that can be used when recycling ABS and polycarbonate, ABS/polycarbonate mixes and polystyrene. Such plastics are common in the electronics industry where identification of materials at disassembly is now more critical as the EU WEEE Directive on recycling of electronic equipment is enforced over the next few years.

✎	*Hiep Nguyen, Gerald Schniedermeier, Yolanda Grievenow, Den Suoss, 3M Deutschland, Germany*	312
⚙	*3M Deutschland, Germany*	312
▬	*Plastics*	
♺	*• Materials labelling* *• Encourages recycling*	327 327
✪	*iF Ecology Design Award, 1999*	332

Earthshell Packaging®

Earthshell Packaging® is a heat-moulded foam laminate derived from plant starches from annual agricultural crops, such as potatoes, corn, wheat, rice and tapioca. These starches are mixed with limestone, fibre and water into a batter-like consistency. The resultant 'batter' is injected into moulds, which are heated and pressurized, releasing the water as steam and allowing the foam laminate to take its final rigid shape. Plates, bowls, hinged-lid sandwich and salad containers are grease- and stain-resistant, and are suitable for cold and hot food and microwave cooking. Life Cycle Inventory (LCI) analysis was applied by leading US consultants Franklin Associates, a pioneer in lifecycle analysis (LCA), in order to minimize the environmental footprint of the packaging. Throughout the lifecycle Earthshell Packaging® shows a smaller footprint than traditional synthetic polymer foam packaging as it consumes less energy, uses less fossil fuel, produces lower greenhouse/air/water emissions and reduces its landfill volume requirement.

	Earthshell Corporation, USA	3
	Earthshell Corporation, USA	3
	Starch, fibre, limestone	
	• *Renewable and geosphere materials* • *Biodegradable* • *Reduced energy of manufacture* • *Reduced emissions*	3 3 3 3

Veltins Steinie

Just 50 years ago most beer bottles were reused by the (local) brewery that produced them, but as beer manufacturing went global the distribution and retailing chains changed beyond recognition, resulting in many bottles making a one-way trip to the landfill. Although 'bottle banks' placed strategically at supermarkets and local sites have encouraged consumers to recycle their empties, and many breweries operate return systems for their pubs and bars, vast amounts of virgin glass is still thrown away. Veltins saw the challenge of re-designing their bottles to improve reuse efficiency as a means to re-brand their beer. This reusable bottle provides a distinctive ergonomic profile and doesn't need a conventional paper label as information is embossed on the base of the bottle.

	Formgestaltung, Alexander Schnell-Waltenberger, Germany	306
	Brauerei C. & A. Veltins, Germany	306
	Glass	
	• *Reduction in materials used* • *Reusable product*	325 329
	iF Design Award, 2003	332

Cull-Un Pack

This is a UN-certified packaging design for the transport of hazardous chemicals in glass containers. A strong moulded pulp base and top, made from used cardboard boxes, protects the containers, which are enclosed in a corrugated cardboard outer. It meets stringent safety tests, including a 1.9 m (6.2 ft) drop, which had previously been met only by using expanded polystyrene packaging. All the materials are from recycled sources and the packs can be delivered flat, saving valuable delivery space.

	Robert Cullen & Sons, UK	320
	Robert Cullen & Sons, UK	320
	100% recycled paper and board	
	• *Recycled, renewable materials*	325

Transfert

Take a close look at the equipment you use in your daily working life, then ask whether it suffers from innovation lethargy because it has become a default industry standard that no one questions despite its obvious drawbacks. This is the process that led to the development of Transfert, a new modular stainless-steel trolley system for restaurants and corporate or municipal kitchens. Existing trollies conform to dimensions set by the GastroNormes international standard containers, which vary in depth from 20–65 mm (¾–2½ in). These containers slide on L-shaped runners set in a fixed frame that usually accommodates ten trays, but the trolley structure is difficult to access to clean properly. Transfert accepts the GastroNormes containers but makes a radical departure from the usual trolley. It is made of a series of individual stacking modules made of a continuous single 10 mm (⅖ in) diameter stainless-steel wire bent in three dimensions without any welding. This modular solution and the unique geometry make Transfert more easily hand-washable and also the first machine-washable trolley. It can be stacked in a stationary or mobile unit to provide flexible work practices and minimize workers' stress and strains. Transfert is the first piece of professional kitchen equipment to be awarded a prestigious Etoile du Design award in 2004.

✎	*François Tesnière & Anne-Charlotte Goût, 3bornes Architectes, France*	310
⚙	*TSA Inox, France*	322
▬	*Stainless steel 204Cu*	
♻	*• Durable • Modular design • Improved health & safety • Improved ergonomics*	325 328 328 328

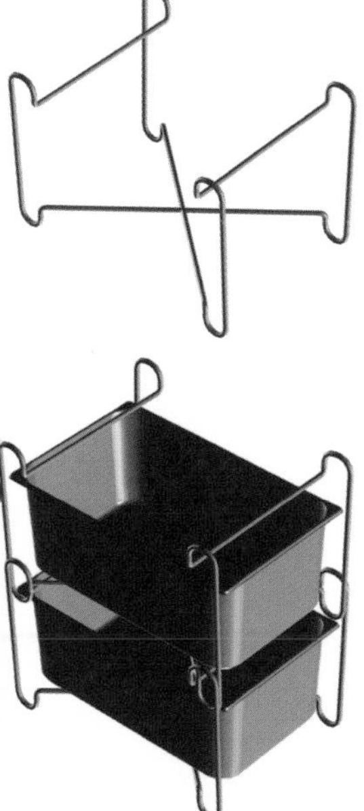

Potatopak

Large-scale commercial manufacturing of polystyrene began by Dow Chemicals in the USA and I G Farben in Germany in the 1930s but it wasn't until the rise of global fast-food corporations and supermarkets that polystyrene began getting some negative publicity. Polystyrene containers tend to have short lives, ending up in landfill sites where there they will remain for many generations. Potatopak products offer a biodegradable alternative. Using starch obtained from low-grade potatoes, the plant can produce a wide variety of food containers with similar handling and rigidity characteristics of synthetic plastics. Labels are not required as the customer brand identity and other product information can be embossed on the bottom of the container. Although not intended to be consumed, the containers are harmless if eaten by mistake.

✎	*Potatopak NZ Ltd, New Zealand*	320
⚙	*Potatopak NZ Ltd, New Zealand*	320
▬	*Potato starch*	
♻	*• Renewable and biodegradable material • Reduced embodied energy • Safe, non-toxic*	325 326 326

MAN Buses

Since the 1990s MAN Nutzfahrzeuge have been testing working prototypes using natural gas (CNG), liquefied petroleum gas (LPG), hydrogen fuel cells and a biofuel called rapeseed oil methylester (RME) as alternatives to diesel. An articulated bus powered by CNG develops 310 bhp, and in combination with closed-loop catalytic converters, conforms to Euro 3 emissions levels proposed by Germany, which are less than or equal to 2 g/kWh carbon monoxide, 0.6 g/kWh hydrocarbons, 5 g/kWh nitrous oxides and 0.1 g/kWh of particulate matter. Recent developments include diesel-hybrid drive buses. These use electric motors with a high performance electric storage unit of ultra-capacitors, which have a higher capacity than batteries, capturing regenerative energy during braking. This means quiet, emission-free exhaust when moving off from bus stops as the electric motors are engaged. MAN's environmental commitment is reflected in their accreditation to management standards for EMAS and ISO 14001.

Designer	*MAN Nutzfahrzeuge AG, Germany*	318
Manufacturer	*MAN Nutzfahrzeuge AG, Germany*	318
Materials	*Various*	
Eco-features	*• Reduced emissions* *• Reduced noise*	328 328

Mercedes-Benz Citaro

Since commencing production in 1997, the Mercedes-Benz Citaro has proven its eco-efficient credentials, and by February 2009 reached sales of 20,000 units. The Natural-gas-drive Citaro and the diesel-powered Citaro were awarded Blue Angel eco-labels in 2003 and 2007. The original two models of the Citaro series have now expanded to 12 models – single or dual-unit buses – the most recent being the Mercedes-Benz Citagro G BlueTec Hybrid articulated bus with its four electric wheel hub motors and the world's largest lithium ion vehicle battery that captures energy from the diesel generator and during regenerative braking. The battery generates a huge 130kW of power but is relatively light at just 350 kg (772 lb) and the diesel engine has been downsized from a unit weighing 1,000 kg (2,205 lb) to just 450 kg (992 lb). These weight savings and sophisticated power control systems enable significant fuel savings of up to 30%.

Designer	*Mercedes-Benz/ Daimler AG, Germany*	318
Manufacturer	*Mercedes-Benz/ Daimler AG, Germany*	318
Materials	*Various materials*	
Eco-features	*• Reduced emissions* *• Hybrid renewable/ fossil fuel energy system*	328 328

Solo

Solo takes midi-bus design to new lows – that is, it provides a low-level platform to enable wheelchair users to mount from pavement to bus by an extendible automatic ramp. A low centre of gravity also produces less roll and a more comfortable ride for all. And in the absence of any steps, buses can pick up and drop off passengers more quickly, enabling them to keep accurately to their specified timetables and reduce emissions while idling.

✎	*Optare Group Ltd, UK*	319
⚙	*Optare Group Ltd, UK*	319
▬	*Various*	
↻	• *Alternative transport for improved choice of mobility* • *Improved functionality for passenger loading/unloading*	327 328

Buses fitted with the GM-Allison Hybrid EP System

An improvement in fuel economy of 50% over conventional powered units is achieved by using the GM-Allison Hybrid EP system developed by Allison Transmission, a subsidiary of General Motors. The system comprises two electric motors for stop-start, recovery of energy from regenerative braking and normal running, giving significant reductions in emissions of carbon monoxide, hydrocarbons, particulate matter and nitrous oxide (the latter reduced 50%). The GM-Allison technology represents an opportunity to create cleaner public transport systems.

✎	*Allison Transmission / General Motors, USA*	312/ 315
⚙	*Allison Transmission / General Motors, USA*	312/ 315
▬	*Various*	
↻	• *Fuel economy* • *Reduced emissions*	329 328

Flexitec

Traffic-calming systems installed using conventional techniques require considerable manpower and cause disruption to traffic during installation. Flexitec, a hard-wearing modular system of kerbs, blocks and ramps, manufactured from recycled rubber, is installed by bolting each module to the existing road surface. It reduces road congestion during installation and can be used for permanent or temporary traffic calming.

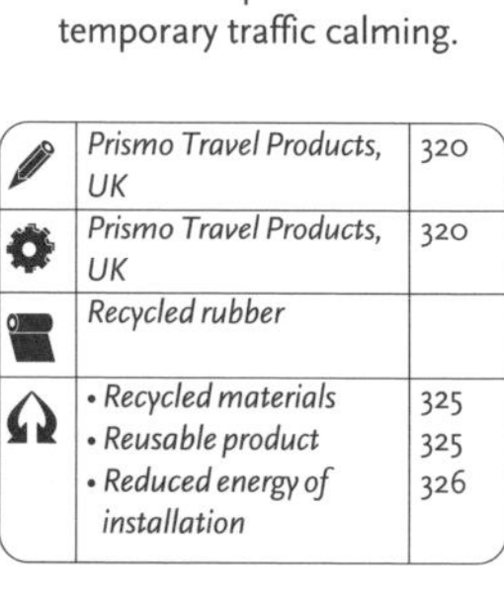

✎	*Prismo Travel Products, UK*	320
⚙	*Prismo Travel Products, UK*	320
▬	*Recycled rubber*	
♻	*• Recycled materials* *• Reusable product* *• Reduced energy of installation*	325 325 326

Metropolitan express train

In contrast to most modern trains, the interior of this new train system is made of entirely natural or recyclable materials. With moulded laminated wood shells and leather upholstery, this design makes significant reference to the original 1956 Herman Miller Model No. 670 and 671 lounge chairs designed by Charles and Ray Eames. For those who travel regularly by rail the little touches, such as a padded head pillow, will be much appreciated.

✎	*gmp – Architekten, Germany*	30
⚙	*Deutsche Bahn AG and Metropolitan Express Train GmbH, Germany*	31
▬	*Leather, plywood, stainless steel*	
♻	*• Renewable and recyclable materials*	32
✪	*iF Design Award, 2000*	33

Checkpoint and Checktag

Loose wheel nuts can lead to accidents, with loss of life and possible spillage of pollutants and toxins into the environment. Checkpoint and Checktag are two types of plastic cap that are pushed over a nut once it has been tightened to the correct torque. The arrows on each cap should be aligned unless the nuts have worked themselves loose. A quick visual check is all that is needed to identify a rogue nut.

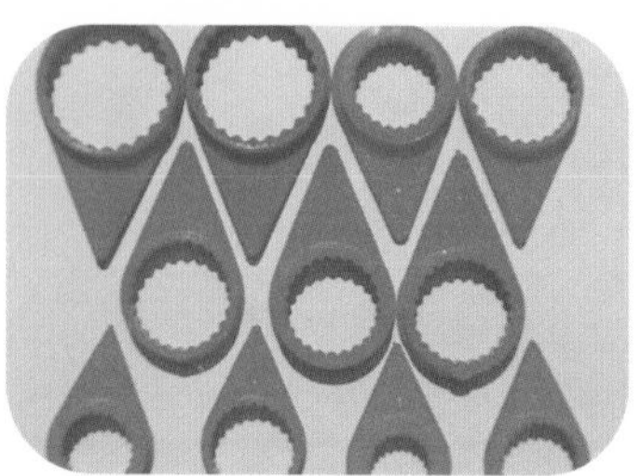

✎	*Mike Marczynski, UK*	308
⚙	*Business Lines, UK*	313
▬	*Plastic*	
♻	*• Improved road safety* *• Reduced pollution*	328 328

Citadis & Flexibility outlook T3000

Enthoven Associates sees accessibility and liveability as the key drivers in addressing sustainable mobility issues. The city environment and its infrastructure are threatened by aerial and water-based pollution caused by current transport modes. Furthermore, the erosion of access to public spaces, gentrification and loss of urban diversity challenge socio-cultural functions. Private transport, i.e. the car, shows a growing incompatibility with accessibility and liveability. Low-emission vehicles will not deliver a sustainable solution. Public transport systems have to embrace strategic and spatial functionality, and so address a wider range of sustainability issues, offering a de facto collective space to compensate for public space eroded by privatization and building for commercial gain. Transport systems have a long history of driving urban development. Lately, many cities in Europe, recalling the efficiencies of 19-century systems, have reintroduced trams. The modern tram is, however, a long way from its noisy, clattering ancestors. It must be different in order to lure modern commuters out of their cars. Enthoven Associates have been involved in designing tram systems for Amsterdam, Brussels and Jerusalem. Their designs reveal a passion to create high-quality interiors, universal access for special-needs commuters and numerous safety features. Over 450 Flexity Outlook vehicles are in successful revenue service in the cities of Brussels (Belgium), Linz and Innsbruck (Austria), Lodz (Poland), Eskisehir (Turkey), Geneva (Switzerland), Marseille (France), and Valencia and Alicante (Spain), and will soon be put into operation in Palermo (Italy), Augsburg and Krefeld (Germany).

✎	*Enthoven Associates Design Consultants, Belgium*	305
⚙	*Various clients including Bombardier Transportation, Brussels and Alstom, Rotterdam*	
▬	*Various materials*	
♻	*• Alternative, sustainable, transport* *• Eco-efficient mass product-service-system (PSS)*	327 325

CIVIA train

In order to reduce the substantial environmental burdens of millions of urban commuters going to work in their cars, there have to be significant improvements in public transport systems.
A consortium, including CAF, Siemens and Alstom Bombardier, worked with their client Renfe to produce the Suburban Train, CIVIA. Bright, light, airy interiors, generous gangways and wide opening doors all contribute to a sense of internal space. Doorway ramps enable access for those with mobility difficulties. Six TFT screens in each carriage inform passengers as to the destinations and timetabling and a CCTV system provides improved passenger security. Colourful, robust synthetic resins and alloys cope with everyday wear and tear while offering comfort for short-haul suburban journeys. The CIVIA can operate with one to four carriages, providing flexible operation. A range of modular carriage designs (A1 and A2 driver's cab, A3 low-floor coach, A4 normal-floor coach) and two types of motor-driven bogies enable operators to mix a train to suit their particular passenger and route needs. Electronic systems monitor energy consumption for internal climate control and obtain the best eco-efficient performance for the passenger load and track conditions.

✎	*A consortium for Renfe, Spain*	320
⚙	*Renfe, Spain*	320
▬	*Various materials*	
♺	• *High-quality public transport product*	327

GE Evolution Hybrid locomotive

Freight by rail is still an important mode of transport for countries that span large continental land masses, and is likely to become even more important as air freight becomes economically and environmentally more costly. General Electric's Ecoimagination R&D unit engineered the world's first hybrid freight train where the diesel power unit generates electrical energy and energy is reclaimed during dynamic braking, when the traction electric motors become alternators when decelerating or travelling downhill. Stored in massive sodium nickel chloride batteries, this electrical power is fed to the traction motor by a computerized system enabling diesel fuel consumption savings of up to 15%. This might not sound like much, but with the huge volumes of rail freight transported in the USA, and with volatile global oil prices, these savings are significant.

✎	*General Electric, USA*	315
⚙	*General Electric, USA*	315
▬	*Various synthetic materials and metals, sodium nickel chloride batteries*	
♺	• *High-quality public transport product* • *Improved fuel efficiency* • *Reduced reliance on fossil fuels and improved energy security*	327 328 328

FanWing

High speed, noise and huge fuel bills are the hallmarks of today's fixed-wing aircraft. The FanWing, tested as a working model prototype in 2005, is an aircraft with near-vertical take-off capabilities that serves as a quiet, slow but fuel-efficient, load-carrying transporter.

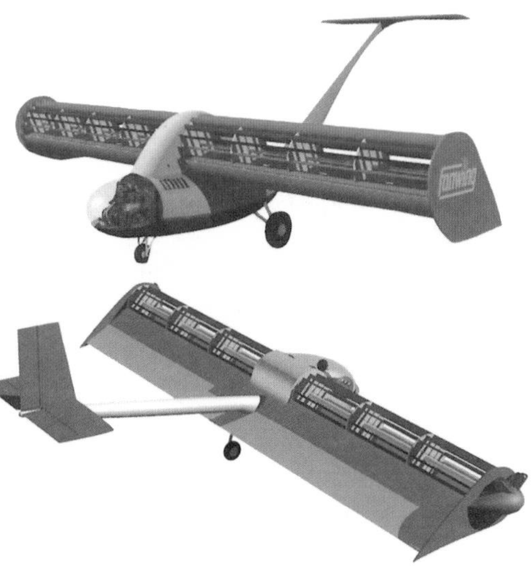

In an intriguing innovation, the designers have introduced a large rotor along the entire leading edge of the wing. The engine directly powers the rotor, which is capable of producing both lift and thrust as the cross-flow fan pulls air in at the front and accelerates it over the trailing edge of the wing. Wind-tunnel testing reveals 15 kg (33 lb) of lift per horsepower, equivalent to a payload capacity of 1 to 1.5 tonnes for a 100-horsepower power unit. These efficiencies make it ideal for air freight, short haul passenger flights and aerial surveillance. By 2007 a low speed urban surveillance drone prototype was developed, with a wingspan of 2.4 m (7.9 ft), weighing in at just 5.5 kg (12 lb) with a payload capacity of 2 kg (4.4 lb). This again proves the value of the concept, although FanWing are still looking to commericalize the concept for passenger and freight flight.

	Patrick Peebles, FanWing, UK	309
	Model prototype	
	Various	
	• Fuel economy *• Simple, low-cost construction*	327 326

Helios

The Helios is an enlarged version of the Centurion 'flying wing'. It has a wingspan of 75 m (247 ft), which is two-and-a-half times that of the Pathfinder flying wing and longer than that of a Boeing 747 jet. An experiment of the early 2000s, the Helios still holds the world altitude record for a propeller-driven aircraft, flying at almost 97,000 ft under solar power. This aircraft is known as an uninhabited aerial vehicle (UAV) and is suitable for remote sensing and reconnaissance with a multiplicity of applications. Perhaps its true potential is signalling that solar technologies could make Helios a reality for future generations who want to fly, but won't have the option of fossil fuels.

	NASA, Dryden Flight Research Center, USA, with AeroVironment, Inc., USA	312
	NASA, Dryden Flight Research Center, USA, with AeroVironment, Inc., USA	312
	Photovoltaic modules, lightweight metals and composites	
	• Zero emissions *• Renewable energy*	325 328

Trent turbofan engines

Rolls Royce is now a leading player in manufacturing fuel-efficient, lightweight and low-cost jet engines, the Trent series (500, 600, 700, 800, 900, 1000, XWB), now fitted to Airbus A330, A340 and A380, as well as the Boeing 777 and variants of the 787. Weight reduction was achieved with titanium fan blades comprising three sheets in close proximity, which were subjected to heat, causing flow of material and bridging between the layers to form a honeycomb structure. Other innovations include 'growing' metal by controlled cooling of the molten alloy in the mould to align all the molecules in one direction, forming an extremely strong single crystal. The turbine blades can operate at almost twice the melting point of normal crystalline metal. A three-shaft design is also more efficient and easier to maintain and upgrade.

✎	*Rolls-Royce International Ltd, UK*	320
⚙	*Rolls-Royce International Ltd, UK*	320
▬	*Various*	
Ω	*• Fuel economy* *• Reduced air pollution*	329 328

Quiet Green Transport

Revolutionary Aerospace Systems Concepts (RASC) is a branch of NASA concerned with conceptual explorations of new mobility and communication systems for air and space travel. Project FY01 included the concept of Quiet Green Transport, by Langley Research Center, whose dual objectives were to substantially mitigate noise and emissions from commercial aviation. Thirty new configurations for aeroplanes were created and several variants identified for further development. A common feature of these concepts was the use of liquid hydrogen fuel, distributed by electric propulsion through ducted fans over the wing.

✎	*Mark D. Guynn, Systems Analysis Branch, Langley Research Center, USA*	306
⚙	*Conceptual prototype*	
▬	*Various materials*	
Ω	*• Reduced emissions* *• Reduced noise*	328 328

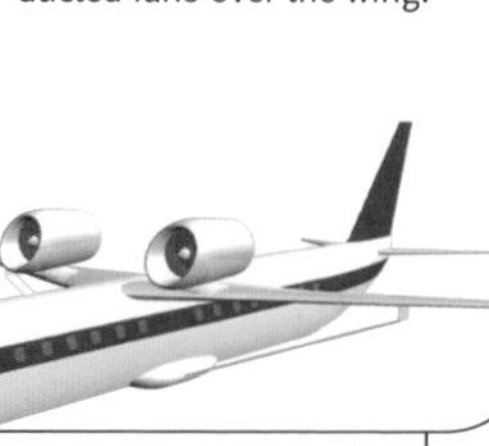

Boeing 787

As pressure increases to curb greenhouse gas emissions and reduce noise levels, the new 'lean green' agenda of the R&D programmes of aeroplane manufacturers is beginning to bear fruit. In January 2005 Boeing's concept 250-seater 7E7 Dreamliner was given an official model designation number, and the first 787 Dreamliner rolled out in July 2007. By January 2009 the fifth 787 Dreamliner was undergoing test flights using the General Electric GEnx engines. The new generation of 787 Dreamliners are capable of carrying 210–330 passengers, with ranges from 2,500–8,500 nautical miles (4,600–15,750 km). Airplanes will use an estimated 20% less fuel than today's similar size[d] airplanes and will travel a[t] speeds up to 0.85 Mach. This is in part achieved by as much as 50% of th[e] primary structure, the fuselage and wing, being made of strong, lightwei[ght] composite materials.

✎	*Boeing, USA*
⚙	*Boeing, USA*
▬	*Various, including lightweight materials*
Ω	*• Improved fuel economy* *• Improved passenger comfort*

Airbus 380

Singapore Airlines took delivery of the first production A380 in October 2007 and currently has six aircraft in operation on routes from Singapore to Sydney, London and Tokyo. By early 2009 more than 80 airports around the world had welcomed the A380, and there were 13 Airbus 380s in operation and 200 more on order. This massive 525-passenger aircraft with a 79.8 m (262 ft) wingspan is an important development in global air travel. The sheer scale of the plane has mirrored the phenomenal challenge to the European aeronautic industry with Airbus's main manufacturing plants in France and Germany, plus significant sub-assembly plants elsewhere in Europe.

Airbus has developed a comprehensive approach to minimizing environmental impact throughout the lifecycle of the product. In 2007 Airbus became the first company to be certified by an innovative integrated site and product oriented environmental management system (SPOEMS) at ISO 14001, which now covers all of the company's European operations. Airbus also initiated the Process for Advanced Management of End of Life of Aircraft (PAMELA) project with its partners to ensure that adequate procedures were developed for de-commissioning, dismantling and recycling aircraft.
The key phase of the life cycle of the Airbus 380 is its operational phase. Radical design strategies throughout the aircraft mean that it incorporates as much as 25% lightweight, high-strength composites – like carbon-fibre – to reduce overall weight. Light-weighting coupled with Trent 900 or GP7000 engines means that less than 3 litres of aviation fuel are consumed per passenger per 100 km (64 miles) at full passenger capacity, giving an equivalent 75 g CO_2/km per passenger, a figure that is 25% lower than the best production cars in Europe for a single occupancy car (typically 100 g CO_2/km).

Airbus claim the A380 is the quietest long-range aircraft on the market; nonetheless the company has signed up to the ambitious 2020 Advisory Council for Aeronautics Research in Europe vision that includes an overall 50% reduction in noise, fuel consumption and CO_2 emissions, and an 80% reduction in NOx emissions. In principle, the Airbus 380 might make an important contribution to reducing the number of flights at key international airports because of its 40% greater passenger capacity compared to other large aircraft.
It is a significant initiative in an industry that needs to improve its eco-efficiency record.

✎	*Airbus, France and Germany*	312
⚙	*Airbus, France and Germany*	312
	Various synthetic polymers, metals, composites, electronic systems	
	• *Improved fuel efficiency by light-weighting, aerodynamics and other measures*	329
	• *Lifecycle approach*	325
	• *Designed for disassembly*	326
	• *Certified to environmental management standard ISO 14001*	330

Solar Sailor Trimaran

This 600-passenger hybrid diesel/solar-powered ferry is ideal for inter-island or harbour and bay transportation. The central hull, 36.4m (120 ft) long, houses the conventional diesel engine power system, while the two outriggers house the electric drive powered by the huge rotating Solar Wing mounted over the wheelhouse, which also acts like a sail. Stable and manoeuvrable, the trimaran configuration in combination with the Parallel Hybrid Technology gives a variety of power options to the captain: sun/wind power enable speeds of up to 11 kph (6 knots), while the low-sulphur diesel engines can achieve between 13–26 kph (7–14 knots). There's currently an operational smaller version of the Solar Sailor stationed in Sydney Harbour, but expect to see more working the harbours of the world in the future.

✎	*Solar Sailor Holdings Ltd, Australia*	321
⚙	*Solar Sailor Holdings Ltd, Australia*	321
▬	*Various synthetics, diesel motor and drive, electric motor/batteries/ photovoltaic panels*	
♺	• *Hybrid power system part-using renewable energy* • *Reduction of fossil fuel consumption*	325 324

Solarshuttle 66 (Helio) and RA82

Kopf are pioneers in developing solar-powered ferries for inland waterways. The Solarshuttle 66, otherwise known as the Helio, is a scaled-up version of a ferry, which has operated between Gaienhofen, Germany, and Steckborn, Switzerland, since 1998. With a maximum speed of 24 km/h (15 mph), the Helio can operate for up to eight hours from the bank of 24 batteries without needing a recharge from the photovoltaic panels. The larger RA82 has a capacity of 120 passengers and is in service in Hamburg and Hannover. Low operating costs and negligible environmental impacts could popularize it in urban areas served by waterways and in ecologically sensitive areas.

✎	*Dr Herbert Stark, Kopf Solarschiff, Germany*	317
⚙	*Kopf Solarschiff, Germany*	317
▮	*Stainless steel, teakwood, photovoltaics, batteries*	
♺	*• Zero emissions* *• Solar power*	328 329

RA

RA is a zero-emissions, solar-powered boat, which is ideal for freshwater transport where the pollution of conventional diesel or petrol motor boats is damaging to water quality. Built to a high specification using Burmese teak and stainless steel, it contains raw materials that are extremely durable, low-maintenance and 100% recyclable. Greenpeace, the international NGO, assisted in obtaining the construction materials. An added benefit of the solar generation and electric motor system is its quietness of operation, making it a more fitting companion for aquatic wildlife.

✎	*Kopf Solarschiff, Germany*	317
⚙	*Kopf Solarschiff, Germany*	317
▮	*Stainless steel, teakwood, photovoltaics, batteries*	
♺	*• Zero emissions* *• Solar power* *• Durable*	328 329 325

Sofanco

Stone is an extremely durable natural material. Oscar Tusquets Blanca has captured the strength of this material but rendered it in a fluid, organic form to create a design of great potential longevity, albeit requiring moderate energy input during manufacturing.

✎	*Oscar Tusquets Blanca, Spain*	304
⚙	*Escofet, Spain*	315
	Stainless steel, reinforced cast stone	
	• Abundant geosphere materials	325

Eco Boulevard

Situated in Vallecas, a typical suburban development on the outskirts of Madrid, the Eco Boulevard is an ambitious landscape system with two complementary objectives: to beneficially modify the local bio-climate while generating positive social activity. The design strategy to achieve these dual aims is borrowed from the horticulture industry using technology normally deployed in greenhouses to create evapo-transpirative cooling. Three pavilions, poetically called 'trees of air', are shaded lightweight cylindrical structures with internal rows of plants that cool the air. Arranged in a boulevard that maximizes pedestrian movement, each pavilion self-sufficiently generates energy which is gathered with a photovoltaic array, and excess energy is sold back to the grid to pay for maintenance of the plants installed on the inside of the structure. These pavilions are seen as a temporary solution to create an approximation of a woodland glade and act as a catalyst to social exchange and dialogue. The goal is that the communities in these new suburban areas will adopt these spaces over time, and when the trees have reached a mature size the structures can be disassembled for recycling and re-located for another project in another town. Here the concept of street furniture is extended to encompass the enhancement of social and natural capital in areas that have undergone rapid urbanization.

✎	*Ecosistema Urbano, Spain*	30
⚙	*Ecosistema Urbano, Spain*	30
	Lightweight metal structure, photovoltaic cells, plants	
	• Renewable energy	32
	• Bio-climate enhancement	32
	• Recyclable structure at end-of-life	32
	• Social regeneration	32

Sussex

With its ranks of straight benches public seating all too often enforces social formality. Robin Day, a British designer who brought a playful modernism to dreary 1950s Britain, continues to show his lightness of touch with this modular seating system. Sussex is based upon a single modular plastic unit attached to a metal frame. It enables subtle curvilinear forms and generates an informal, convivial environment that is much more conducive to social interaction. There is potential for Magis to experiment with the use of a recycled fraction in the plastic and metal components without detriment to the positive social and aesthetic impacts of this system.

✎	*Robin Day, UK*	305
⚙	*Magis SpA, Italy*	318
▮	*Plastic, metal*	
♲	*• Modular* *• Ease of repair*	325 329

Outdoor bench

Baccarne produce a range of outdoor and public seating exclusively from recycled plastics, a mix of polypropylene, polyvinyl chloride and polyethylene obtained from post-production waste streams such as window-frame manufacturing. Planks and sheeting provide basic yet tough functional furniture.

✎	*Baccarne Design, Belgium*	312
⚙	*Baccarne Design, Belgium*	312
▮	*Recycled polypropylene, polyvinyl chloride and polyethylene*	
♲	*• Recycled materials*	325

Giulietta

The bench, in a public or private setting, provides a moment's respite, a counterbalance to the rush of modern life. Re-working the classical typology of the English wooden park bench, Rizzatto has created an animated bench in bright, bold polyethylene. Rotationally moulded in a single piece, the Giulietta possesses an air of permanence, toughness and durability. Equally suited to indoor and outdoor use, this bench is weatherproof and rot-proof, so requires no maintenance and is easily recycled.

✎	*Paolo Rizzatto, Italy*	309
⚙	*Serralunga, Italy*	321
▮	*Polyethylene*	
♲	*• Single material* *• Recyclable*	325 325

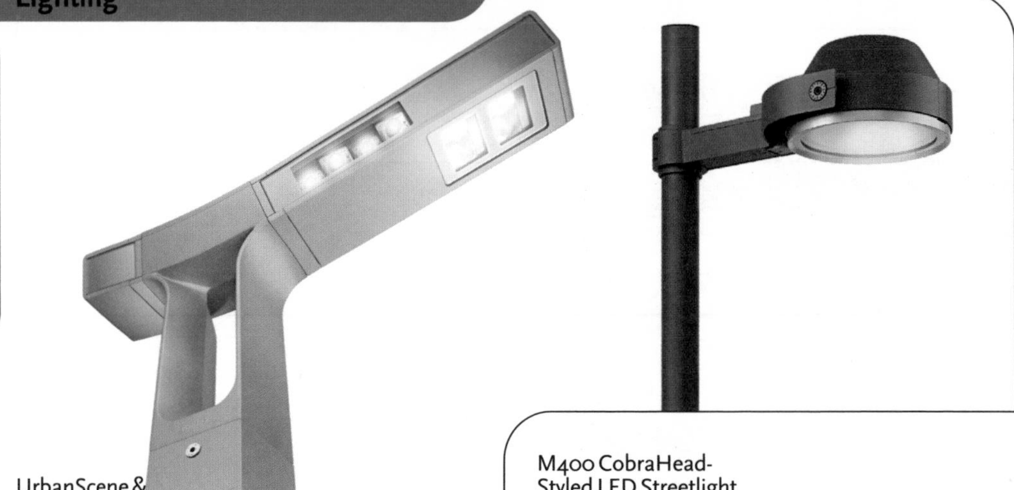

UrbanScene & UrbanLine

If you have ever seen a map of the world at night it is a powerful reminder of the energy consumed to light our urban environment for reasons of visibility, safety, aesthetics and for advertising. Philips, like many companies that compete in the public sector market for urban lighting solutions, is continually striving to offer flexible, easy to maintain low energy lighting solutions. Two of their most recent innovations include UrbanScene, a luminaire system attaching to masts, and UrbanLine, a range of flexible LED-based lighting solution for residential streets, circulation routes and pathways. UrbanScene's high-energy efficiency lamps and electronic gear are designed to be unobtrusive and flexible, with directional lighting control and full recyclability. UrbanLine provides uniform neutral white (4000 Kelvins) or warm white (3000 Kelvins) using an array of LEDs that last over 50,000 hours. The net savings are a 51% reduction in energy when compared to traditional street lighting solutions. Flexible luminaire configurations enable designers and architects to specify arrangements to suit the requirements of people, place and client.

✎	*Philips Lighting, the Netherlands*	319
⚙	*Philips Lighting, the Netherlands*	319
▤	*LEDs, electronic gear, various synthetic polymers and metals*	
♻	*• Reduction of energy consumption*	326
	• Flexible lighting solutions	329
	• Improved recyclability	325

M400 CobraHead-Styled LED Streetlight

Satellite photos reveal the earth at night to be brightly lit in the major urban and mega-metropolitan centres. It is increasingly difficult to escape this 'light pollution' and clearly see the stars. In the USA it is estimated that almost 30% of outdoor lighting doesn't light its target but is lost into outer space. In line with the ambitions of the Dark Skies initiative LEDtronics has introduced an LED streetlight head to replace traditional incandescent streetlights. The M400 CobraHead-Styled LED Streetlight contains 400 warm incandescent-white (3,200K) LEDs arranged to optimize directional light to the target. With only a 19W power drawer and a lifetime of over 100,000 hours for each LED, this represents a massive energy saving against traditional incandescents and means that street lighting is a more feasible option for local renewable energy networks. Old 'Cobra'-style lamps can be retrofitted with the LED units or complete new die-cast aluminium M400 luminaires can be fitted as replacements. Standard voltage is 120VAC but 12V, 24V, 28V and 240V versions are manufactured.

✎	*LEDtronics, USA*	31[illegible]
⚙	*LEDtronics, USA*	31[illegible]
▤	*Die-cast aluminium, LEDs, electonic gear*	
♻	*• Energy conservation*	32

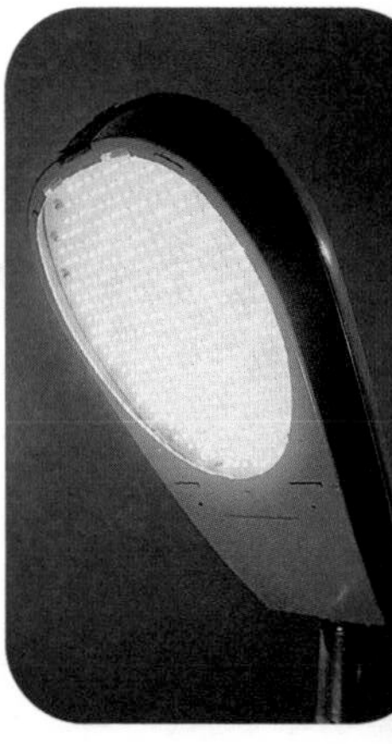

Metronomis

A range of street lighting by Philips is specially designed for energy-saving lamps and low maintenance. Modular components are durable and vandal-proof and permit different design permutations according to customers' preferences.

	Philips Lighting, the Netherlands	319
	Philips Lighting, the Netherlands	319
	Metal, glass, lamp	
	• Energy efficiency *• Modular design* *• Design for disassembly*	329 328 326

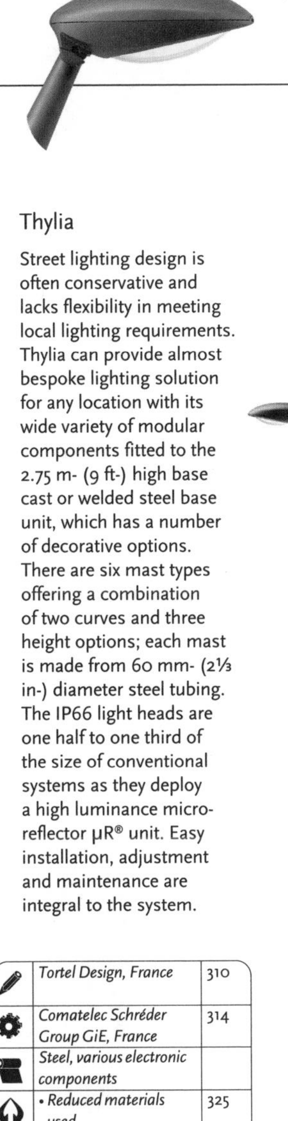

Thylia

Street lighting design is often conservative and lacks flexibility in meeting local lighting requirements. Thylia can provide almost bespoke lighting solution for any location with its wide variety of modular components fitted to the 2.75 m- (9 ft-) high base cast or welded steel base unit, which has a number of decorative options. There are six mast types offering a combination of two curves and three height options; each mast is made from 60 mm- (2⅓ in-) diameter steel tubing. The IP66 light heads are one half to one third of the size of conventional systems as they deploy a high luminance micro-reflector µR® unit. Easy installation, adjustment and maintenance are integral to the system.

	Tortel Design, France	310
	Comatelec Schréder Group GiE, France	314
	Steel, various electronic components	
	• Reduced materials used *• Modular, multifunctional design* *• Improved public street illumination*	325 328 329
	iF Design Award, 2003	332

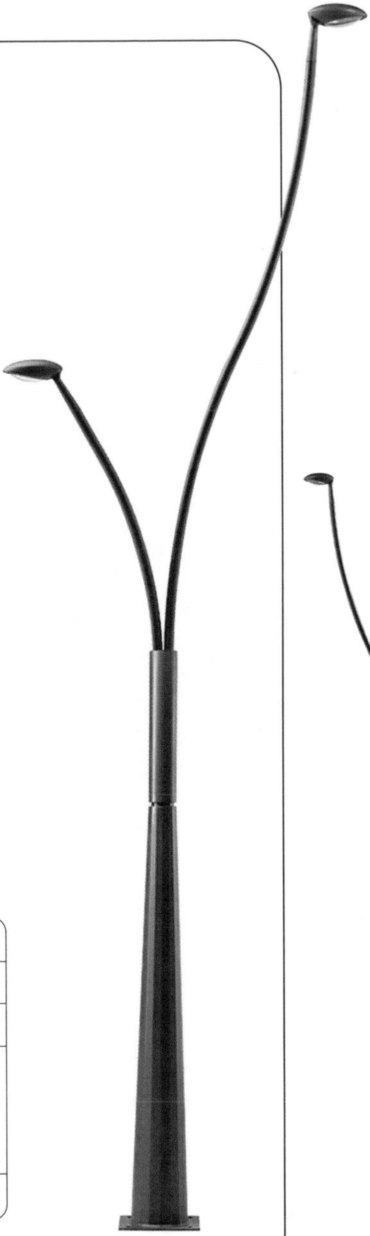

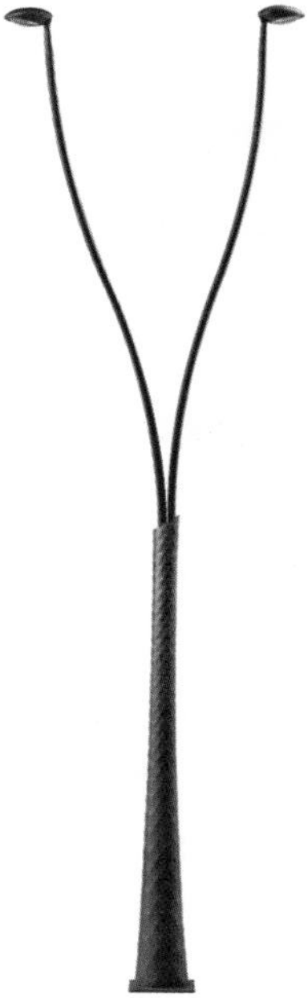

Boase

This housing project was first conceived in 2001 by Force4, a group of young Danish architects. Their aim to provide affordable housing on stilts surrounded by dense planted vegetation, which helps clean up polluted sites, is now being gradually realised. Boase (from the Danish *bo*, to live, and *oase*, oasis) aims for sustainability and accessibility. A survey of Denmark found 14,000 post-industrial urban sites that developers and local authorities would not consider for housing. Usually developers will transport polluted soil from a building site, but with phytoremediation – using plant roots and micro-organisms to render the pollutants safe – and the suspending of a development above ground, the soil can

remain, stilt houses can be erected and fast-growing trees can be planted. The City of Copenhagen offered the blighted area Nørrebro as an experimental site for the first Boase. Suspended access paths made of technical textile that insulates and collects solar energy and sectional supports/bridges made of strong fibreglass-reinforced plastic link the apartments and house units, which are made of the same material, in a minimum-waste production line. A lifesize mock-up of the module was completed in 2005, showing the feasibility of factory production. It marked the completion of Boase's Four Focus Area's Research phase: cleaning of a contaminated site, developing of a solar-membrane, developing of an energy-accumulating façade, and industrial optimization of the private dwelling. In the light of the UN World Research Institute's prediction that human alteration of the earth's surface will increase from one-third today to two-thirds by 2100, Boase offers a way of reducing use of bio-productive land for urban developments.

✎	*Force4 and KHRAS, Denmark*	306
⚙	*Prototype*	
▮	*Fibreglass-reinforced plastic, timber, trees, technosphere materials and photovoltaics*	
♻	• *Conservation of land resources* • *Affordable city homes* • *Cleans contaminated land* • *Solar power* • *Modular, minimum-waste construction*	330 327 330 329 328

Island Wood Center

A building reflects the philosophical diet of an organization, in the same way that 'we are what we eat'. None more so than the Island Wood Center, which is part of a range of sustainable design facilities on a 103-hectare (255-acre) campus. The organization brings a holistic approach to education, integrating scientific inquiry, technology, the arts, energy conservation and community living in order to encourage individual and community stewardship and a deeper understanding of the link between biological and cultural diversity. A reclaimed beam supports the roof trusses in a marriage of old and new, bringing attention to the superb engineering and aesthetic qualities of wood.

✎	*Mithun Architects, USA*	308
⚙	*Mithun Architects, USA*	308
▮	*Hardwoods and softwoods*	
♻	• *Renewable materials* • *Energy conservation* • *Educational resource*	325 329 328
✪	*IDRA award, 2002–2003*	

BRE, Environmental Office

Designed to use 30% less energy than UK 'best practice' in 1996, this office was certified Excellent under the Building Research Establishment (BRE) Environmental Assessment Method, and is still exemplar today. Cooling is achieved by natural automatic ventilation at night combined with ground water pumped through the concrete floors and ceilings, with an efficiency of 1 kWh output for pumping to an equivalent 12–16 kWh cooling energy input. Timber and steel are the primary materials for the structure and originate predominantly from recycled sources. Thanks to a combination of the thermal mass of the building, natural cooling and automated monitoring systems, the building regulates its own climate.

	Feilden Clegg Bradley Architects, UK, Buro Happold, and Max Fordham & Partners	306
	Contractors for the BRE	
	Various	
	• Energy conservation *• Recycled materials*	329 325

Marché International Support Office and GLASSX

This facility for Marché Restaurants is the first 'zero-energy' office building in Switzerland, achieving the Minergie P-ECO Label standard with just 7.8 kWh/m² per annum. This is a holistic approach to solar architecture including building orientation with a fully glazed south-facing façade utilising special anti-glare GLASSXcrystal translucent glazing to control solar gain, and a roof of photovoltaic panels harvesting up to 40,000 kWh per annum. The GLASSXcrystal panels comprise a unique triple glazed sandwich with each layer fulfilling a different function. The outermost pane is prismatic, reflecting the high altitude solar radiation in summer and transmitting the light in winter when the sun is below 35 degrees of the horizon. A central pane system contains a salt hydrate Phase Change Material in a polycarbonate box which 'melts' as it stores heat. And the interior toughened glass pane can be optionally coated with silk screen printing. This all equates to a U-value of 0.48 W/m²K, contributing significantly to the efficiency and comfort of the building envelope's climate. Space heating and air circulation is further controlled by using a ground source heat pump coupled with heat recovery ventilation. Internal 'green walls' provide fresh filtered moist air.

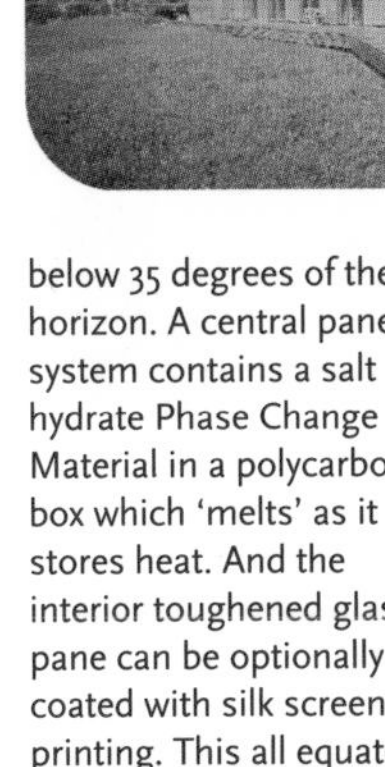

	Beat Kämpfen, Switzerland	307
	Various contractors with GLASSX AG, Switzerland	316
	Timber, steel, GLASSX panels, photovoltaic roof, ground heat pump	
	• Aims for zero energy *• Solar, renewable, energy powered* *• Lifecycle thinking and design* *• Recyclable at end of life*	326 329 325 325
	Schweizer Solarpris 2007	

Microflat

Spiralling property prices in European cities like London have forced key workers such as nurses, teachers and postal staff to the outer suburbs where they commute huge distances to their jobs. The Microflat was conceived as an affordable but compact apartment just 32.5 sq. m (350 sq. ft), two-thirds of a typical one-bedroom flat. The spatial arrangement looks more like that of a boat interior, with sliding panels and oblique partitions. The angled balcony creates a sense of privacy for the residents but also floods the living area with light. Unusually, the architects-cum-property developers suggest that prospective owners will be vetted and those with an income over £30,000 per annum will not be eligible. Only first-time buyers need apply. Another caveat, which has been successfully tried in Germany, is that properties can only be sold on to other key workers. As architectural circles revive ideas around the compact diverse city, Microflats offer an option to reclaim gap urban sites and bring life back to blighted areas. It remains a conceptual design but one that still has great relevance to help encourage cities to get closer and denser as a means to improving the sense of community and make for more walkable neighbourhoods where local shops can thrive.

✎	*Piercy Conner Architects, UK*	309
⚙	*Conceptual design, 2002, for London and Manchester, UK*	
▬	*Various materials during pre-fabrication*	
♻	• *Reduced resource use* • *Affordable and key worker housing*	325 327

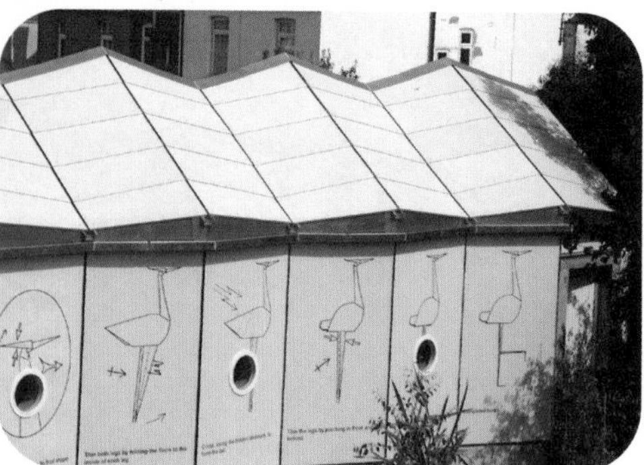

Cardboard Building, Westborough Primary School, UK, 1999–2002

Origami and the intrinsic strengths of folded paper structures inspired the design for this intriguing and imaginative school building whose aim was to use 90% recycled and recyclable materials in its construction.
This was a research and collaborative project between the architects and other professionals, children and staff at Westborough Primary School and the research partners, The Cory Environmental Trust and the Department of Environment, Transport and Roads. The project team included engineering consultants Buro Happold, three cardboard manufacturers and contractor, CG Franklin. All panels were pre-fabricated off-site.
The walls and roof are 166 mm (6½ in) thick timber-edged insulated panels made of layers of laminated cardboard sheet (Paper Marc) and cardboard honeycomb (Quinton & Kaines). Recycled interior products included rubber floor tiles, wastepaper structural board, polyurethane core board, pinboard, and Tetrapak board. At the end of its estimated 20-year life, most of the materials can be recycled. Many of the structural materials have good acoustical and insulation values.

✎	*Cottrell & Vermeulen, UK*	305
⚙	*Various contractors*	
▬	*Recycled cardboard, paper, plastics and rubber*	
♻	• *Recycled and recyclable materials* • *Low-embodied energy materials and construction* • *Design for disassembly*	325 325 326

Solar Settlement

Germany's advanced Renewable Energies Law along with the Passive House and Plus Energy House directives, are a direct incentive for the emergence of sophisticated housing such as Solar Settlement, Freiburg. There is political encouragement for developers to create dwellings that generate electricity locally and gain a net income from putting energy back into the national supply grid. This enlightened attitude helps architects make real practical progress to create self-sustaining neighbourhoods.

Solar Settlement, a mixture of terraced houses and penthouses, is actually part of a larger development that includes a mixed office/ housing block known as the Sun Ship. Together these buildings presently constitute the largest solar village in Europe. The Plus Energy initiative actually means each house is a mini powerstation. This is achieved by lowering the energy required to warm and cool the house, and by generating energy in excess of requirements. The specification means each house is highly insulated, wind-proofed, fitted with a heat recovery ventilation system, faces south for passive solar gain, and is fitted with triple-glazed argon filled, infra-red filtering façades. Cooler north-facing façades have few, small windows and even more insulation. Each roof is a massive array of photovoltaic panels, while acting as the primary rainfall barrier. Heat energy, for space heating and hot water, is provided via a centralized system of solar collectors in the Sun Ship, and any extra energy required on particularly cold days is provided by a wood-chip fuelled boiler. Integral to the design of the settlement is a car sharing system and good connectivity to the existing tramway network with a stop just across the street. This is truly a viable blueprint for future housing developers.

✎	*Rolf Disch, Germany*	305
⚙	*Various contractors*	
▮	*Photovoltaic arrays, solar collectors, rainwater collection systems and various construction materials*	
♺	*• Renewable solar-powered dwellings* *• Zero-energy housing* *• Carbon reduction transportation plan*	329 326 327

Valdemingómez Recycling Plant

Corporations and municipal authorities should increase recycling volumes in the future, but how inherently sustainable are the facilities themselves? Valdemingómez represents a new generation of recycling plants that consider their own environmental impacts, a critical undertaking for a plant in the future southeast Regional Park of Madrid. A cross-section of the 29,000 sq. m (312,000 sq. ft) building reveals a stepped profile with raw waste entering at the highest point and treated waste or waste separated into recyclable mono-materials emerging at the lowest point. This 'gravitational' recycling, in two complementary plants (recycling and composting), is hidden from view by an enormous living structure, a technological hillside of the sedum turf roof (65% of total roof area), which literally blooms in spring. External cladding, comprising recycled polycarbonate, wraps itself around a bolted lightweight steel frame. Both these elements can be disassembled and recycled at end of life – about 20 years. This inspiring example of municipal architecture reflects key points in the micro-manifesto of the collaborating architects, emphasizing hybrid models and interactions between natural and lightweight, energetically active, sophisticated, artificial materials. Valdemingómez, built for the price of a chicken farm, sets high standards of holistic architecture for future recycling plants.

	Ábalos & Herreros with Ángel Jaramillo, Spain	304
	For Vertresa-RWE Process, Auntamiento de Madrid, Spain	
	Recycled polycarbonate, steel, sedum turf roof	
	• *Recycled materials* • *Design for disassembly* • *Water and energy conservation systems*	325 326 329

Solar Office, Duxford International Business Park

This is the first generation of speculative solar-powered office buildings in the UK. It incorporates 900 sq. m (9,660 sq. ft) of photovoltaic cells into the south-facing glass façade inclined at 60 degrees. This array has generated 113,000 kWh per annum for the first two years, a substantial contribution to the energy consumption needs of the building. The solar-powered system is complemented by a natural stack ventilation system with sun-shading louvres, and both systems are controlled and monitored by computer. This 'best-practice' three-storey building gives substantive data to encourage the next generation of solar office buildings.

	Akeler Developments Ltd, UK	304
	Akeler Developments Ltd, UK	304
	Various, including photovoltaic array, monitoring systems	
	• *Energy conservation* • *Solar power*	329 329

›obley Schools ›stainable ›velopment

›obley Schools energy ›nagement system is ›st-bed to extend the ›tainable energy ›iatives of a local ›hority in response to ›al Agenda 21. A holistic ›roach led to a wood- ›l boiler, using locally ›vested coppice ›ndwood, which was ›sen on the grounds ›t it was the most ›tainable system. ›e coppice suppliers › paid according to the ›at output of the wood (supplied as chips) rather than the quantity, encouraging quality supplies. Insulation is to very high standards coupled with computerized monitoring of the under-floor heating and internal environment of the building that work in tandem with passive design features, including solar shading and natural ventilation. The net effect is a very energy-efficient public building using local resources.

	Hereford & Worcester County Council, UK	307
	Various contractors	
	Biomass fuel from coppice	
	• *Energy conservation* • *Renewable energy*	329 329

Fred

Although the concept is not new, Fred is a portable building with some special features. The basic room unit is 3 x 3 x 3 m (27 cu. m, 953 cu. ft) but the floor area can be doubled to 18 sq. m (194 sq. ft) by taking advantage of sliding wall and roof elements, which are electronically controlled. Each unit is equipped with a kitchen, toilet and shower and an area available for multipurpose use, but the basic utility services have to be connected. A fully glazed wall provides excellent natural light, and thick insulation in the walls and roof minimizes energy consumption.

	Johannes & Oskar Leo Kaufmann, Austria	307
	Johannes & Oskar Leo Kaufmann, Austria	307
	Timber, metal, glass	
	• *Multi-use space* • *Low-embodied-energy of fabrication, transport and construction*	328 325/326

Oxstalls Campus, University of Gloucester, UK

Low energy consumption and responsive regulation of the internal environment were the priorities for these new university buildings on the mixed brownfield and parkland campus site. The development comprised the Learning Centre, including 300 workstations, teaching rooms and a 200-seat lecture theatre, and the Sports Science building containing laboratories, teaching spaces and staff offices. The buildings are linked by a glazed 'bridge' over a water feature. To meet an overall energy consumption target of 110kWh per square metre per year for the Learning Centre (one-third of current best practice in the UK) a 'Termodeck' thermal mass floor system with special diffusers was used for heat storage and redistribution of warm/cold air according to the season. At the top of the atrium incoming cool air is warmed by heat extracted from rising (warm) stale air by using a thermal wheel. Exhaust air from the Termodeck heats the buffer zone and entrance areas. Lighting in the library is provided naturally by north-facing windows and by high-frequency T5 fluorescent fittings controlled by occupant and illuminance sensors. Similar strategies are used in the Sports Science building and photovoltaic panels are installed on a 'waveform' roof, providing 65% of energy requirements.

✎	*Feilden Clegg Bradley Studios LLP, UK*	305
⚙	*Various contractors, UK, 2002–2003*	
▮	*Termodeck thermal mass floor system, photovoltaics, various materials*	
♻	*• Energy conservation by intelligent management system* *• Solar power*	329 329
✪	*UK Civic Trust's Sustainability Award 2003*	

Norddeutsche Landesbank

A complex, seemly chaotic, arrangement of glass and steel buildings sits in a transition zone between the old residential area and historic centre of Hannover. This is a self-contained mini-metropolis, a landscape of transparent walkways, atria, and meeting places for 1,500 employees for Norddeutsche Landesbank. The key principles of the design focused on reducing energy use, increasing natural daylight and maximizing worker comfort. Insulation was the initial priority to reduce energy loss by exceeding the 1997 German insulation regulations by 10%. Appropriate air circulation designs by consultants Transsolar Energitechnik harnessed natural ventilation and the chimney-stack effect rather than using air-conditioning to deal with summer temperatures. Balustrade panels fitted with air shutters control incoming air; double-glazed windows fitted with highly reflective aluminium reduce incoming solar gain whi increasing natural light, and radiant slab cooling the floors uses polyethyl tubes embedded in conc in which to pump cool, geothermally heat-exchanged water at nigh This entire system is controlled by office work in their own part of the building so the comfort level is personalized. Th architects demonstrate that an eco-tech strategy delivers reductions in energy expenditure and high aesthetic and architectural standards.

✎	*Behnisch Architekten, Germany*	3
⚙	*Various contractors, Germany, 2002*	
▮	*Steel, aluminium, glass, concrete*	
♻	*• Improved natural lighting* *• Energy conservation*	3 3

Four Horizons 'Autonomous' House

Four Horizons is an on-going design exercise in refining the concept of an autonomous house to suit the particular demands of its location. Lindsay Johnston, a Brit who settled in Australia, designed the original house in 1994, self-built it in 1995–96, added several nearby studio houses and continues to evolve new autonomous technologies. Lacking all the usual utilities, the site is located at the north end of Watagans National Park on the eastern seaboard of Australia in a forested area. Temperatures range from 4°C to 38°C, rainfall averages 110 cm (43 in) per annum and prevailing winds are from the south to south east. Key features had to address the temperature range, bush fires, wood-eating termites and the need for it to be an economical self-build. A significant aesthetic and structural feature of the house is the double 'fly' roof with curved inner trusses that is completely separate from the living modules underneath. This feature is revived from some of the early pioneer buildings in Australia. The roof frame comprises 50 x 50 mm (2 x 2 in) steel tube and 20 mm- (¾ in-) diameter bar covered in silver coloured 0.55 mm BHP Zincalume corrugated steel. This protects the two 9 m x 9 m (29 ft 6 in x 29 ft 6 in) dwelling modules, one including living, dining and kitchen activities and the other bedrooms, bathrooms, laundry and study. A curved thermal mass wall of solid concrete blockwork runs through the middle of each module to receive winter morning sun, and the breezeway permits through-flow of cooling north-east summer winds. South, east and west elevations are protected from excessive solar gain by recycled polyester wool insulation and the north side is clad with local blue gum planks, although use of timber is kept to a minimum throughout the building to preserve existing forests. Exposed concrete floors were treated with Livos hard-wearing finishes. The roof captures between 382,000 and 585,000 litres (84,000–128,000 gals) of water per year into Aquaplate steel tanks. Annual consumption is 220,000 litres (48,000 gals) per annum , so there is a six-month water reserve. Grey water is recycled for use in the walled vegetable garden. Energy is supplied by a combination of a BP Solar battery storing photovoltaic energy, 'Solahart' solar hot water, LPG gas and a back-up diesel generator. While not totally autonomous, the house uses half the energy of a conventional house in New South Wales. Measures to counter the threat from bushfires include cleared vegetation zones, concrete slab ground construction, 'Mini-orb' roofs and steel structure and non-combustible concrete block. This is 'Factor 2' sustainable design and the lessons learnt contribute to improving building specifications for new builds.

✎	*Lindsay Johnston Architecture, Australia*	308
⚙	*Lindsay Johnston Architecture, Australia*	308
	Concrete, steel, recycled polyester insulation, blue gum timber, photovoltaics	
	• *Low energy* • *Passive and active solar heating* • *Autonomous water systems*	329 329 329

Frame house

Modesty and adaptability are the dual visions that guided the environmental agenda for this family home in east London. This bespoke, modular, prefabricated house, built in four months, comprises load-bearing Glulam (glue-laminated) post-and-beam construction of tough Siberian larch. As the family grows, this open internal structure ensures that rooms can easily be expanded or sub-divided. Striking features include the cedar roof and wall cladding wrap and the passive glazed areas. These combine with holistic attention to the use of natural materials throughout, including massive flax insulation with Pavatherm fibreboards, and slate flooring, lime plaster and porous paints enabling the walls to breathe. Space heating is achieved with a combination of a timber pellet boiler coupled with under-floor heating.

✎	*Marcus Lee, FLACQ Architects, UK, with Arup, UK*	3
⚙	*Various contractors*	
▰	*Glue-laminated Siberian larch, flax insulation, timber pellet boiler, solar panels, other natural materials.*	
♻	*• Extensive use of renewable and abundant lithosphere materials* *• Adaptable spaces* *• Renewable energy space heating and hot water system*	32 32 32

Symbiont Friederich

In the small German town of Merzig a three-storey 1960s house gets an inspirational fourth floor with an act of architectural symbiosis. The couple occupying the existing top floor of the building wanted more space, so what better option that to use the existing roof structure. Accessible by a light steel staircase, the cantilevered 'living box' of dark-grey painted chipboard is attached to a lightweight timber frame insulated to 200 mm (7.9 in) depth. A short glass passageway links to a 'winter-garden' box similarly insulated but clad in zinc sheeting with a dramatic green firewall of precast concrete. These two new components provide an additional 50 m² (538 sq. ft) of living space, including a new roof terrace. Making the most of existing roof structures is a design strategy that eloquently increases the density of cities and has the potential of generating an improved quality of life with additional green spaces.

✎	*FloSundK Architektur + Urbanistik, Germany*	306
⚙	*Various contractors*	
▰	*Timber frame, precast concrete firewall, zinc, wooden boards, glass, steel*	
♻	*• Lightweight structures* *• Uses existing urban space*	325 325

Hope House

Hope House, built in 1995, is a home, office, energy generator and leisure zone. Passive solar design combined with photovoltaic generation is sufficient to maintain an ambient internal climate and to run an electric car for up to 8,500 km (5,300 miles) per year, resulting in a net saving of about 4.13 tonnes of carbon dioxide per year. Mains water usage is minimized by using rainwater for the toilets and laundry room. Between 2004 and 2006 bio-fuel devices (a log-burning stove and wood pellet boiler) replaced the gas boiler, and a micro-wind turbine was fitted. This project was the inspiration for the BedZED low-energy housing development near Croydon, south London and remains an experimental, evolving house where architect Bill Dunster can test new technologies as they emerge.

	Bill Dunster Architects, Mark Lovell and Oscar Faber, UK	304
	Various contractors	
	Various	
	• *Integrated energy-efficient control system for home and domestic transport* • *Water conservation system* • *Solar power*	329 329 329
	Design Sense award, 1999	

SU-SI

Many people associate mobile or trailer homes with holiday parks and dubious lifestyles. Not so this customizable 21st-century modular home system, which can be erected on site within a few hours and is easily disassembled and reused in another location. The factory-produced modules measure 12.5 m x 3.5 m x 3 m (41 ft x 11 ft 6 in x 9 ft 10 in), each one interlocking with the next to create versatile domestic, office or exhibition spaces.

	Johannes & Oskar Leo Kaufmann, Austria	307
	Johannes & Oskar Leo Kaufmann, Austria	307
	Timber, metal, composites, glass	
	• *Reusable buildings* • *Low-embodied-energy of fabrication, transport and construction*	325 325/326
	iF Design Award, 2000	332

Airtecture

Weighing just 6 tonnes and easily packed on to a road vehicle for transport, Festo's portable building comprises a protected floor space of over 357 sq. m (3,810 sq. ft). This is achieved by supporting an inflatable cross-beamed roof on two rows of inflatable, Y-shaped columns. Stiffness is given to thin cavity wall panels by tensioning them with pneumatic muscles, which contract to oppose the effect of the wind.
Air is the main insulator to assist with internal climate control.

✎	*Festo & Co., Germany*	315
⚙	*Festo & Co., Germany*	315
▮	*Various*	
♻	• *Reduction of resource consumption* • *Reusable and portable building* • *Multi-use single space*	326 327 328

Ecover factory, Oostmalle, Belgium Project

Growth of Ecover's business in the early 1990s required an expansion of the existing factory near Antwerp, Belgium. Using an ecological grading system devised by the University of Technology, Eindhoven, building materials were selected for their minimal environmental impact. Structural timber was obtained from sustainably managed forests and bricks from a clay-based residue from the coal industry provided high-insulation material.
A huge multi-ridged turf roof covers the 5,300 sq. m (57,050 sq. ft) building, providing excellent insulation, controlling storm-water run-off and helping integrate the factory into the local landscape. In line with the company's philosophy of balancing commerce with social and environmental concerns, the factory has been developed to enhance conditions for the workforce: many roof-lights create natural lighting and there are solar-powered showers.

✎	*University of Technology, Eindhoven (Building Initiative Environmental Standards), the Netherlands, with Ecover, Belgium*	310/ 314
⚙	*Various contractors*	
▮	*Various natural materials, turf roof, bricks from clay-residue*	
♻	• *Energy conservation* • *Local materials from sustainable sources* • *Natural lighting*	329 325 329

BedZED Housing

BedZED is a pioneering mixed-use and mixed-tenure development of 100 homes, workspace for 100 people, and public areas, constructed in 2002 on an old sewage works – a 'brownfield' site – south of London. It remains one of the largest eco-villages developed in the UK and is a benchmark scheme. Architect Bill Dunster and environmental consultants BioRegional have, in collaboration with the client, the Peabody Trust, adopted a holistic view of the needs of the mixed-use community, including a green transport system that was actually built into the planning permission and ratified by the local authority. BedZED homes require only 10% of the heating energy of a typical new home. Further energy is saved by reducing the need to travel between living, work, shopping and recreational facilities. Reduced transport impacts are also encouraged by promoting good networking with existing public transport services and by providing decent bicycle storage facilities, attractive pedestrian links and on-site charging points for electric vehicles. (There is a ten-year target to produce enough solar electricity on-site to power 40 electric vehicles.) Materials were selected from natural, renewable or recycled sources, within the local area where possible. Each dwelling is an energy-efficient design using passive solar gain and a high insulation specification, including triple-glazed windows. A central combined heat- and power-generation facility utilizes tree waste and is combined with on-site generation from photovoltaics to help the transition towards a 'carbon neutral' development. Water conservation is encouraged with 18% of consumption coming from stored rainwater and recycled water, and through water-efficient appliances. Based upon BedZED's success, BioRegional created a toolkit for carbon neutral developments and influenced the UK government in its 'zero carbon' thinking.

✎	*Bill Dunster Architects and BioRegional, UK*	304
⚙	*Various contractors for the Peabody Group, UK*	
▬	*Various, including locally sourced*	
♺	• *Energy neutral development* • *'Carbon neutral' aspiration*	329 329

The Seashell House

Simplicity and relaxed living are the intertwining themes driving this eclectic, yet unified, modern spatiality that is completely harmonized with revitalized vernacular elements. This predominantly wooden structure, a commodious 237 m^2 (2,551 sq. ft) of living space, wraps around a central concrete column housing the fireplace and supporting a natural rubber-clad spiral staircase. The protective gastropod form gives the house its Finnish name (*Kotilo*) and was an inspiration for the 180 prefabricated timber panels supported in part by a steel frame. Larch shingles, which will grey as they age, clad the exterior of the building, while light-coloured traditional Finnish aspen shingles clothe the interior surfaces. A massive 200 mm (7.9 in) thick layer of Rockwool insulation combined with triple, argon-filled glazing and a heat recovery ventilation system, all ensure a cosy atmosphere in winter. Most remarkable is the variation in spatial experiences generated by a combination of differential natural lighting, the extensive use of natural materials, and the wedge-shaped cut-out in the middle of the spiral. In this dwelling the rhythms of the day and season can be readily experienced.

	Olavi Koponen, Finland	307
	Olavi Koponen and family, Finland	307
	Various timbers (larch, aspen), steel, concrete, natural rubber, Rockwool insulation, triple-glazed glass	
	• *Sustainably managed forest timber*	326
	• *Predominantly timber construction*	326
	• *Natural lighting*	329

House of Steel and Wood

Vernacular architecture offers a rich palette of aesthetics and embodied knowledge based upon bio-climatic conditions, but often raises negative socio-cultural perceptions when copied too literally. Not so with Ecosistema Urbano's House of Steel and Wood which re-envisions and enriches future possibilities by borrowing stilts from the raised granaries of this region in northern Spain and its fenestration from the urban office block with vast windowed surfaces. The double height façade combines local sustainably managed timbers, North Pine and Douglas Pine, with precise geometric composition, while providing thermal regulation by encouraging air circulation. Orientation to the compass ensures the best passive thermal gain and cooling, and orientation to the landscape complements what already exists. What emerges is a home with a strong personality that mirrors the client himself, a local rugby hero.

✎	*Belinda Tato and Jose Luis Vallejo, Ecosistema Urbano, Spain*	305
⚙	*Contractor, Espinareu, Spain*	
	Wood, steel, glass	
	• *Passive solar gain and cooling*	329
	• *Local, renewable materials*	325
	• *Small physical footprint*	330

Project 19, Project 23

In the 1990s Germany saw the emergence of an eco-tech aesthetic expressing new forms of solar architecture. Thomas Spiegelhalter, an architect and town planner, cleverly mixed high-tech photovoltaics with passive systems and created a modern eco-tech design. Both Project 19, a new settlement at Rieselfeld near Freiburg, and Project 23, solar row houses at Ihringen, were intended to demonstrate low-cost, low-energy-technology houses with a plan area of 250–300 sq. m (2,700–3,200 sq. ft) requiring less than 30 kWh/ sq. m per annum to operate. Both projects deployed prefabrication, Internet-based collaboration, lifecycle assessment, and energy performance simulation. Most importantly, the local, social, cultural and landscape features were meshed with climatic data to create an exciting solar vernacular architecture.

So the curved north-facing roofs of Project 19 and materials selection reflect traditions yet experiment with new materials. Passive solar gain through deployment of south-east-facing glasshouses and use of cooling northerly air with natural stack ventilation ensures a maintenance-free system for comfort in winter or summer. Spiegelhalter's vision was holistic, extending to soft landscaping, rainwater harvesting systems, use of recycled materials and the need to create flexible spaces for living and working. To reduce costs the Rieselfeld project was based on a segmented assembly method from basement to fourth floor. Service systems for water, PV electricity, heating, telecommunications and sewage with integrated heat recovery, were centralized in each house. Prefabricated, insulated panels of reinforced concrete or wood–steel modules were carefully jointed to prevent thermal bridging to maintain a U-value equal to 0.1W/m2K. Double- and triple-pane low-e glazing with Argon gas maintains U values of between 0.7 to 1.5 W/m2K while maximizing transmission of visible light for day lighting. Rooms are orientated to provide natural lighting and ventilation, and simple controls enable people to select their level of comfort. Twenty-kilowatt photovoltaic arrays generate sufficient electricity to meet 70–78% of the entire low temperature heat, cooling and electricity demand, with excess power at daytime peak hours exported to the national grid. The PV panels also provide shading to the south-facing façades to prevent excessive solar gain. These projects demonstrate the eco-efficiency levels attainable for temperate European houses and set valuable new standards for lowering the total carbon dioxide emissions of future housing stock. The eclectic mix of technology, technosphere materials and quirky features create low-cost houses with individuality.

✎	*Thomas Spiegelhalter, USA and Germany*	309
⚙	*Various*	
	Concrete, steel, wood, mineral insulation, glass, photovoltaics	
	• *Low-energy housing* • *Affordable housing*	329 327

Fab Tree Hab

Working with nature, this conceptual design is a living structure supported by a reusable scaffold prefabricated using Computer Numeric Controlled (CNC) machines. Conceived as a 'local biota and living craft structure' this concept emerged from the Massachusetts Institute of Technology's Human Ecology Design Team, whose designers now work within a design group called Terreform 1. These dwellings are an integral part of the ecological community, providing synergistic and symbiotic opportunities for humans and other living organisms to co-exist. They are comprised of 100% living nutrients, providing effective services to the ecosystem (including carbon capture, new habitat provision, edible goods on your doorstep) and suggest new systems of farming and house 'manufacturing'. This is design 'imagineering' at its best. Now the challenge is to turn the dream into reality...

✎	*Mitchell Joachim, Lara Greden and Javier Arbona, Terreform 1, USA*	310
⚙	*Concept design*	
▮	*Living components, inert reusable scaffold*	
♻	• *A carbon capture device* • *Synergistic, symbiotic human/nature housing*	324 325

Villa Eila, Mali, Guinea

This is a private residence for Eila Kivekäs, founder of the Indigo Association, responsible for promoting vocational training for women in Guinea, West Africa. Built in 1995, it is a powerful example of 'intermediate' or 'appropriate' technology using local resources and skills to develop a arenas for the natural materials to speak quietly. Villa Eila is more than a vernacular building – although it does embrace all the virtues of thinking locally and being true to local materials, it is a statement of self-help and self-belief in local, rather than Western industrial solutions.

distinctive aesthetic. All the bricks and floor tiles were made of local soil mixed with 5% cement, manually pressed and then dried in the sun. Roof tiles were made in a similar fashion but sisal fibre was added to a higher cement–soil mix to form a stronger composite. Woven bamboo screens provide dappled light and shade to protect internal walls from heat absorption. The mono-pitch roof covers a variety of simple rectilinear or circular rooms, the simple geometry creating mini-

✎	*Heikkinen-Komonen Architects, Finland*	306
⚙	*Various*	
▮	*Soil, clay, cement, sisal, bamboo*	
♻	• *Local renewable or geosphere materials* • *Low-embodied-energy materials and construction*	325 325

Loftcube

Look out over the rooftops of any major city and acres of unused space come into view. Werner Aisslinger's Loftcube aims to colonize these sunny plots of prime urban spaces. Two prototypes, a 'home' and an 'office' version, were launched at Designmai in Berlin in 2003 on the roof of the premises of Universal Music Deutschland. Today the dream has become a reality, with Loftcube units now being produced, with 55 sq. m (588 sq. ft) of gross living space.
This prefabricated, modular structure utilizes honeycomb-type wooden modules protected with white laminate plystyrol with durable Bankirai wood for exterior features. The structural 6.6 x 6.6 x 3 m (21 ft 8 in x 21 ft 8 in x 9 ft 10 in) body rests on supporting legs and offers customizable window spaces (transparent, translucent, or wooden slats) according to clients' needs. Interior panels and partitioning, made of DuPont Corian® Glacier White, is also based on a modular structure and offers considerable flexibility. Flooring is the uniform, tough quartz-crystal composite of DuPont Zodiac® in wet areas or those needing frequent cleaning. In the office version DuPont Antron® Excel nylon-based fibre is used in tough-wearing 'Nandou' carpet from Vorwerk Teppichwerke, but a fluffy, warmer look is achieved in the home version by using a loopware carpet, Antron® Supergloss. Aisslinger's belief that his 'flying building' would be appealing to young urbanites, business people and those wanting to be in the heart of the city is being realised, and installation of a Loftcube on a flat rooftop can be accomplished with a crane in 2–4 days.

✎	*Werner Aisslinger, Studio Aisslinger, Germany*	310
⚙	*Loftcube GmbH, Germany*	318
▬	*Wood, plystyrol, DuPont Corian/ Zodiac/ Antron and other materials*	
♻	*• Modular, prefabricated* *• Design for disassembly* *• Recyclable technosphere materials* *• Conserves land resources*	328 326 325 329

Minibox

Designers Holz Box have tested Minibox on the citizens of Innsbruck in Austria. At the heart of the home – which is not for the claustrophobic – is the multifunctional kitchen/ dining/work table, with a wood-burning stove neatly inserted underneath.
A single and double sleeping platform, top-lit by a roof-light, occupy the top third of the transportable prefabricated 2.6 m- (8 ft 6 in-) cube unit. Every object has multiple functions, so the ply benches have steel storage shelves underneath and a storage unit doubles as steps to the beds.
Another storage area contains a shower and camping-style toilet.
Every detail of the larch and softwood laminate/ plywood panels, stainless-steel stove pipe and metal shelving contributes to a modern, functionalist aesthetic. Four units fit on a standard EU lorry so Minibox is adaptable to a variety of needs from temporary housing, autonomous extensions, holiday chalets or roof-top urban living.

✎	*Holz Box ZT Gmbh, Austria*	307
⚙	*Holz Box ZT Gmbh, Austria*	307
▬	*Wood laminates and plywoods, stainless steel, various metals, rooflight*	
♻	*• Reduced resource consumption* *• Spatial economy*	326 328

Cape Schanck House

Carefully sited among the scrubland of native tea trees on the Mornington Peninsula south of Melbourne, the shell of this house is wrought from detailed studies of the wind strength, direction and turbulence combined with patterns of seasonal sunlight and rain. The aerodynamic external skin and continuous internal skin represent a sensitive, spatially intelligent, bio-climatic response to the site. This response can be read in a number of eclectic individual features. A rain tank in the centre of the living area cools ambient air in summer, collects and stores rainwater, and structurally supports the roof load. Made of 6 mm (0.2 in) mild steel, this tank keeps water close to 21°C in summer and provides acoustic entertainment after rainfall. Excess water drains to another tank used for flushing toilets, irrigating the garden and washing wetsuits. Wind scoops in the south elevation trap cooling summer winds. Vertical louvers for the glazed panels in the rear bedroom provide shade from the afternoon sun. ECOply structural plywood clads the sleeping areas, and provides a sharp contrast with the Alucobond aluminium composite panel and fully glazed surfaces. Awareness of place is in the details: even the terrace paving is made of bespoke shapes that mimic the patterns found in local lava formations. Sensitive, intelligent and dynamic, this house reveals ideas of a 'better' life.

✎	*Paul Morgan and a team of architects/ designers at Paul Morgan Architects, Australia*	309
⚙	*Owner-builder, Paul Morgan, Australia*	309
▬	*Aluocobond, glass, timber, plywood*	
♻	*• Passive design to harness site's natural benefits*	330
	• Extensive use of eco-materials	324

LockClad terracotta rainscreen

Combining the aesthetics and durability of fired clay tiles with ease of installation, this rainwater cladding on aluminium rails is a cost-effective method of protecting the exterior of a building from the elements. Each clay tile is locked in place on an extruded aluminium rail, LockRail, which meets all UK and Ireland wind loadings. This minimal-maintenance, lightweight cladding permits extra insulation materials to be applied to the outer skin of the building's structure, improving energy conservation. Natural ventilation behind the clay tiles and protection from the sun reduce temperature variations in the load-bearing structure.

✎	*Hanson Building Systems, UK*	316
⚙	*Hanson Building Systems, UK*	316
	Clay, aluminium	
	• *Durable, recyclable materials* • *Improved energy conservation for buildings*	325 329

Criss Cross

Making a welcome change from the ubiquitous rectangular paving block, the Criss Cross paving system comprises four different forms that can be interlocked in regular or random patterns. Glindower Ziegelei still fires these blocks in a kiln dating from 1870. Natural variation in the clay minerals yields a range of colours and textures.

✎	*Ecke: Design, Germany*	305
⚙	*Glindower Ziegelei GmbH, Germany*	316
	Clay minerals	
	• *Abundant geosphere materials*	325
	iF Design Award, 2000	332

eco-X

If you are seeking an alternative to natural stone or synthetic petroleum-based products for countertops, tabletops, or similar uses, then eco-X might be an alternative. It is a cement-based composite material containing up to 70% post-industrial and post-consumer waste, including a glass fraction. There are 12 standard colours, with no limit to the customized colours that can be blended. Dimensional specifications are varied with thicknesses ranging from standard 19 mm (.75 in) up to 50.8 mm (2 in) depth with a density of about 0.7 kg per sq. m (17 lbs per sq. ft).

✎	*Meld USA, USA*	318
⚙	*Meld USA, USA*	318
	70% recyclate including glass, 30% cement and reinforcing fibre.	
	• *High fraction of recycled materials* • *Suitable to gain US LEED credit points*	325 330

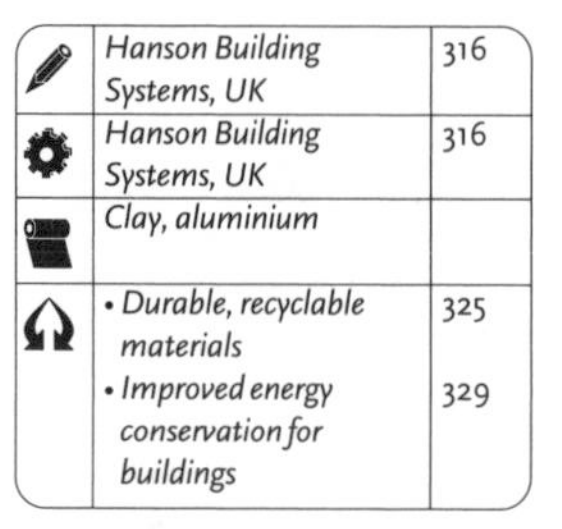

swall®

›ost-and-beam structural
d is created by filling wall
ms with reinforced
ncrete. Wall forms are
nufactured using K-X®
ycled wood waste chips.
e entire wall structure,
own as Faswall®,
nprises up to 85% K-X
gregate (from waste
od) bound with Portland
ment (containing up
15% fly ash content by
ume). A finished Faswall
ows good R-values
ermal insulation) of
ween 18 to 24 and it is an excellent sound barrier and substrate for dry-wall or direct finishes. Standard blockmaking equipment permits local manufacturing of Faswall® components.

✎	*ShelterWorks Ltd, USA*	321
⚙	*ShelterWorks Ltd, USA*	321
	Waste wood and fly ash, Portland cement	
	• *Partially recycled and renewable content*	325

FERMACELL gypsum fibre board

This product is made of 80% recycled gypsum and 20% recycled cellulose fibres from waste paper, and is available in standard and custom size boards. It uses no additives or preservatives and at end of life can be recycled again, making it a real cradle to cradle product. The whole lifecycle approach means that it is also manufactured using minimal energy and water resources. The fine texture of the board enables the elimination of plastering, although a very smooth finish or Fine Surface Treatment can be applied as a skim coat. FERMACELL is also suitable for flooring for soundproofing, under-floor heating systems and providing a warm layer over polystyrene foam.

✎	*Xella International, Germany*	323
⚙	*Xella International, Germany*	323
	Recycled gypsum and paper fibre	
	• *Recycled and recyclable materials* • *Lower embodied energy than virgin materials*	325 325

Eco-shake®

Made of 100%-recycled materials, reinforced vinyl and cellulose fibre, eco-shake® shingles are available in four colour shades designed to mimic weathered wooden shakes. The shakes qualify under strict fire-rating, wind and rain resistance and impact tests.

✎	*Re-New Wood, USA*	320
⚙	*Re-New Wood, USA*	320
	Recycled wood, recycled plastics	
	• *Recycled materials*	325

Parallam®, Timberstrand®, Microllam®

TrusJoist produces a range of patented engineered timbers made by drying short or long veneer 'strands' or sheets, bonding them with adhesives or resins and subjecting them to high pressure and/or heat. They produce three 'timbers', Parallam® PSL, Timberstrand® LSL and Microllam® LVL, and a special composite structural timber floor joist, the Silent Floor® Joist. They claim improved strength and avoidance of defects such as cracking and warping for all their timbers. Furthermore, thanks to the raw veneer ingredients, they can use virtually the whole diameter of a sawn log and small-diameter second-growth trees. This results in a considerable saving on raw materials to produce the same amount of structural timber as sawn wood. For example, the Silent Floor® Joist system uses one tree to every two to three trees for a conventional sawn-wood joist flooring system. Microllam® LVL uses 30% more of the timber from each tree and, being stronger than solid timber, provides almost double the structural value per unit volume of raw material than sawn wood. However, quite a lot of energy is needed to make these composite timbers, so detailed examination of the embodied energy of TrusJoist versus sawn timber should be made on a case-by-case basis.

✎	*Weyerhaeuser, USA*	322
⚙	*Weyerhaeuser, USA*	322
▬	*American softwoods and hardwoods, waterproof adhesives, polyurethane resin*	
♻	*• Efficient use of raw materials*	326

EnviroGLAS

EnviroGLAS Terrazzo, known as EnviroTRAZ is a flooring surface that will last for 40 years or more and is a good replacement for marble terrazzo, allowing conservation of natural geological resources. Comprised of 75% waste glass and porcelain bound by a thin set epoxy binder that is VOC-free after curing, the flooring is poured in situ so it provides a seamless, easy-to-maintain surface ideal for corporate, public or domestic environments. EnviroSLAB is a terrazzo style countertop in a standard size, and

✎	*EnviroGLAS, USA*	315
⚙	*EnviroGLAS, USA*	315
▬	*EnviroTRAZ is 75% post-industrial (pre-consumer) and post-consumer recycled glass, 25% epoxy binder*	
♻	*• Recycled and recyclable materials* *• VOC-free product* *• Qualifies for US LEED credit points*	325 330 330

EnviroPLANK are tiles that facilitate the laying of patterned flooring and are suitable for smaller floor areas. There is an endless variation of colours and recycled constituents to ensure huge visual diversity.

tternut Glass

:hitectural Systems
duce a wide range of
'l panel, ceiling, screen
d table top options for
erior design, including
unique Butternut Glass,
ich is made of an ultra-
n veneer of salvaged
:ternut wood (part of
black walnut family)
ninated within layers of
ss. Light filters through
paper-thin veneer to
ow the natural growth
terns, worm holes and
eaks associated with
timber from this tree.
tural colours can be
mbined with coloured
erlayers to meet
stomized specifications,
d a wide range of clear
nealed, tempered, acid
hed or ultra clear glass
tions are also available.

Architectural Systems, Inc., USA	312
Architectural Systems, Inc., USA	312
Glass, salvaged wood veneers	
• *Abundant geosphere materials*	325
• *Reclaimed material*	325
• *Qualifies for US LEED credit points*	330

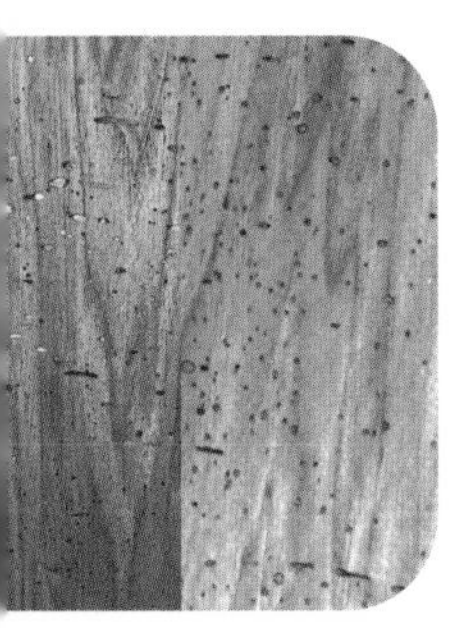

2Zones2®

The 1974 'Marche en Avant' (walk ahead) regulations changed catering practices in France by encouraging nuclearization of activities

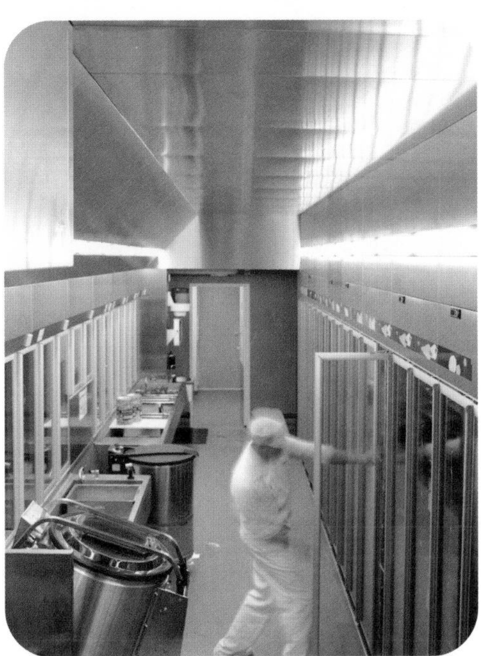

to avoid cross-contamination of raw, cooked or waste food. This led to the development of mono-functional rooms, resulting in a significant expansion in floor space required. The 1989 Hazard Analysis and Critical Control Point (HACCP) regulation enforced new cleaning and disinfecting procedures, which meant increased investment and time was required. Recently 3bornes Architectes re-examined the parameters needed to maximize health and safety while balancing business and eco-efficiency. Their vision was simple: divide any activity according to its temperature and hygiene requirements (cold/clean, cold/soiled, warm/clean, warm/soiled), create a series of parallel climate-controlled zones and utilize standard GastroNorme (GN) containers and a new trolley, Transfert, to transfer food between zones, which are separated by the temperature-controlled cabinets that can be accessed from adjacent zones. Food enters the reception and is stored in GN containers in the first cabinet; the chef retrieves the container and places it in a steel canal – in effect a multifunctional stainless-steel work surface plumbed with water and waste drainage below. Food is unpacked, vegetables and fruit are peeled and disinfected and rinsed above the canal. At the end of the canal prepared food is placed in another GN container and the trolley is wheeled into the cabinet adjacent to the transformation zone (for cutting, mixing, cooking, assembly) on a similar canal system. The prepared food is then placed in the cabinet accessible from the distribution zone. 2Zones2 is a modular stainless-steel building with in-built ventilation, lighting, plumbing and waste disposal, producing a 50% reduction in floor space to a traditional kitchen layout. A final innovation is the use of directed temperature-controlled air streams, such as down onto each canal, to force bacteria towards the soiled areas that are easily disinfected. Local chilling makes better working temperatures for the staff and substantial energy savings. 2Zones2 kitchens have been installed for various institutions, including the Association National des Directeurs de Restaurants Municipaux.

	François Tesnière & Anne-Charlotte Goût, 3bornes Architectes, France	310
	Bourgeois SC with Halton (ventilation), Electricité de France (electrics)	313
	Stainless steel, electrics, ventilation & refridgeration units	
	• *Energy efficiency* • *Improved health & safety* • *Improved ergonomics* • *Modular design* • *Durable*	328 328 328 328 325

UPM ProFi® Deck

Surplus self-adhesive paper and plastic labels are blended with polypropylene and turned into new lightweight yet rigid deck planks. With a hollow core that can also be used for cabling, the weight of one square metre of deck planks is just 2.5 kg (5.5 lb). Standard lengths are 3.2, 4, and 4.8 m (10.5, 13, and 15.7 ft). UPM ProFi® Deck is in fact a complete decking system with support rails, cover strips, planks and T-clips that overlap the edge groove to hold down each successive plank. Easy to install (and remove and re-install), the polymer recyclate is weatherproof, UV resistant, hard-wearing and low maintenance. The grooved texture on the surface of the plank provides good frictional grip and, of course, doesn't splinter.

✎	*UPM, Finland*	322
⚙	*UPM, Finland*	322
▪	*Pre-consumer waste paper and plastic, polypropylene*	
♻	• *Lightweight, durable* • *Recycled materials* • *Low-embodied energy* • *Recyclable at end of life*	325 325 325 329

Paperform V2

These striking sculptural 3D modular wallpaper tiles offer interior designers the option to add dynamic elements that transform the character of a room or space. You can create your own pattern by changing the orientation and colour of the tiles. Made of 100% recycled waste paper, the 12 inch (305 mm) square tiles are lightweight, just 0.04 inch (1.0 mm) thick, can be attached permanently with wallpaper glue or temporarily by double-sided tape. MIO also produces two other Paperform textures, the Ripple and the Flow, which offer other exciting possibilities for bringing new life to wall and ceiling surfaces. There's also a special Acoustic Weave tile for improving the sound qualities within a space.

✎	*Jaime Salm, MIO, USA*	309
⚙	*Jaime Salm, MIO, USA*	309
▪	*100% post- and pre-consumer waste paper*	
♻	• *Recycled materials* • *Recyclable at end of life*	325 329

proSolve370e

Described as 'decorative, depolluting architectural tiles', proSolve370e made its debut at the 11th International Architecture Exhibition, Venice Biennale 2008. The prototype comprises 55 photocatalytic tiles in a 10 m² suspended screen to demonstrate its potential application as a post- or pre-construction element of a building façade. Each tile is coated with superfine Titanium oxide in SigmaCoating's Siloxan NOx, a nano-technology coating that can neutralize nitrogen oxides (NOx), a product of the combustion of fossil fuels, and Volatile Organic Chemicals (VOCs), a wide range of chemicals offgasing from modern materials. As this neutralization occurs in the presence of ultraviolet light and humidity, the tiles are designed to maximize surface area exposed at different orientations of the sun. The organic shaped non-orthogonal grid is derived from a five-fold symmetrical pattern made up of only two constituent parts. The random pattern is intended to be a visual provocation and intervention into the urbanscape.

✎	*Daniel Schwaag and Allison Dring at elegant embellishments, Germany*	305
⚙	*Prototype*	
▮	*Photocatalytic tiles, Titanium oxide*	
♺	*• Attenuation of air pollution* *• Interventions in interstitial space of urban landscape, raising awareness*	328 330

GrüneWand GreenWall System

There is much written on the adverse effects of spending long hours in office environments where automated systems such as HVAC control atmospheric conditions. The GreenWall System designed by Indoorlandscaping addresses this through a natural climate controlling system. The modular, pre-cultivated vegetation system is composed of three layers. A base of polystyrene panels, 20 mm (¾ in) deep, is attached to a tubular steel frame. Plants raised in greenhouse conditions are connected to the base through a middle layer of synthetic growth medium made of fully hardened phenolic resin foam substrate mats. Typically a GreenWall is capable of passive vaporization resulting in an adiabatic cooling capacity of 50–110 W/m2 at an evaporation rate of 1.8–3.8 l/m²/day. They have been installed in a range of buildings from banks to restaurants, and the ecosystem mix of plants in each installation depends upon the nature of the space. This is a visually compelling, germ-free and low-maintenance permaculture system that improves air quality and is capable of generating a 'feel-good' experience for a building's occupants.

✎	*Indoorlandscaping, Germany*	307
⚙	*Site-specific installation, Indoorlandscaping, Germany*	307
▮	*Steel, polystyrene, plants, water*	
♺	*• Air purification* *• Air cooling* *• Improved psychological effect for users of building*	328 328 327
✪	*• Focus Green Design Award 2008*	

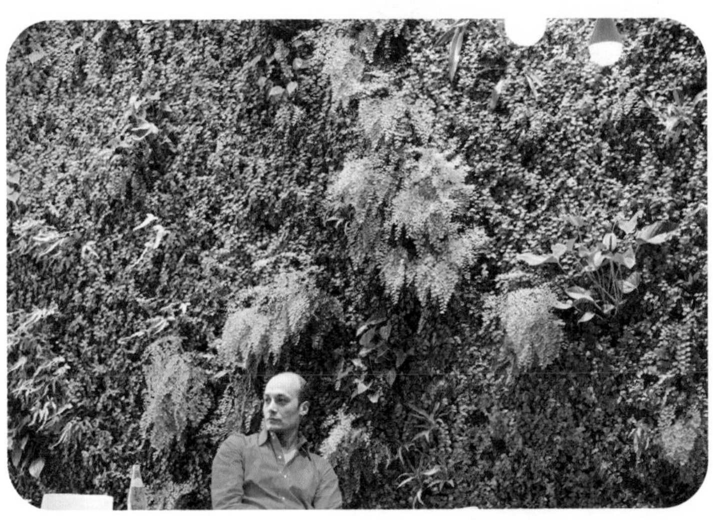

SunPipe

Natural daylight provides a more relaxing spectrum of light than artificial sources, and reduces energy consumption in work spaces. SunPipe is a system of conveying natural sunlight from rooftops into buildings. A transparent dome of UV-protected poly-carbonate is held on the roof by an ABS/acrylic universal flashing, and below the dome is a tube made of Reflectalite 600, silverized coated aluminium sheeting with 96% reflectance. Four standard-diameter tubes, 330 mm up to 600 mm (13–24 in), and a range of elbow joints permit light to be directed into the required space. A vertically oriented SunPipe of 330 mm can deliver 890 Lux in full summer sun and 430 Lux in overcast conditions in the temperate British climate, which is sufficient to provide natural daylight to an area of approximately 14 sq. m (150 sq. ft).

✎	*Terry Payne, Monodraught Ltd, UK*	318
⚙	*Monodraught Ltd, UK*	318
▮	*ABS/acrylic, polycarbonate, aluminium*	
♻	*• Natural lighting*	329

power glass® walls, balustrades

LED technology is driving diverse creative ideas for illumination. Two projects by Glas Platz demonstrate the unique features of their transparent panels of glass, conductive transparent materials and LED electronic systems. A complete power glass® façade was installed in Tour Europe. The spatial dynamics within the building are transformed by the wall of light. Ano project at the Galeries Lafayette, Paris, is the installation of balustrad over three floors. It indicates the graphical and communications potential of power glass systems, which can pro economical-to-run light and dramatic new spati interpretations for inter designers.

✎	*Glas Platz, Germany*	316
⚙	*Glas Platz, Germany*	316
▮	*Glass, conductive materials, LEDs and electronics circuits*	
♻	*• Low-energy lighting and wall systems*	329

ShetkaStone™

This material is engineered using a patented process to convert waste paper fibres into tough, durable, work surfaces, tabletops and counters. Fibres originating from newsprint, cardboard, office paper and other cloth or plant fibres are sourced within 100 miles of the production facility. ShetkaStone™ is scratch, stain and water resistant, and class 'A' fire rated. It is available in thicknesses between 0.5–1.125 in (12.7–28.6 mm) and is a homogenous material with a density of 7 lb/sq. ft (0.3 kg/sq. m), which can be used like any hard wood.

✎	*ShetkaWorks, USA*	321
⚙	*ShetkaWorks, USA*	321
▮	*100% post-industrial or post-consumer waste paper fibre*	
♻	*• Recycled materials* *• Recyclable* *• Locally sourced* *• Zero-waste manufacturing system*	325 325 325 327

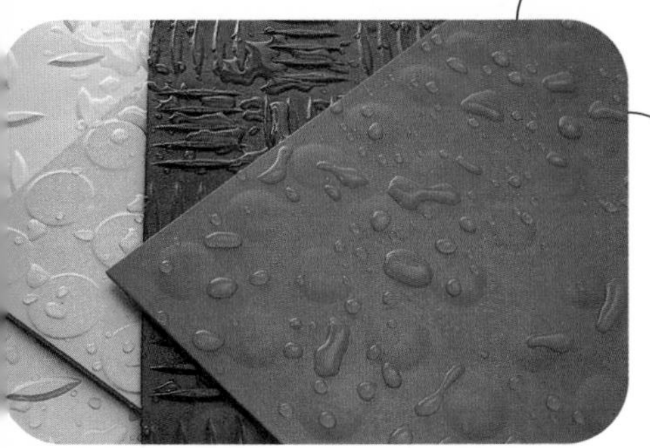

Dalsouple

Dalsouple manufactures standard and bespoke rubber flooring tiles in a huge variety of colours and surface textures that are 100% recyclable.

	Dalsouple, UK	314
	Dalsouple, UK	314
	Synthetic and natural rubbers	
	• Recyclable *• Clean, chlorine-free production process*	325 326

All Dalsouple rubber is free from PVC, CFCs, formaldehyde and plasticizers. Production waste is virtually all recycled within the manufacturing plant and emissions meet local statutory requirements. Service partners to Dalsouple include Uzin Adhesives, who offer water-based and solvent-free adhesives including polyurethane and epoxy resins.

Ecoplan/ecoment

Nora manufactures a diverse range of rubber flooring, but 'ecoplan' and 'ecoment' are the only two made with up to 75% factory and post-installation waste. Granite and marbled-effect patterns are available. All products are free of PVC, plasticizers, formaldehyde, asbestos, cadmium and CFCs and production follows stringent waste-management procedures, minimal packaging and a zero-emissions environment for the workforce.

	Nora Systems, Germany	319
	Nora Systems, Germany	319
	Rubber, recycled rubber	
	• Recycled and recyclable materials *• Clean production*	325 326

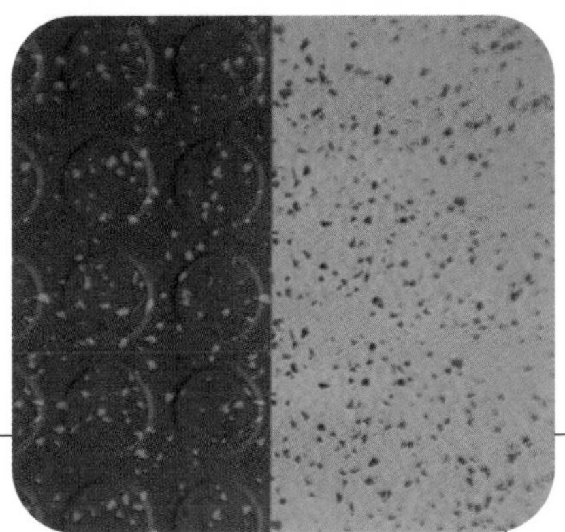

Eco-surfaces

ECORE International is North America's leading recycler of waste tyre rubber, annually recycling over 60 million pounds, equivalent to 2,000 trailer loads and saving over 1 million barrels of oil. A diverse range of tiles, available as squares or on rolls, is available in indoor and outdoor specifications and a vast range of colours. The latest addition to the product range is ECOstars™, a range of neutral colours suitable for corporate or public environments.

	ECORE International, USA	314
	ECORE International, USA	314
	Post-consumer tyre rubber, pre-consumer waste and organic fillers ColorMill EPDM, water-based polyurethane polymer.	
	• Recycled materials *• Recyclable* *• Low-VOC emissions* *• Contributes to points for the US LEED certification for sustainable building*	325 325 328 330

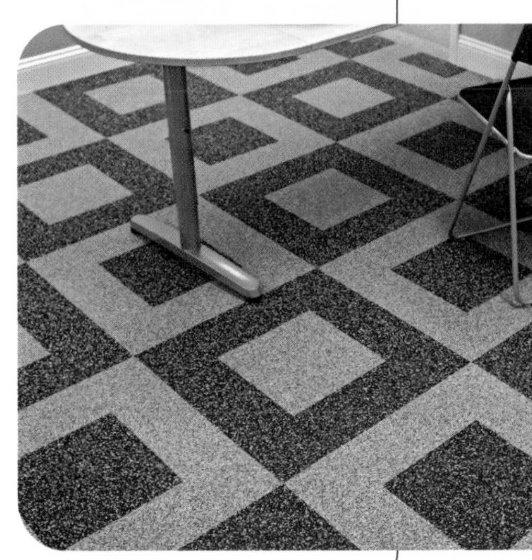

Re:Source Floor Care

Interface was an early US advocate of adopting sustainable thinking into a company's business strategy and practices. Chief Executive Ray Anderson was an early adopter of the Natural Step philosophy and today the company has its own 'Seven Steps to Sustainability'. Re:Source Floor Care is a service system (SS), although it is also available as a product-service-system (PSS) if Interface flooring products (such as Entropy™ carpet tiles) are purchased. The core focus is to create bespoke cleaning solutions for clients' carpets on the basis that no two stains and traffic wear patterns are alike, and that maintenance requirements vary with clients. A range of non-abrasive, low-moisture, easy-to-remove products includes Intercept® anti-microbial and IMAGE care products. Any replacement carpet tiles can be stuck down with Factor 4 Spray Adhesive System requiring less than 25% of the volume of a conventional adhesive. Re:Source service providers provide the first pan-European network within the floorcovering industry so the benefits of lower environmental floor care impact can reach further afield.

Designer	*Interface, Inc., USA*	317
Manufacturer	*Interface, Inc., USA*	317
Materials	*Various cleaning and adhesive materials*	
Eco	*• Reduction in use of consumables* *• Improved product lifespan*	329 329

Silencio 6

This is a special 6 mm thick (about ¼ in) fibreboard composed of 100%-waste softwood fibre, which provides good attenuation against impact sound and insulation as an underlay for wooden and laminate floating floors.

Designer	*Hunton Fiber, Norway*	316
Manufacturer	*Hunton Fiber, Norway*	316
Materials	*Recycled softwood fibre*	
Eco	*• Waste materials* *• Recycled, recyclable materials*	326 325

PLYBOO®

The basic component of all this company's products is strips of bamboo measuring 0.5 x 1.9 x 183 cm (⅛ x ¾ x 72 in), which are extremely durable. These strips can be bent, woven and laminated as required for flooring, interior-decoration fittings and furniture. Four-, two- and one-ply laminates are available.

Designer	*Smith & Fong Company, USA*	321
Manufacturer	*Smith & Fong Company, USA*	321
Materials	*Bamboo, adhesives*	
Eco	*• Renewable material*	325

Acousticel Acoustic Underlay (A10)

This underlay is suitable for reducing noise transmission for wood, concrete and asphalt floors. It has a top layer of high-grade felt (jute and hair mixture) fixed to a layer of recycled rubber crumb. Supplied in 1.37 x 11 m (4 ft 6 in x 36 ft) rolls it is easy to lay and is recommended on uneven floors.

Designer	*Sound Service (Oxford), UK*	321
Manufacturer	*Sound Service (Oxford), UK*	321
Materials	*Felt, rubber*	
Eco	*• Recycled and renewable materials* *• Reduced noise*	325 328

Entropy™

As the name suggests, Entropy™ carpet tiles embody principles of energy movement in natural systems. Tiles are made of DuPont Antron Lumena® yarn, a recyclable nylon using the tufted tip-sheared production process. All tiles are designed for installation in random patterns in any direction. This means that Entropy™ provides near-zero installation waste, ages gracefully and has an extended lifecycle because worn tiles can be swapped or replaced. Using one dye lot ensures even colouring. Worn-out tiles are taken back and recycled. This product embraces Interface's 'Seven Steps to Sustainability' – elimination of waste, benign emission, renewable energy, closing the loop, resource-efficient transport, sensitivity hook-up (creating synergistic communities to work with the company) and redesign of commerce. This is further reflected in a new scheme Interface have introduced, with the Climate Neutral Network, called Cool Carpet™ – buy 15 sq. m (18 sq. yds) of carpet, plant a tree. Interface has calculated that new tree planting absorbs the same amount of carbon dioxide as is emitted by their carpets. Paying for new forestry plantings with one's purchase may become a regular feature of retailing.

✎	*Interface, Inc., USA*	317
⚙	*Interface, Inc., USA*	317
▬	*DuPont Antron Lumena® yarn*	
♲	• *Recyclable product* • *Reduction in emissions* • *Ease of repair*	329 328 329

Loomtex

Lloyd Loom is renowned for its patented process for creating furniture from metal wire and twisted paper. The technology has been further developed to create a highly durable woven paper carpet with 100% cotton welt. A wide range of patterns is available in widths up to 2.5 m (8 ft 2 in) and any length can be made to special order.

✎	*Lloyd Loom of Spalding, UK*	318
⚙	*Lloyd Loom of Spalding, UK*	318
▬	*Paper, cotton*	
♲	• *Renewable materials* • *Durable material*	326 325

Composite lumber

This decking is a composite material using oak fibre and recycled polyethylene with foaming compounds and additives. Containing over 90% recycled materials by weight, it is very moisture-resistant and durable. It also weathers like conventional wood but without any associated rotting.

✎	*U.S. Plastic Lumber, USA*	322
⚙	*U.S. Plastic Lumber, USA*	322
▬	*Composite wood*	
♲	• *Recycled materials*	325

Dal-lastic

Dalsouple's huge range of coloured, textured floorcoverings for domestic and heavy industrial use is well known. Dal-lastic is a new variant, a thick chunky tile suitable for outdoor use and made from 100% recycled rubber fibre. It has a springy impact and sound-absorbent surface that is self-draining and feels warm underfoot. Tiles are loose-laid but jointing is invisible. Plain, flecked or terrazzo patterning plus a large range of colours gives plenty of choice for patios, terraces, roof gardens and other landscape features.

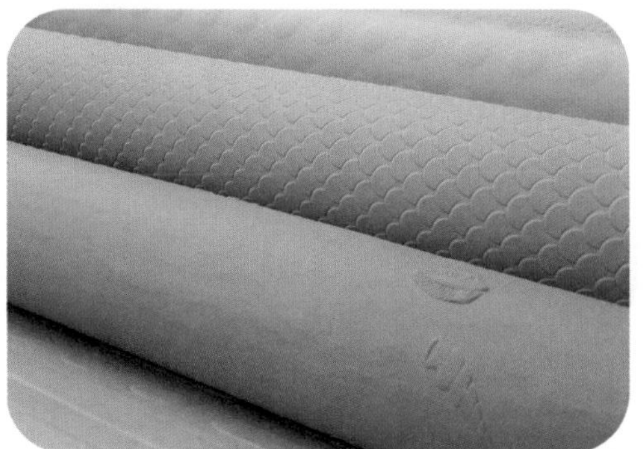

✎	*Dalsouple, UK*	314
⚙	*Dalsouple, UK*	314
▬	*100% recycled rubber*	
♲	•*Recycled materials*	325

Super Duralay

Over 60,000 used car tyres are processed each week at a new Duralay plant to provide the raw material for a range of rubber crumb underlays suitable for carpets and wooden flooring. Super Duralay is rated for heavy domestic use but other grades are suitable for contract usage. Bacloc, a woven backing of paper and synthetic thread, gives extra strength to the underlay. Other grades use a mixed backing of jute and plastic.

Duralay, UK	314
Duralay, UK	314
Recycled tyres, latex, Bacloc or polyjute	
• Recycled materials *• Conservation of landfill space*	325 329

Coconutrug®

While most people associate coconut fibre with uninspiring doormats, Deanna Cornellini has worked with the natural characteristics of the fibre to produce an inspiring range of textured rugs dyed with non-toxic colours. As well as standard ranges, including warm spectrum colours, blues, purples and greens, there are a number of limited edition runs, such as 'Signs', 'Views', and 'Horizons', with juxtaposed colour panels and hand-stitched lines. Ultra-pure fibre is hand-woven on traditional looms to produce distinctive finishes to each product line. These tough rugs suitable for hard-weari areas, from children's playrooms to offices. E carpet arrives rolled up in a jute bag. Cleaning easily achieved by brus with a broom or using carpet beater – good o fashioned but time-tes methods that require c human energy.

Deanna Cornellini, Italy
G. T. Design, Italy
Coconut fibre, non-toxic colouring dyes
• Renewable materials *• Safe, non-toxic*

Heart woods

Heart woods, from the slower growing core of the tree, usually have closer, and tougher, grains. The Goodwin Heart Pine Company specializes in antique Longleaf heart pine, heart cypress and other native American timbers. Once covering 41% of the landmass of the American deep South, some 90 million acres, the Longleaf's slow growing trees, many of them 500 years old, now provide only 2% forest cover. The company ensures that their antique heart pine is reclaimed and revived so it can find new life.

Goodwin Heart Pine Company, USA	316
Goodwin Heart Pine Company, USA	316
Wood from US sources	
• Use of native and reclaimed hardwoods	325

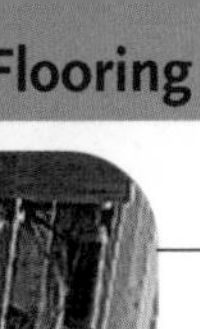

Teragren

Teragren has manufactured bamboo flooring, panels and veneers since 1994 using sustainably harvested Optimum 5.5 Moso bamboo grown in China. Harvesting takes place on a 5 to 6 year cycle, producing older, durable bamboo, and leaving the cut plants to regenerate for the next crop. All products are made in China in factories certified to ISO 9001 and ISO 14001 standards, and are laminated using low-formaldehyde or formaldehyde-free glues then finished with water-based solvent-free coatings. The company is committed to operating a sustainable business, and is a member of the US Green Building Council, Business for Social Responsibility, and follows The Natural Step programme.

Designer	*Teragren, USA*	322
Manufacturer	*Made in China for Teragren, USA*	322
Material	*Bamboo*	
Eco	*• Rapidly renewable material* *• Sustainably managed plantations* *• Low or zero emissions* *• Contributes to points for the US LEED certification for sustainable building*	325 326 328 330

Stratica

Stratica is a laminated flooring product comprising a very tough, durable layer of chlorine-free, ionomer coating, DuPont Surlyn®, a printed layer, a backing layer to the print and a final bottom layer. Surlyn® is the finishing material on golf balls. Stratica is naturally flexible but doesn't use plasticizers and is free of volatile organic compounds (VOCs). Over 45 different 'natural' surfaces can be mimicked in the printing process, from stone to marbles, granites, terrazzos and woods, plus over 20 solid colours. Abrasion resistance is very high and maintenance costs are low. With certification to ISO 14001, recycling pre-consumer waste, preventing pollution and saving energy in production are high priorities.

Designer	*Amtico International, UK*	312
Manufacturer	*Amtico International, UK*	312
Material	*Dupont Surlyn®, mineral-filled ethylene copolymer*	
Eco	*• Clean production*	326

EcoDomo

Recycled leather from the furniture, shoe and car industries is combined with natural rubber and a natural binding agent from tree bark to make tiles for luxury floor or wall finishes. They have the feel and quality of real leather and come in eight different colours and four textures.

Designer	*EcoDomo, USA*	314
Manufacturer	*EcoDomo, USA*	314
Material	*65% post-industrial/ pre-consumer recycled leather, 20% natural rubber, Acacia tree bark*	
Eco	*• Recycled and rapidly renewable materials* *• Low VOC emissions* *• Contribute to US LEED certification credits for sustainable building*	325 328 330

Logamax U112-19

This slim wall-mounted boiler unit offers an output capacity of 20kW. It uses an efficient ceramic burner to provide more complete combustion of the gas fuel, resulting in emission levels well under those specified in the German Blue Angel eco-label scheme.

	Buderus Heiztechnik GmbH, Germany	313
	Buderus Heiztechnik GmbH, Germany	313
	Steel, various metals including aluminium-silicon alloy, ceramics (burner unit)	
	• Energy efficient *• Blue Angel eco-label*	329 330

Farm 2000 'HT' boilers

A range of high-temperature (HT) boilers has been designed to accommodate typical biomass fuels available on the farm, such as circular or 1-tonne rectangular straw bales, as well as woodchips, cardboard and other combustible wastes. Heat outputs vary from 20 kW to 300 kW depending on the boiler and the equivalent electricity costs per kWh are between 25% and 33% of those of kerosene oil or natural gas. An upper refractory arch encourages complete burning of gases, improves the overall efficiency and minimizes atmospheric emissions. Annual or short-rotation crops for biomass fuels absorb carbon dioxide that is released on burning, so this cycle is neutral and makes no net contribution to the greenhouse effect.

	Teisen Products, UK	321
	Teisen Products, UK	321
	Steel	
	• Reduced emissions, carbon-dioxide neutral, heating system *• Renewable, local biomass fuels*	328 329

Logano G144 ECO

This free-standing boiler unit offers an output capacity between 13 kW and 34 kW depending on the exact model, making it suitable for heating single or multiple dwellings. A key feature is the efficient ceramic burner, which provides more complete combustion of the natural or liquid gas fuel, reducing emission levels of nitrous oxides and carbon monoxide below the levels of the German Blue Angel eco-label scheme.

	Buderus Heiztechnik GmbH, Germany	313
	Buderus Heiztechnik GmbH, Germany	313
	Cast iron, various metals including aluminium-silicon alloy, ceramics (burner unit)	
	• Energy efficient *• Blue Angel eco-label*	329 330

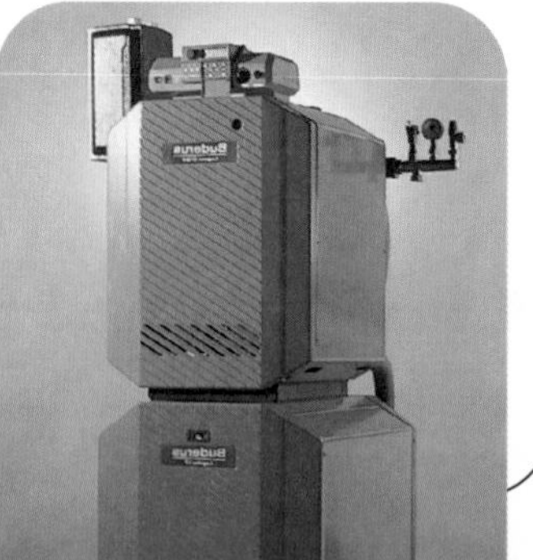

Filsol Stamex Collector

Water heating in buildings in temperate climates can be readily supplemented by installation of a solar collector on or in the roof. Sunlight enters the acrylic collector, which transmits 89% of incident light, and heats a 'Stamax' absorber plate made of a specially coated (18% chromium, 10% nickel) stainless steel.

✎	*Filsol Solar, UK*	315
⚙	*Filsol Solar, UK*	315
▮	*Stainless steel, aluminium, alloys, acrylic, poly-isocyanurate foam*	
♻	*• Solar power (passive)*	329

Colourless oxides of chromium, nickel and iron provide an absorption of 93% of the incident energy, transferring it to an aqueous antifreeze mixture running in channels in the absorber plate. This hot mixture is pumped when its temperature is higher than the water in the tank, where heat exchange also occurs.

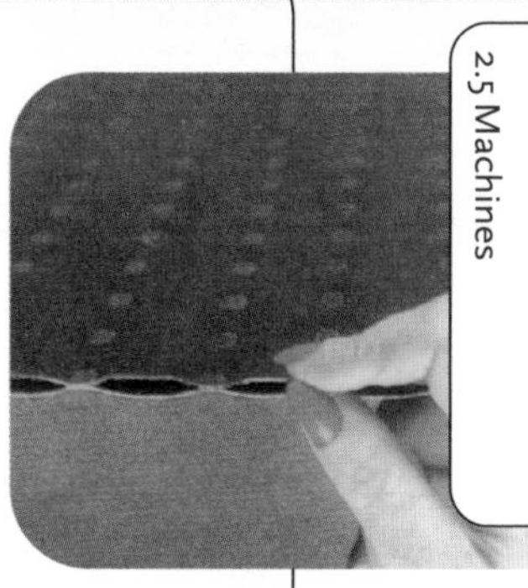

C21e roof tiles

The C21e range from Solar Century, one of the leading installers of integral, photovoltaic roof tiles in the UK, fits conventional roof batten systems and is available in tiles or slate, suitable for a minimum roof pitch of 22.5 degrees or 30 degrees respectively. The tiles hook together with push-fit connectors and are screwed onto the battery. Up to 2000 kWh per annum – enough to meet the needs of an average household – can be generated by this system, with excess electricity sold back to the national grid. This requires approximately 42–48 C21e tiles or slates and creates a carbon offset of about 1,030–1,177 kg per annum.

✎	*Solar Century, UK*	321
⚙	*Solar Century, UK*	321
▮	*Photovoltaics, glazing, other synthetic materials*	
♻	*• Solar power generation*	329

momax®

late solar collectors a reduction in ncy from 60% at an ting temperature of down to about 40% temperature doubles. o with vacuum-tube collectors, which ain efficiencies of over at temperatures of A semi-conductor ber plate sits within acuated glass tube. pecial liquid-filled heat s in intimate contact he absorber plate heat from the auses the liquid to rate to the top of a enser unit. Between pe and the condenser spring made of -memory metal, which limits heat transfer through the pipe when pre-set temperatures are reached, so preventing overheating. Water surrounding the condenser absorbs heat as it flows. Energy conversion even on cloudy days is very efficient, with an overall annual efficiency of over 70%. The reduction of gas or electricity heating costs for an average household is about 40%. All Thermomax manufacturing plants comply with ISO 9001.

✎	*Thermo Technologies, USA*	322
⚙	*Thermo Technologies, USA*	322
▮	*Low-iron soda glass, copper, shape-memory metal*	
♻	*• High-efficiency solar-powered hot-water system*	329

GFX

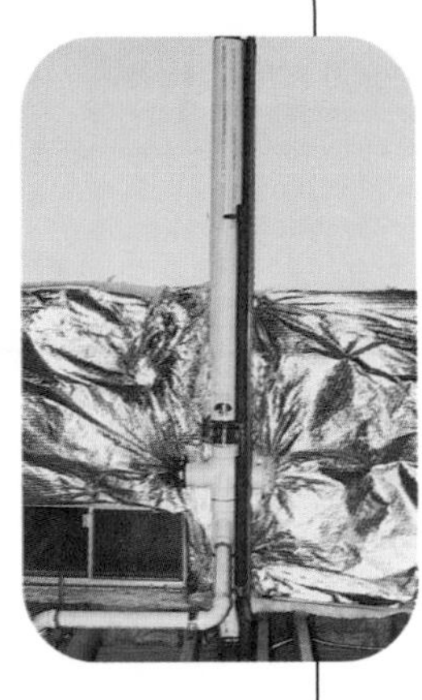

The heat within waste water from domestic showers, baths and sinks can be reclaimed to heat the incoming cold-water supply to the hot-water tank. GFX is an insulated spiral coil of copper tubing carrying the cold supply, which is in intimate contact with a falling-film heat exchanger through which the waste hot water travels by gravity. The system is capable of saving up to 2kW of power from each shower.

WaterFilm Energy Inc, USA	322
WaterFilm Energy Inc, USA	322
Copper, insulation	
• Energy conservation	329

Paradigma CPC Star Azzurro

The modular design of this solar collector, in four sizes of 14 to 21 tubes, allows for easy separation of and almost 100% recycling of components. Materials have been kept to a minimum, giving a lightweight structure with high efficiency in low sunlight and at temperatures below freezing.

Büro für Produktgestaltung, Germany	304
Paradigma, Germany	319
Aluminium, glass, other synthetic materials	
• Design for disassembly and recycling	326
iF Ecology Design Award, 2000	332

Topolino

Traditional wood-burning stoves are stoked with timber in a haphazard fashion, causing rapid, uncontrolled combustion with significant heat loss up the chimney stack. GAAN's range of wood-burning heaters encourage optimal combustion because wood is stacked vertically and burns from the top, like the wick of a candle, producing a fuel efficiency of 85%. As the warm combustion gases rise they are forced through a double swan-necked constriction where heat is absorbed into the surrounding materials. Immediate space heating is provided by radiated heat from the toughened glass door, while the remaining heat passes into the surrounding cast stone, granite or steatite body panels, where 60% of the total combustion energy is stored and emitted over the next six to eight hours. Emissions are significantly lower than required under existing EU and Swiss regulations.

GAAN GmbH, Switzerland	306
Tonwerk Lausen AG, Switzerland	322
Steel, glass, granite	
• Improved energy generation *• Durable construction*	329 329

SeaGen Tidal Turbine

These submarine windmills, driven by tidal or ocean currents, can capture large quantities of energy and are, arguably, less contentious interventions than their above ground cousins. Axial flow rotors of 15–20 m (49–65.6 ft) in diameter are mounted on twin lateral wings, each with a unique blade pitching system that allows 180ffl rotation to capture bi-directional flow, driving a generator via a gearbox system. Trials of the SeaGen in the Strangford Narrows, Northern Ireland in December 2008 generated energy at the maximum designed capacity of 1.2 MW during the operational period of 22 hours per day, generating enough energy to power up to 1,000 homes and setting a world record. Now that the technology is proven, we can expect to see more of these being put into operation: seven SeaGen units are planned for installation off the coast of Anglesey, North Wales in 2011–12.

✎	*Marine Current Turbines, UK*	318
⚙	*Marine Current Turbines, UK*	318
	Various metals, synthetic polymers, electronic and electric systems	
	• *Renewable energy generation* • *Reduction of visible intrusion in the landscape*	329 330

Multibrid wind turbines

These wind energy converters are designed to work offshore exposed to high-speed, salt-laden winds, so all components are sealed to prevent ingress of water. Unique aspects of the design include slow rotational movement to ensure that the unit can operate without maintenance for the first three years. Specifications for each turbine varies according to client and site needs, with precise engineering to get the best out of the rotor's aerodynamics.

Designer	*Bartsch Design GmbH, Germany*	30[illegible]
Manufacturer	*aerodyn Energiesysteme GmbH, Germany*	312
Materials	*Various*	
Environmental	*• Renewable energy*	329

Rutland 913 Windcharger

Weighing just 13 kg (28 lb), this wind generator has a 20-year pedigree and has been well proven in a wide variety of climates by yachtsmen and scientific researchers and for military and telecommunications operations. Continuous electrical generation starts at wind speeds of 5 knots (5.75 mph). Durable marine-grade materials are combined with quality engineering, units being manufactured to ISO 9001.

Designer	*Marlec Engineering Company Ltd, UK*	318
Manufacturer	*Marlec Engineering Company Ltd, UK*	318
Materials	*Stainless steel/ aluminium, glass-reinforced polymer*	295
Environmental	*• Renewable wind power*	329

Underwater 100 Micro Hydro Generator

This water turbine is designed to be in a fixed location where a moving body of water passing by can generate up to 5 amps continuous charge for 12/24 volt systems, or can be used at sea on a moving vessel. In a fast-running stream, just 400 mm (15.7 in) deep, the UW100 can generate up to 2.4 kW per day. Durable marine-grade materials are used with double 'O' seals and hydraulic fluid in the alternator body to provide maintenance-free turbines.

Designer	*Ampair Ltd, UK*	312
Manufacturer	*Ampair Ltd, UK*	312
Materials	*Marine-grade metals and plastics*	
Environmental	*• Renewable, water-driven, power*	329

Solar-powered service station canopy

This was an early trial by BP Amoco using photovoltaic installations on canopies to generate the station's electricity. Solar power has not become standard in all petrol service stations, but BP has developed its solar capability by acquiring significant capacity in solar panel manufacturing under its BP Solar division, as well as renewable technology resources with its BP Alternative Energy division. How the energy transition is managed by oil companies is critical to climate change events.

	BP, UK and international	313
	BP, UK and international	313
	Photovoltaic panels	
	• Renewable, solar power *• Provision of operational energy needs of station*	329 329

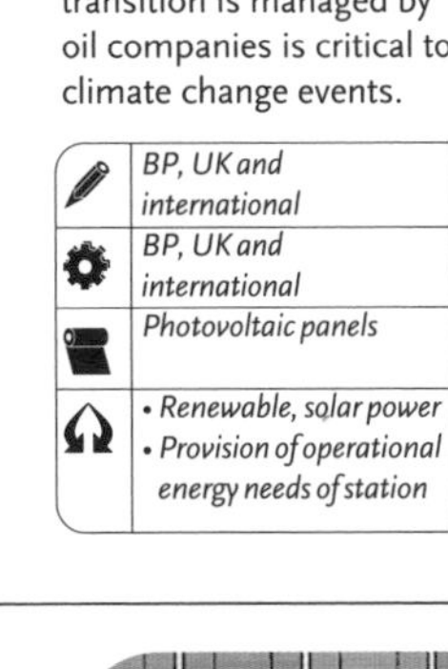

Solar Cells

Up to 17.4% of incident sunlight is converted into electrical power with these Sunways monocrystalline and multicrystalline solar cells, a very efficient ratio compared with conventional solar cells. Various versions are manufactured, including those offering up to 30% transparency. The transmitted light is white, yet there is a range of external colours for the cells, employing a process of texturing that avoids the use of chemicals. Now solar cells can be integrated into any aperture intended to introduce light into a building, such as windows and roof lights, thereby reducing overall construction costs.

	Roland Burkhardt, Sunways, Germany	304
	Sunways, Germany	321
	Silicon mono or multicrystalline wafers	
	• Dual-function *• Solar power*	329 329

LifeStraw® Personal

More than a sixth of the world's population does not have access to safe drinking water, and each year more than 1.8 million people die from drinking dirty water. The LifeStraw® Personal, a compact portable water purifier weighing only 140 g (5 oz) with its inbuilt filters, is, literally, a lifesaver. Enclosed in the high-impact polystyrene tube is a complex filter made of elements of halogenated resin Elutes, a strong base anion exchange resin, and granular activated, silver-impregnated carbon. These ingredients kill and remove 99.9% of waterborne bacteria and 98.2% of waterborne viruses, and remove particulate matter from 125 microns down to 15 microns. Over 700 litres (185 gallons) of water can be purified with one filter, giving an equivalent personal usage of 2 litres (½ gallon) of clean water per day. The designers at Vestergaard Frandsen are working towards achieving the Millennium Development Goal of reducing by one-half the proportion of people without sustainable access to safe drinking water by the year 2015.

✎	*Vestergaard Frandsen, Denmark*	310
⚙	*Vestergaard Frandsen, Denmark*	310
▬	*High impact polystyrene, various filter ingredients*	
♺	*• Creates potable, safe water at point of use*	329
✪	*Saatchi & Saatchi Award for World Changing Ideas, 2008; Index Award 2005*	

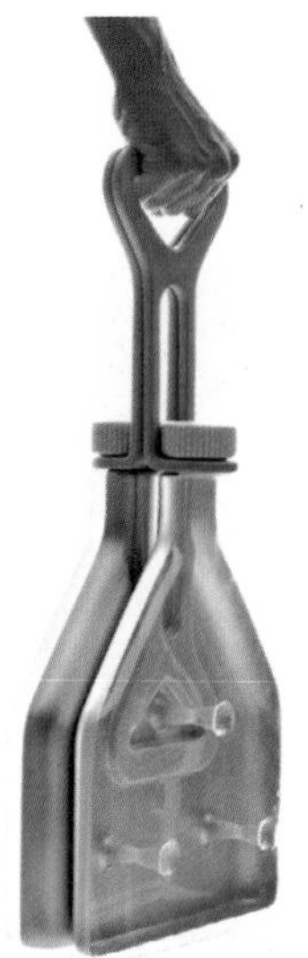

Solar Bottle

The renowned Italian designer Alberto Meda, winner of two Compasso d'Oro prizes for lighting products for Luceplan, turns his attention to the urgent and growing global challenge of providing access to safe drinking water. A traditional DIY solution is to place old PET soda bottles on a roof to sterilize the water by UV-A sunlight and high temperatures. This system is called the Solar Water Disinfection (SODIS) system and has been shown to be effective in treating a wide range of bacteria and viruses, but is less effective against protozoa, such as Giardia and Cryptosporidium. Meda and his colleague Francisco Paz re-examined SODIS and discovered that the diameter of soda bottles meant that sunlight has to penetrate up to 6–10 cm (2.3–4 in) to reach water at the centre of the bottle. Their solution was to flatten the bottle and coat one side with a reflective surface, thereby increasing the intensity of the UV-A and temperature sterilization through infrared absorption. The volume of treated water has been increased to 4 litres (1 gallon), but the solar bottle is just 60 cm (23.6 in) deep ensuring good sunlight penetration even of water with a high turbidity. Bottles stack vertically to save space and a clip in/out handle acts as a prop to get the best angle of inclination to maximize incident sunlight. Clear instructions are embedded on the bottle's external surface.

✎	*Alberto Meda and Francisco Gomez Paz, Italy*	308
⚙	*Prototype, seeking manufacturer*	
▬	*Eastman's Eastar PETG 6763 copolyester*	
♺	*• Improved functionality and ergonomics* *• Robust, durable design* *• Light-weighting per volume of treated water*	328 328 330
✪	*Index Award Home category, 2007*	

Q Drum

Developed in 1993, the Q Drum is now used in ten countries throughout Africa. This tough, durable rotationally moulded Linear Low Density Polyethylene (LLDPE) plastic roller enables 75 litres (19.8 gallons) of water to be easily transported – that's 75 kg (165.3 lb) in weight – without causing the inevitable human damage associated with lifting and carrying substantive weights on the head and shoulders. Demonstrating a simple ingenuity, this kind of product can transform daily life from burdensome chores to bearable tasks.

✎	*P.J. and J.P.S. Hendrikse, South Africa*	307
⚙	*Q Drum (Pty) Ltd, South Africa*	320
▬	*Linear Low Density Polyethylene (LLDPE)*	
♻	*• Encourages use of safe water sources by facilitating transportation*	329

Oxfam bucket

Jerrycans holding about 22 or 45 litres (5 or 10 gals) are the normal means of carrying water for aid or disaster relief work by agencies such as Oxfam. But rigid jerrycans take up a lot of valuable space on aid work planes, so Core Plastics developed a stackable bucket with a removable lid, which incorporates a filler hole/ spout with snap-on top. An indentation in the base helps reduce the risk of spinal injuries when the bucket is carried on the head. The bucket design improves the efficacy of relief efforts.

✎	*Core Plastics, UK for Oxfam, UK*	319
⚙	*Oxfam, UK*	319
▬	*Lightweight, UV-resistant plastic*	
♻	*• Reduction in transport energy*	327

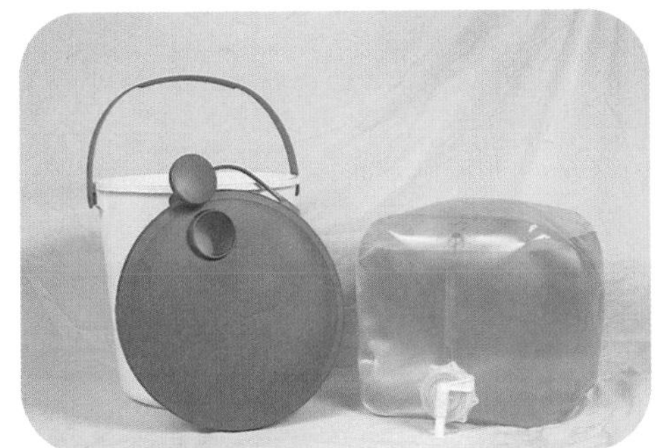

Watercone®

Paradoxically for a planet whose surface is covered by 70% oceanic water and has two major ice caps, the availability of fresh drinking water is extremely limited in many parts of the world. Potable reserves are unequally and unequitably distributed, and almost 40% of the world's population doesn't have access to adequate supplies of safe drinking water. This self-help solar still provides a functional, low-cost solution to in-situ provision of safe drinking water. It can provide 1–1.5 litres of water per day, a basic daily water supply. Contaminated, dirty water or seawater is placed in the central reservoir of the base section. Sunlight causes evaporation from the water source, which condenses on the underside of the vacuum-formed polycarbonate cone and trickles down inside the cone to the collection trough. A screw-thread apex on the cone allows it to be used to empty fresh condensate and water from the trough into any suitable receptacle. Watercones are easily stacked for transportation and are ideal for disaster relief as well as serving poor communities. With a duration of 3–5 years the US $50 investment pays back well within the lifetime of the product, so it is attractive to not-for-profits and NGOs for its economic effectiveness and eco-efficiency. The UV-resistant polycarbonate of the cone is robust, non-toxic, non-flammable and 100% recyclable, and the base section is made of 100% recycled polycarbonate.

Designer	*Stephan Augustin, Augustin Produktentwicklung, Germany*	304
Manufacturer	*Disc-O-Bed GmbH, Germany*	314
Material	*Polycarbonate*	
Benefits	*• Water generation* *• Improved health*	329 328
Award	*iF Design Award, 2003*	332

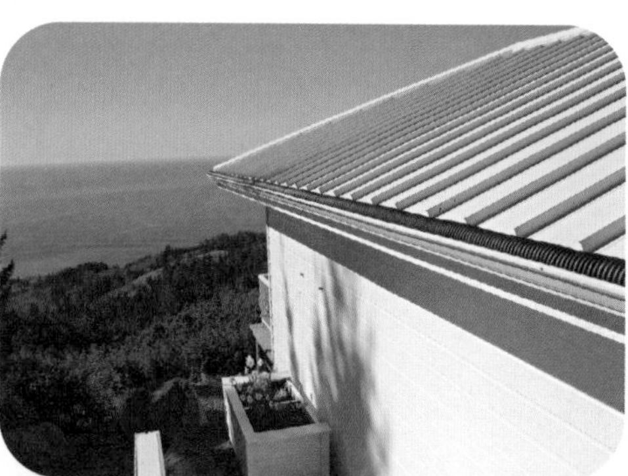

RainTube®

RainTube® is a hydraulically engineered solution that prevents clogging and overflowing gutters while enhancing rainwater harvesting even during high intensity rainfall of up to 100 inches (254 cm) per hour. The flexible tube fits snugly into existing gutters adopting its specific shape once installed. Grooves and holes permit percolation of water but prevent larger particles and debris, so provide a first filter system for rainwater harvesting making it easier to use downstream filters to get pure water. Made from 100% recycled, post-consumer, food grade plastics that normally end up in landfills, and manufactured using renewable energy, this product is also recyclable at end-of-life. This lifecycle approach qualifies RainTube® for a Cradle2Cradle Gold award, certified by William McDonough Braungart Chemistry, USA, and means that it also contributes to points in the US LEED eco-building certification scheme.

Designer	*RainTube Technologies, USA*	320
Manufacturer	*RainTube Technologies, USA*	320
Material	*Recycled post-consumer HDPE plastic*	
Benefits	*• Improved water capture* *• Decreased maintenance compared to traditional systems* *• 100% recycled and recyclable materials i.e. zero-waste product*	329 328 325

Axor Starck two-handled basin mixer

This functional, easy-to-clean mixer tap limits water output to 7.2 litres (1.6 gal) a minute, eliminates limescaling and has a special stop valve.

✎	*Philippe Starck, France*	309
⚙	*Hans Grohe GmbH, Germany*	316
▬	*Chromium-plated steel*	
♻	*• Water conservation*	329
✿	*iF Design Award, 1999*	332

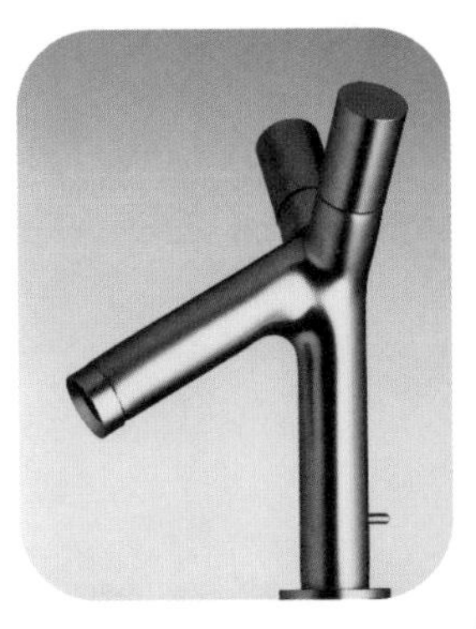

Plush tap

Conservation of resources in public buildings ought to be a high priority but it often needs an innovation to encourage capital investment before tangible results can be achieved. Plush Tap is such an innovation. It allows existing cross-head taps to be converted into push taps by using an adaptor to fit on the old tap body. Water is conserved, as taps cannot be accidentally left running. The advantages of push taps are especially felt where water is metered.

✎	*Flow Control Water Conservation Ltd, UK*	315
⚙	*Flow Control Water Conservation Ltd, UK*	315
▬	*Brass, seals, stainless steel*	
♻	*• Water conservation*	329

Axor Starck showerhead

This free-standing shower unit is sparing in its use of materials as parts of the frame also act as hot- and cold-water pipes to deliver to the overhead and hand-held roses. Pleating the polyester curtain prevents it from clinging to the user. The unit is easily plumbed in, provides excellent access for maintenance and can be repositioned when moving house. Stainless steel would be a preferred substitute for the chromium-plated steel to minimize the impacts of this unit even further.

✎	*Philippe Starck, France*	309
⚙	*Hans Grohe GmbH, Germany*	316
▬	*Enamelled and chromium-plated steel, polyester, polymer base*	
♻	*• Reduction in materials used*	325
✿	*iF Ecology Design Award, 2000*	332

Ifö Cera range

The humble toilet bears the hallmark of a couple of hundred years of traditional industrial design but few people know what happens inside the water cistern. In traditional toilet designs extravagant volumes of water are used to flush even small quantities of human effluent. Today, to meet the need for water conservation, sanitary-ware manufacturers such as Ifö Sanitär have introduced dual-flush cisterns offering two- or four-litre (0.4 or 0.9 gal) water delivery and, more recently, adjustable flushing volumes, from three to eight litres (0.7–1.8 gal) in the Ifö Cera range. Polypropylene or duroplastic seating is available, ergonomically designed to fit a variety of posteriors, and with hygienic surfaces in a typically clean, sculptural, Scandinavian form.

✎	*Ifö Sanitär AB, Sweden*	316
⚙	*Ifö Sanitär AB, Sweden*	316
▦	*Ceramics, polypropylene or duroplastic*	
↻	• *Water conservation* • *Improved ergonomics*	329 328

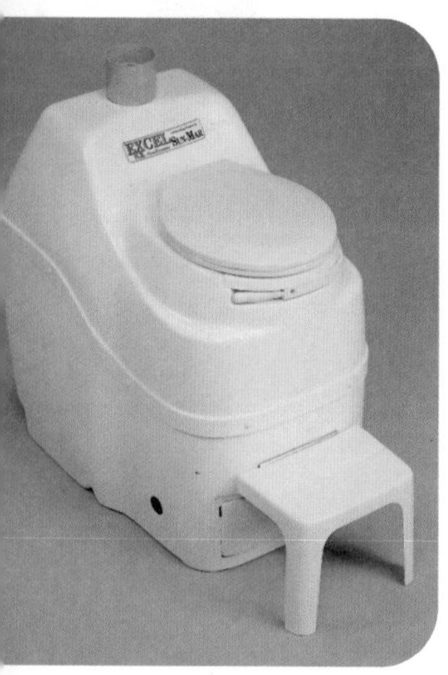

Excel NE

With over 25 years' experience of designing composting, waterless toilets, Sun-Mar Corporation has developed a range of self-contained and central composting toilet systems. Most models are equipped with an electrically driven fan to provide an odour-free atmosphere but the Excel NE is totally non-electric, using a vent chimney instead. The operating principle in all Sun-Mar toilets is identical. A mixture of peat, some topsoil and/or 'Microbe Mix' is added to the Bio-drum™ and a cupful of peat bulking mix is added per person per day. After use the Bio-drum is mechanically turned between four and six revolutions every third day or so to aerate the mixture. Fully degraded compost is removed from a bottom finishing drawer as required. An evaporating chamber at the rear of the drum ensures excess moisture is removed. So confident are the manufacturers in the robust design of their toilets that they offer free parts for three years and a 25 year warranty on the fibreglass body.

✎	*Sun-Mar Corporation, Canada*	321
⚙	*Sun-Mar Corporation, Canada*	321
▦	*Fibreglass*	
↻	• *Water conservation* • *Local composting of waste*	329 329

Naturum

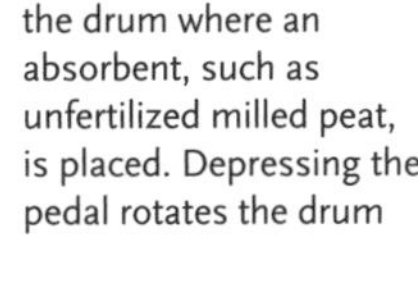

Scandinavia has a well-developed market for composting toilets, so expectations of standards are high. The Naturum is a very practical and efficient rotary-drum composter suitable for installation indoors in any bathroom serving up to five people daily. Whether used in a seated or standing position, the urine is separated by the shape of the bowl. As the compost space is closed with a shutter-seal, a shower hose turns the bowl into a bidet. Solid wastes fall directly through the trap door into the drum where an absorbent, such as unfertilized milled peat, is placed. Depressing the pedal rotates the drum and instantly turns old compost over the new waste. Each time the pedal is depressed more mixing occurs. As the digestion and decomposition process proceeds the compost mass of about 30 litres (6–7 gals) is kept constant and rises to its operating temperature. Any excess mass is shaken out in sterile, odourless particles by the rotation motion into a 10-litre (22-gal) separate container, which is emptied when required – about once a month if used by four people. The front and bowl are fibreglass but inner parts are recyclable polyethylene and any metal parts are high-grade stainless or acid-resistant steel.

Biolan, Finland	313
Biolan, Finland	313
Glass fibre, polyethylene, stainless or acid-resistant steel	
• *Water conservation* • *Local composting of waste* • *No chemical use*	329 329 328

Clivus Multrum composting toilets

This company has been manufacturing composting toilets since 1939. This particular model, made of 100% recycled polyethylene, provides adequate sanitation for a three-bedroomed house. An integral moistening system ensures biomass volume reductions of 95%. Water vapour and carbon dioxide are the only emissions.

Clivus Multrum, Canada	313
Clivus Multrum, Canada	313
Recycled polyethylene	
• *Zero water consumption* • *Recycled materials* • *Local composting of waste* • *Canadian eco-label Environmental Choice EcoLogo M certified*	329 325 329 330

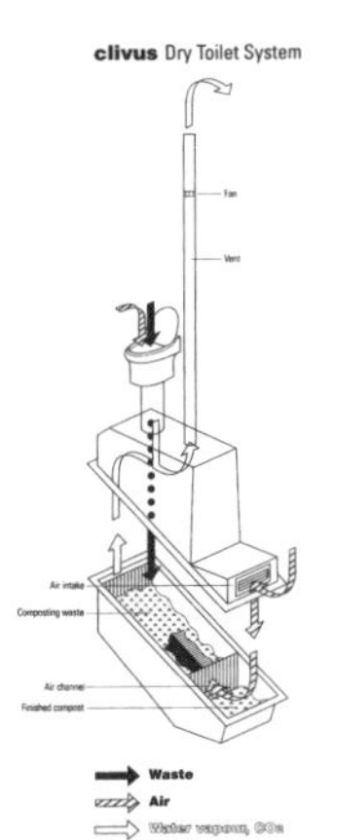

Pureprint

Conventional web-offset printing processes use water with about 10% industrial alcohol, such as IPA, to ensure the plates stay wet so the inks can flow. IPA is highly mobile as it readily evaporates and 'dissolves' in water. It is also a carcinogen and therefore creates a potentially toxic environment for workers. Beacon Press avoids using water or alcohol and instead use silicon rubber to ensure appropriate 'wetting' of the plates and sharper resolution. Nor are any chemicals used in preparation of filmwork, and a strong corporate environmental policy ensures that Beacon Press operates a clean technology printing plant in all aspects, from supply-chain management to car-sharing for employees.

✎	*Originally developed in Japan and USA*	
⚙	*Beacon Press, UK*	312
▬	*Silicon rubber*	
♻	*• Water conservation* *• Avoidance of use of toxic substances*	329 326

'W' High-Efficiency Motor

Brook Crompton is a major supplier of heavy-duty electric motors to UK industry. The 'W' High-Efficiency Motor uses a new type of steel with improved magnetic characteristics, which increases electrical efficiency by 3%. Electric motors account for over 65% of the energy usage by UK industry, so this new motor can potentially reduce carbon dioxide emissions in the UK by up to 2.5 million tonnes per annum as existing motors are replaced. Use of this special steel also allows a reduction of 30% in weight compared with conventional steels.

✎	*Brook Crompton with Sheffield and Cambridge Universities, UK*	313
⚙	*Brook Crompton, UK*	313
▬	*Steel, copper and other materials*	
♻	*• Energy efficiency* *• Reduction in materials used*	329 325

MIMID

Land mines planted during military and civil conflicts during the 20th century kill or maim innocent civilians every day. An estimated 100 million undiscovered mines form a lethal legacy for future generations, so this portable, compact mine detector is a useful addition to the tools available to mine-clearance personnel.

✎	*Gerhard Heufler, Austria*	307
⚙	*Schiebel Elektronische Geräte AG, Austria*	320
▬	*Various*	
♻	*• Improved health and safety*	328

Tensar® range

Steep construction slopes can be reinforced with Tensar® 80RE, a uni-axial grid of polypropylene with elongated apertures, which improves the shear strength of the exposed surface layers. Extruded sheets of special polyethylene are punched with regularly shaped apertures, then stretched under heat to create a high-strength grid. When this geomat is laid on the surface it reduces soil erosion by absorbing the energy of raindrop impact and providing an anchor for plant roots.

	Tensar International, UK	322
	Tensar International, UK	322
	Polypropylene or polyethylene	
	• Protection against soil erosion and slope failure	330

Erosamat Type 1, 1A, 2

Woven mats of natural fibres placed on the surface of the soil absorb raindrop impact and significantly reduce water run-off and consequent soil erosion. Erosamat Type 1 and 1A are made from jute fibres whereas Erosamat Type 2 is a heavier duty geotextile of coir fibres extracted from the husks of coconuts. The latter takes longer to decompose but affords greater protection to soils more at risk from erosion. All types of mat can be seeded to create a dense sward of vegetation, which further bonds the surface of the soil.

	Unknown	
	Various for ABG Ltd, UK	312
	Jute, coir	
	• Renewable materials *• Protection against soil erosion*	325 330

Tartan Outdoor Millennium track

Designing and laying a bespoke athletics track requires large amounts of synthetic material to create the right texture, density and bounce. Athletic Polymer Systems integrates recycled materials into the composite mixture for the Tartan Outdoor Millennium, which has options for embedded (smoother surface for maximum abrasion resistance) or encapsulated (rougher surface for maximum traction) tracks that incorporate a recycle fraction crumb within the site-specific polyurethane mix. These tracks are in use at numerous American universities.

	Athletic Polymer Systems, USA	312
	Athletic Polymer Systems, USA	312
	Polyurethane, plastic recyclate	
	• Recycled materials	325

Dyson Airblade™

Dyson Appliances is known for its engineering and design innovation. The Airblade™ sets a challenge to the commercial hand dryer market by contesting the traditional design using a downward projection of warm air. Instead, air is forced through a narrow continuous slit aperture by a digital pulse motor spinning at 81,000 rpm to create an airstream with a speed of 400 mph. This high velocity wind literally 'scrapes water from the hands'. One of the first high profile public sites to use the technology was the ferris wheel tourist attraction the London Eye, where 13,600 visitors pass through daily. Drying hands faster is a priority in order to accommodate such visitor numbers and the Airblade™ facilitates this with its rapid drying airstream and reduces the energy required by up to 80% on conventional hot-air driers. Hygiene standards are high as touch free infrared sensors activate the machine and a lifetime anti-microbial HEPA filter removes 99.9% of bacteria.

✎	*Dyson Ltd, UK*	314
⚙	*Dyson Ltd, UK*	314
▬	*Polycarbonate-ABS, steel, copper, zinc, ferrite cores, rubbers*	
♻	*• Durable, tough machine*	325

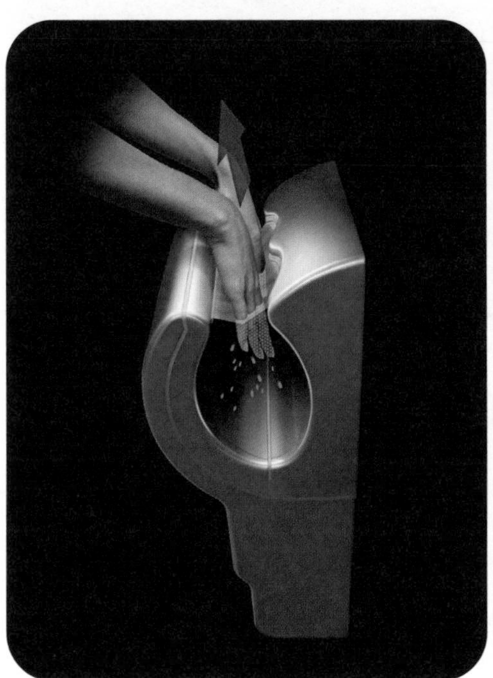

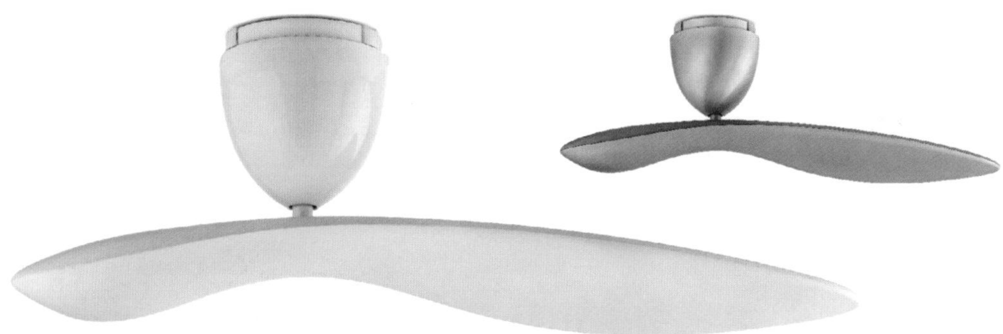

Sycamore Fan

This single blade ceiling fan uses as much as 50% less energy than a conventional four-blade fan. With its aerodynamic Aerofoil shape, this dynamically balanced asymmetric blade only requires 35 W to achieve air flow efficiencies of 115W/CFM at 160 rpm and even at high speeds only produces 38 dB whisper-like sound levels. More air flow can be delivered at lower blade velocities with the Sycamore Fan. Easy to assemble and disassemble, the parts of the fan are robust ABS plastic mouldings or die cast zinc or aluminium, with a zinc balance weight in the asymmetric blade.

✎	*Michael Hort and Danny Gasser, Sycamore Technology, Australia*	31[illegible]
⚙	*Sycamore Technology, Australia*	31[illegible]
▬	*ABS plastic, metals, motor*	
♻	*• Significant reduction in energy consumption* *• Reduction in noise levels* *• Straightforward assembly and disassembly*	32[illegible] 32[illegible] 32[illegible]

CONBAM joints

Bamboo locks up vast amounts of carbon dioxide as many species have exceedingly high growth rates, converting it into carbon-based molecules and releasing oxygen as the plants photosynthesize. Known for its elasticity, hardness, strength and beauty, the use of bamboo should be encouraged as it can be a lucrative cash crop for small holders and larger scale farmers alike. Christoph Tönges' series of stainless-steel joints, of different configurations, enable bamboo poles to be easily fitted together for lightweight constructions from awnings to shelters and even houses. The jointing system also permits fast mounting and demounting, offering flexibility of purpose.

✎	*Christoph Tönges, CONBAM Advanced Bamboo Applications, Germany*	305
⚙	*CONBAM Advanced Bamboo Applications, Germany*	305
	Stainless steel, bamboo	
	• Renewable and recyclable/ compostable materials	325
	• Encourages use of a fast-growing renewable building material	325
	iF product design Gold award, 2007	332

3.0 Materials

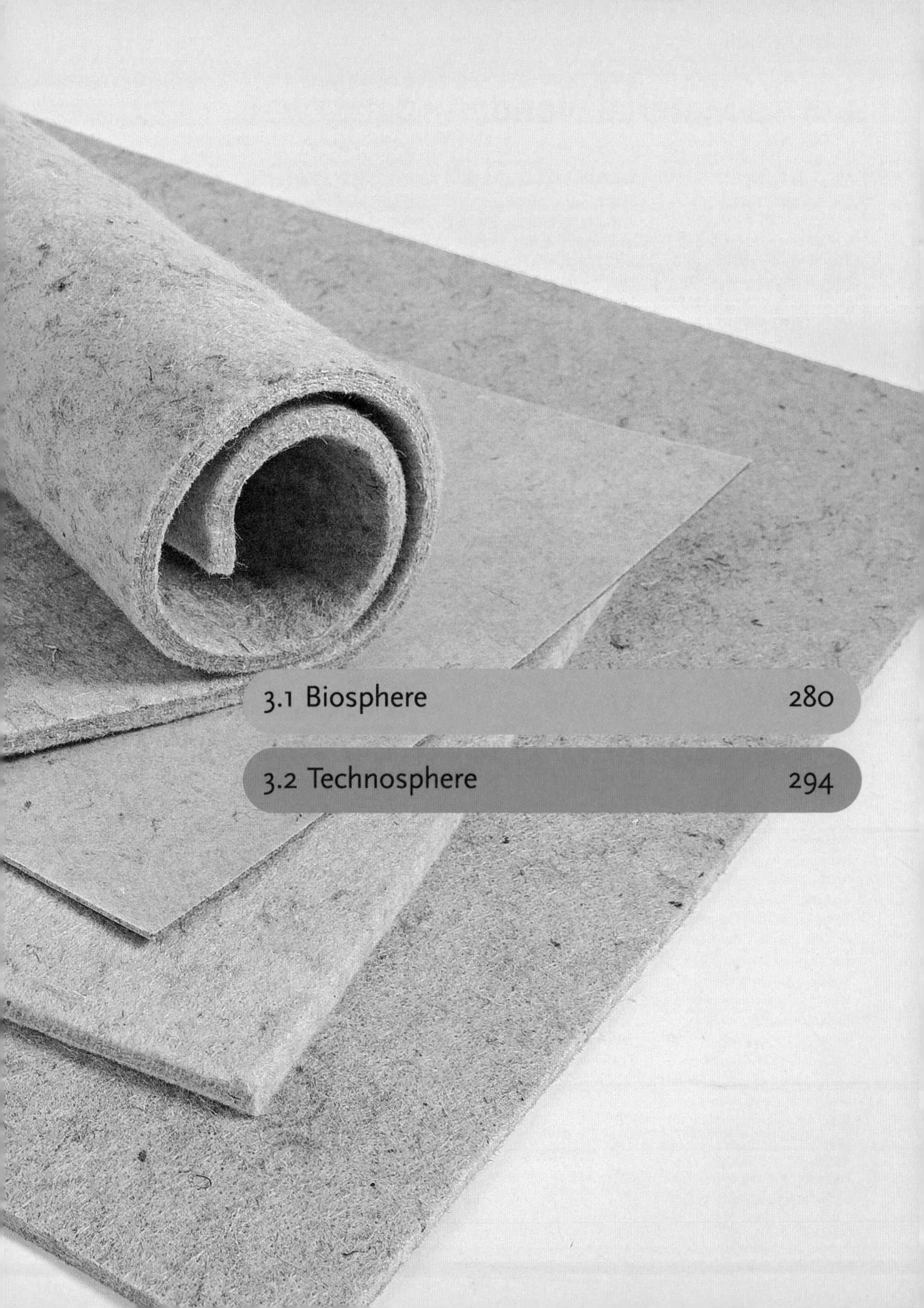

It's a Material World

Today designers, manufacturers and consumers are faced with a mind-boggling array of thousands of materials to choose from, some of which have little or no impact upon the environment while others generate an enormous amount of environmental impact, including depletion of non-renewable resources, toxic or hazardous emissions into the air, water or land, and the generation of large quantities of solid waste. How we choose and use materials today might well determine whether we experience 'ecological recession' in the future.

While designers have traditionally selected materials on the basis of their physical, chemical and aesthetic properties, as well as by cost and availability, other parameters – such as embodied energy, carbon footprint and resource depletion – are proving more and more important. Designers are now obliged to observe legal restrictions on the use of materials from endangered species, as listed in the 1973 Convention on International Trade in Endangered Species (CITES). Various voluntary certification schemes, such as the Forest Stewardship Council (FSC), Programme for the Endorsement of Forest Certification Schemes (PEFC) and SmartWood ensure that materials originate from sustainably managed forests. Designers specifying materials for interior design or architectural projects can now gain additional credit points within certain sustainable building codes, such as the LEED system in the US and BREEAM in the UK. The checklist in Table 1 offers a way of considering the overall potential environmental impacts of a material.

Ecomaterials

An ecomaterial is one that has a minimal impact on the environment but offers maximum performance for the required design task. Ecomaterials from the biosphere are easily recycled by decomposing agents in nature, while ecomaterials from the technosphere are those recycled by man-made processes.

Embodied energy

All materials represent stored energy, captured from the sun or already held in the lithosphere of the earth. Materials also embody the energy used to produce them. One tonne of aluminium takes over a hundred times more energy to produce than one tonne of sawn timber, so the embodied energy of aluminium is comparatively high. Materials extracted directly from nature and requiring little processing tend to be low-embodied-energy materials, while man-made materials tend to possess medium to high embodied energy (Table 2).

In complex products involving many materials, the calculations of embodied energy are more involved. For instance, using lightweight aluminium as opposed to steel in the chassis of a car will ensure greater fuel efficiency and so reduce the total energy use over the lifetime of the product. Selection of high-embodied-energy materials, which are durable and extend product life, may be preferred to lower-embodied-energy materials, which have a short product life. A very important consideration is the embodied energy of the material over the lifespan of the product, which depends on physical durability and cultural trends.

Table 1: A checklist for selecting materials

Material attribute	Low environmental impact	High environmental impact
Resource availability	Renewable and/or abundant	Non-renewable and/or rare
Distance to source (the closer the source the less the transport energy consumed) km	Near	Far
Embodied energy (the total energy embodied within the material from extraction to finished product) MJ per kg	Low	High
Recycled fraction (the proportion of recycled content) per cent	High	Low
Production of emissions (to air, water and/or land)	Zero/Low	High
Production of waste	Zero/Low	High
Production of toxins or hazardous substances	Zero/Low	High
Recyclability, reusability	High	Low
End-of-life waste	Zero/Low	High
Cyclicity (the ease with which the material can be recycled)	High	Low

Materials from the biosphere and lithosphere

Materials derived from the living components of the planet – the biosphere – are renewable and originate from plants, animals and micro-organisms. Biosphere materials include special groups of man-made materials such as compostable biopolymers and biocomposites derived from plant matter. Such materials are readily returned to the cycles of nature. Materials derived from the lithosphere (the geological strata of the earth's crust) fall into two main categories. The first category is widely distributed or abundant materials such as sand, gravel, stone and clay, while the second category includes materials whose distribution is limited, such as fossil fuels, metal ores and precious metals/stones. Materials from the biosphere or lithosphere are often processed by synthesis or concentration to create technosphere materials.

Materials from the technosphere
Technosphere materials are generally non-renewable. Synthetic polymers (plastics, elastomers and resins) derived from oil, a fossil fuel, are technosphere materials. Embodied-energy values tend to be much higher than in biosphere materials. Most technosphere materials are not readily returned to the cycles of nature and some, such as plastics, ceramics (glass, glass/ graphite/ carbon fibres) and composites (ceramic, metal), are inert to microbial decomposition and will never re-enter the biosphere. In a world of finite resources we must be aware of the need to recycle technosphere materials.

Resource depletion
The Living Planet report by the World Wildlife Foundation reveals that, at present rates of consumption, we will need two planets by 2030 to sustain our present way of living. Anthropogenic use of resources is currently 30% more than the earth can sustain. Much of the developed North, the Middle East and, more recently, India and China, have moved into ecological debt, where the nation's consumption footprint exceeds biological capacity. An article in *New Scientist* in 2007 revealed that many of our key minerals, essential to electronics, energy generation and transportation, were in limited supply based on current demands. For example, supplies of zinc were estimated to have just 20–30 years before exhaustion, silver 15–20 years, and lesser known minerals like indium just 5–10 years. Add the phenomenon of peak oil and related resource depletion must be addressed with responsibility by designers, manufacturers and consumers.

Table 2: Embodied energy values for common materials

Material type	Typical embodied energy (MJ per kg)
Biosphere and lithosphere materials	
Ceramic minerals, e.g., stone, gravel	2–4
Wood, bamboo, cork	2–8
Natural rubber (unfilled)	5–6
Cotton, hemp, silk, wool	4–10
Wood composites, e.g., particleboards	6–12
Technosphere materials	
Ceramics – bricks	2–10
Ceramics – glass	20–25
Ceramics – glass fibre	20–150
Ceramics – carbon fibre	800–1,000
Composites – titanium-carbide matrix	600–1,000
Composites – alumina fibre reinforced	450–700
Composites – polymer – thermoplastic – Nylon 6 (PA)	400–600
Composites – polymer – thermoset – epoxy matrix – Kevlar fibre	400–600
Foam – metal – high-density aluminium	300 –350
Foam – polymer – polyurethane	140–160
Metal – ferrous alloys – carbon steel	60–72
Metal – ferrous alloys – cast iron – grey (flake graphite)	34–66
Metal – light alloys – aluminium – cast	235–335
Metal – non-ferrous alloys – copper various alloys	115–180
Metal – non-ferrous alloys – lead various alloys	29–54
Metal – precious metal alloys – gold	5,600–6,000
Polymer – elastomer – butyl rubber	125–145
Polymer – elastomer – polyurethane	90–100
Polymer – thermoplastic – ABS	85–120
Polymer – thermoplastic – nylon	170–180
Polymer – thermoplastic – polyethylene	85–130
Polymer – thermoplastic – polypropylene	90–115
Polymer – thermoset – melamine	120–150
Polymer – thermoset – epoxy	100–150

Adapted from Cambridge Engineering Selector, version 3.0, Granta Design Ltd, UK

Recycling
Nature recycles all its biosphere and lithosphere materials but humans recycle only certain man-made synthetic materials; the rest get dumped in controlled landfills or distributed in our landscapes. Expensive ferrous metal and light alloys often include a recycle fraction of between 70–80%, non-ferrous metals between 10–80%, and precious metal alloys (gold, platinum, silver) between 90–98%. Relatively inexpensive polymers (plastics), on the other hand, exhibit variable recycle fractions of between 0–60%, and, as recent global events have shown, are very susceptible to price fluctuations in financial recession. Specialist technosphere materials, especially composites, for example thermosets and reinforced thermoplastics, often have less than 1% recycle fraction.

Closed loop recycling of materials from the technosphere significantly reduces environmental impacts. Metals made entirely of recycled content and recycled plastics have an embodied energy that is often only half or even as little as 10% of that of virgin metals. Increasing the recycle fraction in more materials, by re-evaluating the idea of 'waste', will bring savings in energy.

Green procurement
Designers can reduce the impacts of materials they use if they specify sources of materials and minimum recycle fractions and if they insist on compliance with certain standards, such as eco-labels or voluntary industry schemes (see Green Organizations, p. 331). Specifiying suppliers or manufacturers that comply with internationally recognized environmental management systems, such as ISO 14001 or EMAS, is also desirable. Labelling of materials and products with a Carbon emissions value is also being tested now, so this could be a future supply chain requirement.

Boards and Composites

Belmadur® plywood

Durability is a material characteristic that designers should seek out wherever possible to ensure that our artefacts endure. Belmadur® is a weatherproof European beechwood laminated wood, from PEFC or FSC sustainable forests, suitable for architectural and design products and furniture. It is twice as hard as oak, as durable as teak, has a high resistance to fungal attack, and is able to take up complex or simple shapes, making this an ideal material to extend the design canon developed by the likes of Thonet, Aalto and Eames. It i37s a versatile material for seating frames, monocoque structures – such as the OTO chair by Peter Karpf – and hybrid designs applying CNC manufacturing techniques. It is closely related to BECKER Incendur®, a moulded plywood with enhanced fire resistant applications.

⚙	*Fritz Becker KG, Germany*	315
♻	*• From certified sustainable sources* *• Excellent durability* *• Multifunctional applications*	326 325 328

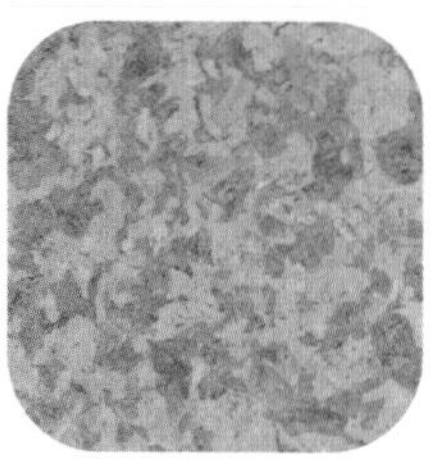

Environ™

Environ™ is possibly the first example of a mass-produced biocomposite using a plant-based resin to bond recycled materials. It is manufactured from recycled paper and soy flour into sheets and floor strips and is claimed to be harder than oak wood and suitable for interior decoration and furniture.

⚙	*Environ Biocomposites, USA*	315
♻	*• Renewable and recycled materials*	325

WISA® plywoods

Birch, spruce and pine from managed Finnish forests are used to manufacture a range of plywoods suitable for interior, exterior and concrete formwork and as laminboard. Special tongue-and-groove panel plywood laminated floorings include Schauman Birchfloor, Sprucefloor and Spruce Dek. The company is certified to ISO 14001.

⚙	*WISA Wood Products, Finland*	323
♻	*• Renewable materials* *• Stewardship sourcing*	325 326

Hemp particle board

Hemp holds great promise as a crop from which new products and applications continue to be derived. Hemp requires little husbandry and needs few if any synthetic inputs such as fertilizers or pesticides, so it is ideal for organic production. This MDF is made from 100% hemp hurds, the short inner fibre that is high in lignin and cellulose. Manufactured with formaldehyde-free resin binder, the material is available in thicknesses ranging from ¼ in to 1 in (6.35 mm to 25.4 mm), in standard 8 ft x 4 ft (2.4 m x 1.2 m) sheets.

⚙	*Hemp Traders, USA*	316
♻	*• Recycled materials* *• Fast growing bio-mass crop* *• Formaldehyde-free*	325 325 325

General-purpose particleboard

Some 14,000 tonnes of waste wheat straw annually go to make this ½ in (1 cm) thick, wheat-based particle-board suitable for furniture, construction and interior design.

⚙	*Prairie Forest Products, USA*	320
♻	*• Renewable material*	325

Kirei board

Stalks of sorghum, a grain and fodder plant, are heat-pressed with a zero-formaldehyde adhesive and layered with poplar wood. The result is a lightweight but strong board with a distinctive texture and visual appeal, suitable for interior architectural design, furniture, exhibition and flooring applications. Sheets vary from 305 to 910 mm (12 to 35 in) in width, by 1820 mm (71⅔ in) in length.

⚙	*Kirei, USA*	317
♻	*• Utilizes agricultural waste* *• VOC-free* *• Formaldehyde-free*	326 325 325

MeadowBoard™

Compressed panels and sheeting of ryegrass straw are suitable for all structural and decorative uses, and furniture production.

⚙	*Meadowood Industries, Inc., USA*	318
♻	*• Renewable material*	325

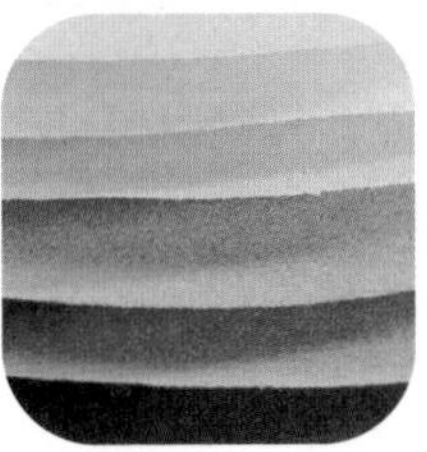

Kraftplex

This versatile, high-density material is pliable and strong, fabricated from 100% sustainably managed softwood stock, using unbleached wood fibres held together without chemical binders or adhesives. It can be permanently shaped, cut, drilled, embossed or stamped, making it especially suitable for furniture, product design and decorative objects. Kraftplex is available as performance or contour types, varying from 1 to 3.2 mm (about 1/24 in to 1/8 in) and 0.8 to 1.6 mm (about 1/32 in to 1/16 in) in thicknesses respectively, and the smooth surface takes oils, waxes, varnishes and paints for a quality finish.

	Well Ausstellungssystem GmbH, Germany	323
	• Renewable materials	325
	• Sustainably harvested	326
	• Environmental management standard ISO 14001	330
	• No chemical binders or adhesives	325

Medite ZF

Medite ZF is the trade name for an interior-grade, MDF manufactured using softwood fibres bonded with formaldehyde-free synthetic resin. Free formaldehyde content of Medite ZF is less than 1 mg/100 g (one part in 100,000), which is equivalent to or less than natural wood, and formaldehyde emissions are well below general ambient outdoor levels. All other Medite MDF boards are manufactured to Class A EN622 Part 1, complying with free formaldehyde content of less than 9 mg/100 g (nine parts per 100,000). The company has applied for FSC certification for Medite.

	Weyerhaeuser, USA	323
	• Renewable materials	325
	• Reduction in toxic ingredients and emissions	328

Resincore™

Resincore is formaldehyde-free particleboard composed of sawdust, phenolic resin and wax.

	Rodman Industries, USA	320
	• Renewable and recycled materials	325

Panda Flooring

Bamboo is harder, tougher and more elastic than many temperate hardwoods, and shows phenomenal growth with reportedly one metre per day for some species. Panda bamboo flooring is made from three-year growth canes of Chinese bamboo, a straight, regular-diameter cane with a very dense fibre and lower carbohydrate content that improves resistance to insect and fungal attack. Canes are harvested from managed forests. They are then treated by steaming, heating, dehydrating and oven-drying followed by anti-insect and mildew treatments. Lengths of bamboo are cut and prepared before laminating into tongue-and-groove boards 920 x 92 x 15 mm (36 x 3½ x ½ in) and receiving UV-resistant lacquer protection. Colours vary from light beige to dark browns.

	Panda Flooring Co., USA	319
	• Renewable resources	325
	• Selective harvesting from managed forests	326
	• Durable	325
	• Ease of maintenance	328

Multiboard

Composite sheeting and boards are made from a diverse range of recycled materials: 55% originates from used PE-coated milk cartons, newsprint and paper, the remaining 45% from the industrial or production waste streams. The company is registered to ISO 9001, ISO 14001 and EMAS.

	Fiskeby Board AB, Sweden	319
	• Recycled materials	325
	• EMAS policy	330

Kronospan®

Kronospan manufactures a diverse range of particle boards, tongue-and-groove panels, MDF, Kronoply, laminate flooring and post-formed panels and surfaces for interior use. Laminated flooring sheets are FSC certified, the formaldehyde-free, panel-type 'Hollywood' qualifies for a Blue Angel eco-label, and the company is certified to ISO 14001. Timber is generally sourced locally.

	Kronospan AG Switzerland	317
	• Renewable materials	325
	• Clean production	326
	• Certification to FSC or Blue Angel	330

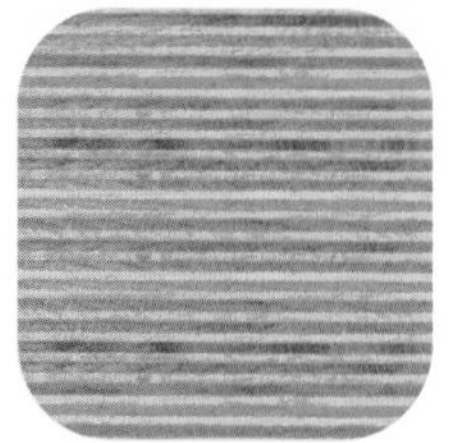

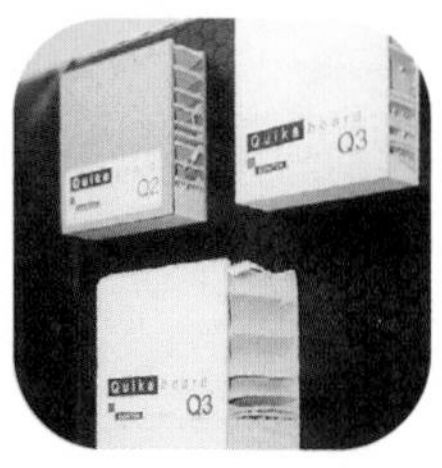

PaperStone Certified

This high-finish architectural and interior design panel or worktop surface has a fibre constituent composed of post-consumer recycled office paper and a petroleum-free phenolic resin derived in part from CNSL (cashew nut shell liquid). The Certified series is currently the only architectural solid surface chain-of-custody certified to FSC standards by the Smartwood program of the Rainforest Alliance, enabling it to qualify for points on the US eco-building LEED standard. The standard panel sizes of 60 x 144 in (1524 mm x 3658 mm) vary in thickness from ¾ in to 1¼ in (19 mm to 32 mm), with a density of 1.4 to 1.45 g/cu. cm (87.4 to 90.5 lb/cu. ft), and come in a range of darker and natural-spectrum colours.

⚙	*Paneltech Intnl., LLC, USA*	319
♻	• *Recycled materials* • *FSC certified*	325 330

Plexwood®

This range of engineered wood veneers has a distinctive surface texture derived from the strip laminating techniques used in its production. Plexwood® contains no formaldehyde and uses VOC-free glue. Its versatility as a design material originates from the diverse range of sizes and specifications, including 6 mm or 14 mm thick (⅕ or ⅔ in) strip, parquet strip or plank flooring, tiles, MDF panel (one/two sided), bendable panel and profile. The colour profiles vary according to the source woods – birch, beech, deal, meranti, ocoumé, pine, poplar and others.

⚙	*Plexwood, the Netherlands*	319
♻	• *From certified PEFC or FSC sustainable sources* • *Formaldehyde- and VOC-free* • *Versatile applications*	326 325 328

Quikaboard

Quikaboard is a versatile range of lightweight honeycomb boards and panels. Reconstituted paper is used for the honeycomb core varying from under 10 mm to over 60 mm (about ½ to 2½ in). Thickness of the boards and panels varies according to honeycomb depth and the thickness and nature of the external facings. Typical facing materials include hardboard, composites, laminates, plywood, chipboard and MDF. Boards and panels can be bent to a range of radii for curvilinear designs. Quikaboard has high strength-to-weight ratio, is durable and is suitable for a wide range of applications, such as signage, point of sale, shopfittings, exhibitions, office interiors and caravan fixtures. An ambitious application of Quikaboard was the waterproofed laminates for the external and roofing panels used at the Cardboard Building, Westborough Primary School, UK, designed by Cottrell & Vermeulen (p. 232).

⚙	*QK Honeycomb Products, UK*	320
♻	• *Recyclable* • *Durable*	325 328

Wellboard

This flexible corrugated sheeting is hot-pressed, without using chemical additives, from 100% wood pulp from sustainably managed forests. Available in rolls of 2000 mm width by 440 mm length (78²/₃ in by 17¹/₃ in), the thickness varies between 1 to 2 mm (about ¹/₂₄ in to ¹/₁₂ in) depending upon the profile. Curvilinear or geometric profiles are available with centres varying between 10.5 mm to 38 mm (⅔ to 1 in), and profile heights of 4.5 mm to 7.5 mm (⅕ to ⅓ in). Wellboard is suited to diverse applications where curved structures are required, including furniture, interior partitions and decoration, and exhibition stands.

⚙	*Well Ausstellungssystem GmbH, Germany*	323
♻	• *Renewable material* • *Sustainably harvested* • *Environmental management standard ISO 14001*	325 326 330
🏆	*iF material Award 2005*	332

Timbers

Accoya® Wood

Wood is known for its tendency to shrink, warp and move with exposure to moisture and varied temperatures, yet today's architectural and design applications demand improved tolerances and specifications. Accoya® wood provides a better engineered technology because the free hydroxyls (OH molecules) that cause warping are replaced by acetic anhydride (itself derived from acetic acid, or vinegar in dilute form). This acetylation of the wood reduces its ability to absorb water, making it more stable. Consequently, Accoya® Wood is more durable and more dimensionally stable than teak or dark red meranti. The treated wood mainly utilizes pine species from sustainably managed forests.

⚙	*Titan Wood Limited, USA*	322
♻	• *Renewable material from sustainable sources* • *Improved durability*	325 328

Bendywood®

A wide range of sections and profiles of Bendywood® are available in oak, beech, ash and maple, offering designers/manufacturers a ready-made material for forming tight- or wide-radius curves for furniture and other applications. Bendywood® is created by subjecting timber to heat and directional pressure, which restructures the cellulose fibres, allowing them to concertina when subjected to bending. This enables bending in a cold dry state to a radius of about ten times the thickness of the section or profile. Tighter bends can also be achieved. Available stock sizes are up to 100 x 120 x 2200 mm (about 4 x 4½ x 86½ in) depending on the type of wood.

⚙	*Candidus Prugger S.a.S., Italy*	313
♻	• *Renewable material*	325

Certified timber

This company supplies high-density prepared boards from sustainably harvested palm trees. The timber is guaranteed 100% chemical-free and is suitable for structural, flooring and furniture applications. A wide range of North American and tropical hardwoods is supplied from certified sources.

⚙	*Eco Timber Intl., USA*	314
♻	• *Renewable materials* • *Non-toxic* • *Certified sources*	325 325 325

Iron Woods®

Diniza, Purpleheart, Greenheart and Macaranduba are tough, exceptionally dense and durable, tropical hardwoods. Promoted as THL Iron Woods®, these sawn and planed woods are certified by the FSC and the Rainforest Alliance's SmartWood® schemes as originating from sustainably managed forests. They do not require any chemical treatments to prolong life.

⚙	*Timber Holdings Intl., USA*	322
♻	• *Renewable materials* • *Certified sources*	325 325

Microllam® and Intrallam®

Layers of aspen wood are bonded with resin to form high-strength timber composites, with minimal wastage from forest roundwood. Microllam™ has thin even layers and Intrallam™ is made of irregular layers and chips.

⚙	*Weyerhaeuser, USA*	323
♻	• *Reduced production waste* • *Efficient use of materials*	326 326

Rubber wood

There are over 7.2 million hectares (17.8 million acres) of cultivated rubber trees worldwide, of which over 5.2 million hectares (12.8 million acres) are in Malaysia, Indonesia and Thailand. Latex yields decline 25 to 30 years after planting. These older trees are now being harvested for rubber wood. In 1990 the the Association of southeast Asian Nations produced about 17 million cu. m (600 million cu. ft) of rubber wood. The timber is suitable for a wide range of applications, such as flooring, kitchen utensils and furniture.

⚙	*Numerous manufacturers in tropical countries*	
♻	• *Renewable material*	325

TimberStrand® LSL

Weyerhaeuser produce a range of engineered timbers composed of strands or sheets of veneer bonded with adhesives or resins at high pressure and heat. These products encourage better resource usage than sawn timber since almost all the sawn log is used in the composite timber. Timberstrand® LSL is a general-purpose structural timber.

⚙	*Weyerhaeuser, USA*	323
♻	• *Renewable material* • *Efficient use of materials*	325 326

Bamboo

Bamboo sourced from Vietnam is the principal material for strip and laminated flooring manufactured by the company, but poles and bamboo for structural purposes can also be supplied.

⚙	*Bamboo Hardwoods, USA*	312
♻	• *Renewable material*	325

Biopolymers

Solanyl®

This biopolymer – derived from the starch by-product created in potato processing – is a truly bio-degradable plastic, available as grains for moulding machines that can be formed at a lower temperature (110° C) than synthetic plastics (170° C). Suitable for everything from fine detail to robust solid forms, Solanyl® is mechanically similar to PE or PS and is suitable for applications in a wide range of sectors for industrial, furniture and product design.

⚙	*Solanyl Biopolymers Inc., Canada*	321
♻	• *Renewable material* • *Reduced energy in production compared to synthetic polymers* • *Compostable*	325 326 325

Biopolymers

Biopolymers and industrial starches are extracted and processed from the corn (maize) plant.

⚙	*Cargill, USA*	313
♻	• *Renewable material* • *Compostable*	325 325

Bioplast®

Biotec specializes in biodegradable plastics using vegetable starch as the raw ingredient. Trade products include Bioplast®. It has similar properties to polystyrene, so it is suitable for making disposable cups for vending machines and catering companies.

⚙	*Biotec, Germany*	313
♻	• *Renewable materials* • *Compostable*	325 325

Biopolymers and resins

This company specializes in the manufacture of starch-based biopolymers and resins suitable for injection moulding. Clean Green is loose-fill packaging that is water-soluble.

⚙	*StarchTech, Inc., USA*	321
♻	• *Renewable materials* • *Compostable*	325 325

Eco-Foam®

Polystyrene chips, often made by injecting CFC gases, can now be substituted with biodegradable chips of foamed starch polymer, Eco-Foam®, where steam is used as the blowing agent. This biopolymer is made of 85% corn starch, so it is biodegradable, water-soluble and reusable. It is also free of static, which makes the packaging process easier.

⚙	*StarchTech, Inc., USA*	321
♻	• *Renewable materials* • *Reusable packaging* • *CFC-free*	325 329 325

Mirel Bioplastics

Mirel Bioplastics are biodegradable plastics derived from plants or bacteria that are water-soluble and easily recycled. PHAs are suitable for medical and food-packaging uses, and BiOH brand polyols.

⚙	*Metabolix, Inc., USA*	318
♻	• *Renewable materials* • *Compostable and recyclable*	325 325

Flo-Pak Loosefill

This loosefill packaging material in a figure-of-eight shape is made by extruding corn, wheat and/or potato starch biopolymer, and is suitable for use in all standard commercial packaging systems. Like all starch-based biopolymers it readily dissolves in water and is biodegradable.

⚙	*FP Intl., USA*	315
♻	• *Renewable materials*	325

PLA

Plastics manufacturers all over the world are examining the commercial viability of making plastics from renewable resources. In 2000 Cargill Dow's 'NatureWorks Technology' created a new bioplastic called polyactide (PLA), derived from the maize plant. A factory in Nebraska produces up to 150,000 tonnes per annum, and PLA has attracted interest from the packaging and computer industries. Questions remain over the lifecycle benefits of PLA and the use of GM maize as source material.

⚙	*Cargill, USA*	313
♻	• *Renewable materials* • *Compostable*	325 325

Green Cell Foam

This bio-based, biodegradable foam, very similar to expanded polystyrene (EPS) in character, is most suitable for single-use packaging. It originates from non-GM high-amylose cornstarch (HACS). The extruded foam is of a corrugated design to optimize cushioning. At the end of its useful life it can be composted, recycled, dissolved or burnt.

⚙	*KTM Industries, Inc., USA*	317
♻	• *Renewable materials* • *Compostable, recyclable* • *Non-GM raw materials*	325 325 330

Eco-Flow

Eco-Flow is an extruded packaging material primarily composed of wheat starch.

⚙	*American Excelsior Co., USA*	312
♻	• *Renewable material* • *Compostable*	325 325

NatureFlex™

Several grades of NatureFlex™ – transparent one-side coated cellulose film – are available, with or without heat-sealable properties. The cellulose is derived from wood pulp from FSC-certified forests. Some films are made with combinations of PLA with starch-based resins. There is also a version of cellulose film, with an appropriate adhesive, enabling easier removal of labels for glass bottle reuse.

⚙	*Innovia Films Inc., USA*	317
♻	• *Renewable material* • *Some certified FSC materials*	325 325

Natural rubber (NR)

Today over 70% of rubber production centres around Malaysia, Indonesia and Thailand. Trees have a productive lifetime of up to 30 years, after which latex production declines. Plantations also act as a sink for absorbing carbon dioxide. NR is used pure or mixed with synthetic rubbers and fillers to manufacture a huge range of products from tyres and tubes, industrial components and medical goods, to footwear and clothing. Special grades of NR produced by Malaysia include SUMAR (non-smelly rubber), ENR (epoxidized NR), DPNR (deproteinized NR), and PA/SP (superior processing rubber).

⚙	*Numerous manufacturers in tropical countries*	
♻	• *Renewable material* • *Versatile natural polymer*	325 325

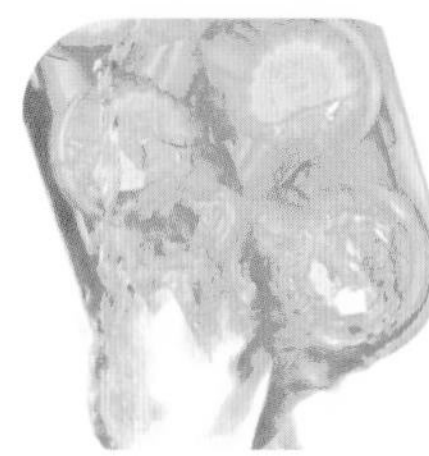

Plantic®

This thermoplastic starch-based polymer is derived from non-GM high-amylose (70%) corn. The biopolymer is extruded as sheet or resin and is ideal for most processing and moulding technologies, especially injection moulding for applications that need high stiffness. It comes in several formulations including EG500, GP100, HF300, WR700, WR701 and R1 for different applications.

⚙	*Plantic Technologies, Australia*	319
♻	• *Renewable materials* • *Non-GM sources* • *Compostable*	325 330 325

BIOGRAPH.ics

Pace Industries is an Environmental Protection Agency (EPA) Green Power Partner, utilizing renewable energy (one third wind, one third hydro and one third methane from landfills) to manufacture their bio and synthetic polymers. BIOGRAPH.ics is based on agricultural starches, not just cornstarch, and polymers. It is manufactured as an opaque, natural or white-coloured sheet and roll product varying in gauge. All grades of BIOGRAPH.ics are suitable for packaging, with good impact resistance, and have good ink receptivity for excellent graphics. It is labelled with recycling symbol 7 indicating it is biodegradable and is certified compostable to ASTM D6400 standards by the US Composting Council.

✿	*Pace Industries, Inc., USA*	319
♻	*• Renewable material*	325
	• Manufactured using 100% renewable energy	326
	• Biodegradable, compostable	325

Tenite Cellulose Acetate/plastic

Eastman manufactures a range of biodegradable polymers from cellulose acetate.

✿	*Eastman Chemical Co., USA*	314
♻	*• Renewable materials*	325
	• Compostable	325

SoyPlus®

This non-petroleum-based polymer is derived from soybeans. Formulations can be made for film and rigid plastic with variable decomposition rates, enabling the polymer to be utilized for diverse applications from food service items to plant propagation systems.

✿	*Soy Works Corporation, USA*	321
♻	*• Agricultural production polymer*	326
	• Compostable	325

Fillers/Insulation

Chanvrisol, Chanvribat

Loose-fill and blanket insulation is made by combining cellulose fibres with hemp fibre. Chanvrisol is loose-fill insulation, Chanvribat is supplied in a roll and has a thermal conductivity of 0.049W/K.

✿	*LCDA, France*	317
♻	*• Renewable materials*	325

Thermafleece™

Thermafleece™ is made from wool fibres that can absorb and desorb water vapour without changing their excellent insulation properties. Manufactured in batts of 50, 75 or 100 mm (2, 3, or 4 in) thickness, 1200 mm (47 in) length and 400 mm or 600 mm (15½ or 23½ in) width, they can be layered according to insulation needs in walls, roofs or floors. Water absorption is 100%, relative humidity is 40% and the thermal conductivity (K-value) is 0.039 W/m.K, similar to other natural-fibre insulation materials. Thermafleece™ requires much less energy to manufacture than glass-fibre insulation. Fire resistance of wool is higher than cellulose and cellular plastic insulants. Naturally derived additives are used for insect and fire proofing so Thermafleece™ does not contain permethrin, pyrethroids, pesticides or formaldehydes. It can be readily reused and recycled.

✿	*Second Nature UK Ltd, UK*	321
♻	*• Renewable material*	325
	• Non-toxic	325

Greensulate™

Ecovative Design won the prestigious PICNIC Green Challenge in Amsterdam in October 2008, which awarded the company Euros 500,000 to develop their innovative new rigid insulation material. Greensulate™ is made by harnessing the power of fungi and converting agricultural and biological waste resources. It produces ten times less CO2 than its competitors, has a comparable R value and is non-flammable. Aimed as a replacement for EPS or Styrofoam, Greensulate™ is suitable for building insulation, product containers, packaging and more.

✿	*Ecovative Design LLC, USA*	314
♻	*• Utilizes agricultural and biological waste*	326
	• Biodegradable, compostable	325
	• Non-toxic	325
🏆	*PICNIC Green Challenge 2008*	

Heraflax

Long and short flax plant fibres are separated; the former are used for weaving linen and the latter are manufactured into insulation battens and quilts. In the Heraflax WP battens and Heraflax WF quilt, the flax fibres are integrated with polyester fibres to form standard 60 mm or 80 mm (2¼ or 3⅓ in) thick products. Both materials are good insulators with a thermal conductivity of 0.42 W/sq. m.

⚙	*Heradesign, Austria (Part of Knauf Insulation, GmbH Germany)*	316
♺	*• Renewable materials*	325
	• Energy-saving product	329

INNO-THERM

Recycled cotton textile fibres provide the stock for this durable, biodegradable, insulation material, which has no hazardous content. INNO-THERM is manufactured as part of a larger charitable project in the UK – the Schools & Homes Energy Education Project – that offers training in energy efficiency and renewable energy technologies. INNO-THERM is an excellent acoustic absorbency material and has a thermal conductivity K of 0.040W/ m.

⚙	*Recovery Insulation Ltd, UK, under license from INNO-THERM LLC, USA (Hickory Springs Manufacturing Company)*	320
♺	*• Recycled materials*	325
	• Non-toxic	325
	• Biodegradable	325

Thermo-Hemp®

Hemp is an ideal crop for all aspiring organic farmers. It does not require the application of any herbicides or insecticides, it is a good weed suppressant, helping to clean the land, and it is a prolific producer of biomass and fibre, growing up to 4 m (13 ft) high in 100 days. Hemp cultivars with minimal active 'drug' chemicals have been grown in Germany since 1996 specifically for the production of this new insulation material. Fibres are extracted from the harvested plants and reworked into panels using 15% polyester for support and 3–5% soda for fireproofing. It is suitable for insulating between stud walls and roofing timbers. Thermo-Hanf® (thermo hemp) conforms to all DIN-Norm standards and has a thermal conductivity of 0.039 W/mk for DIN 52612. It also has in-built resistance to insect attack from the plant's own natural defences.

⚙	*Hock Vertrieb GmbH/ Swabian ROWA, Germany*	316
♺	*• Renewable material*	325
	• Clean production	326

Warmcel 500

Thousands of tonnes of post-consumer newsprint waste are collected by Excel's own kerbside vans and given new life as Warmcel 500, a professionally applied 'blown' system of fibre installation for buildings. Natural, inorganic, non-toxic additives help bind the fibres together and the specification can be altered to suit the requirements to control breathability and R values.

⚙	*Excel Industries Limited, UK*	315
♺	*• Recycled materials*	325

UltraTouch natural cotton fibre

This Class A fire-rated insulation, protected with natural borates, is made of 85% recycled denim jeans and cotton fibres from factory production waste. It is not itchy to use, is non-toxic, and the fibres provide good R values varying between 13 for 3.5 in (89 mm) batts and 30 for 8.0 in (203 mm) batts. It is also eligible for points on the US eco-building LEED certification scheme.

⚙	*BJ Green, LLC, USA*	313
♺	*• Recycled materials*	325
	• Non-toxic	325

Paints/Varnishes

Nutshell®

Nutshell produces a full range of adhesives, paints, herb and resin oils, varnishes and stains with natural pigments.

⚙	*Nutshell Natural Paints, UK*	319
♺	*• Renewable materials*	325
	• Non-toxic	325
	• Clean production	326

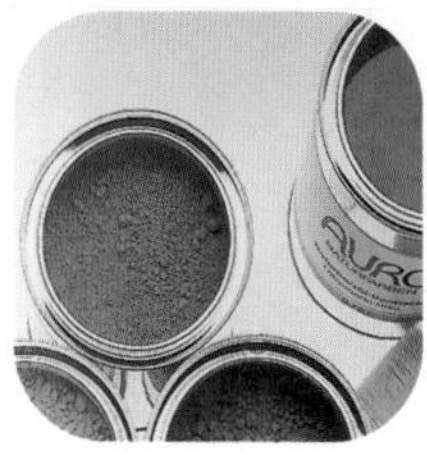

Auro paints, oils, waxes, finishes

Auro manufactures an extensive range of 'organic' paints, oils, waxes, stains and other finishes without the use of fungicides, biocides or petrochemicals including white spirit (an isoaliphate). Oils originate from renewable natural sources such as ethereal oils, balm oil of turpentine or oil from citrus peel, so waste from the manufacturing process is easily recycled and the potential health hazard of the finished products is less than in petrol or isoaliphatic-based manufacturing systems. Emulsion paints for interior use include white chalk and chalk casein paints, which can be tinted using pigments from a range of 330 colours. Exterior-grade gloss paints and stain finishes are suitable for applying to wood, metal, plaster and masonry.

⚙	*Auro Pflanzenchemie AG/Auro GmbH, Germany*	312
♻	*• Non-toxic ingredients* *• Clean production*	325 326

BioShield paints, stains, thinners, waxes

BioShield Paint Company manufactures a diverse range of paints, stains, thinners and waxes from natural ingredients such as oils from linseed, orange peel and soy bean.

⚙	*BioShield Paint Co., USA*	313
♻	*• Renewable materials*	325

Livos

In 1975 Livos developed techniques for dispersing ingredients in natural resins. Its range includes natural-based primers (with linseed oil), hardening floor agents (pine tree resins), transparent glazes (phytochemical oils such as citrus), wood polishes (beeswax), wall glazes (beeswax and madder root) and varnishes. Livos URA Pigment Paint comprises organic beeswax, linseed/stand oil, orange-peel oil and dammar mixed with water, methylcellulose, isoaliphate, ethanol, iron oxide, mineral pigments, borax and boric acid. Colour varies according to the amount of pigment.

⚙	*Livos Pflanzenchemie, Germany*	317
♻	*• High content of natural, renewable materials* *• Low or VOC-free content* *• EU eco-label for some products*	325 325 330

Milk-based paints

Traditional milk-based paints, suitable for interior design, restoration and furniture production, are made by the Old Fashioned Milk Paint Company. These paints follow authentic recipes and are free of synthetics.

⚙	*Old Fashioned Milk Paint Co., USA*	319
♻	*• Renewable materials*	325

The Natural Choice

All paints and finishes in the Natural Choice collection utilize natural oils and solvents from citrus peel or seeds, resins from trees, waxes from trees and bees, inert mineral fillers and earth pigments. Oils are extracted by cold pressing or with low heat and all products are packaged in biodegradable or recyclable containers.

⚙	*BioShield Paint Co., USA*	313
♻	*• Renewable materials* *• Reduced pollution manufacturing* *• Recyclable packaging*	325 326 327

OS Color

Waxes from carnauba and candelilla plants and oils from sunflower, soy bean, linseed and thistle are the raw ingredients of a wide range of natural stains and protective finishes for exterior and interior wood surfaces. OS Color wood stain and preservative is a natural oil-based, microporous, water-repellent treatment for timber exposed to the weather. The natural oils, water-repellent additives and lead-free siccatives (drying agents) form the binder, which comprises almost 85% of the solids content. This binder is mixed with the active (bacteria and fungi) protective ingredients, low-odour solvents (benzole-free, medical-grade white spirit) and pigments (iron oxide, titanium dioxide). Floor treatment, such as the OS Color hardwax-oil, is an oil-based application containing neither biocides nor preservatives. Manufacturing plants are covered by ISO 9001 and ISO 14000.

⚙	*OSMO, UK*	319
♻	*• Reduced toxins (solvents, VOCs, biocides, preservatives and citrus oils)* *• Natural, renewable raw materials*	325 325

Papers/Inks

Paperback papers

Paperback offers the most extensive range of gloss- and matt-coated papers, uncoated offsets, letterheads and speciality grades manufactured from recycled waste paper in the UK. This process uses less than half the energy required to make paper from virgin wood pulp. The company was set up in 1983 and is committed to encouraging use of recycled paper to decrease the disposal of 6 million tonnes of waste paper annually in the UK. Boards range in weight from 225 gsm to 300 gsm with a variety of finishes from smooth white watermarked to natural-coloured micro-fluting. All Context papers and boards contain 75%-de-inked used waste to a NAPM approved grade and Context FSC is made from 75%-de-inked fibre and 25%-FSC-endorsed pulp.

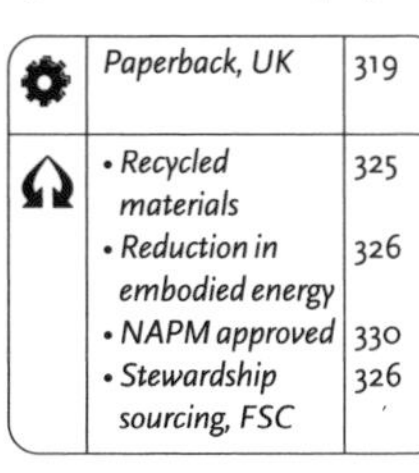

⚙	*Paperback, UK*	319
♻	• *Recycled materials*	325
	• *Reduction in embodied energy*	326
	• *NAPM approved*	330
	• *Stewardship sourcing, FSC*	326

Elephant Dung Paper

Since the 1950s the global elephant population has been decimated, with an estimated decline of over 90%. The imperative to help conserve habitat and to look after the remaining elephant populations is urgent. This paper is manufactured by taking fibrous elephant dung, extracting the fibre and reconstituting it to create a range of colourful, textured, 100% handmade papers. The papers are completely free of bacteria and odour, and are available via a number of worldwide distributors co-ordinated from the USA.

⚙	*Thai Elephant Conservation Center, Thailand with Elephant Dung Paper, USA*	315
♻	• *Non-tree paper*	330
	• *Helps elephant conservation programme*	330

Vanguard Recycled Plus™

This tree-free, bond-quality paper is manufactured from 10% hemp/flax and 75% post-consumer waste paper.

⚙	*Living Tree Paper Co., USA*	317
♻	• *Renewable and recycled materials*	325
	• *Conservation of forest resources*	330

New Leaf Encore 100

New Leaf offer a range of FSC-certified paper stocks including coated, writing, offset and copy papers. The Encore 100 is made of 100% recycled post-consumer paper waste, is acid free and is processed without the use of chlorine, exceeding local state standards for recycled papers. The manufacturing facility uses Green-e® renewable energy sources and the company also uses non-tree virgin fibres from hemp and kenaf. Encore 100 is suitable for general office use as it performs well with a range of photocopier and printing processes.

⚙	*New Leaf Paper USA*	318
♻	• *Renewable post-consumer waste*	325
	• *Chlorine and acid free*	325
	• *FSC-certified non-tree paper*	330

Tree-free paper

A range of papers is made from natural plant fibres, such as cotton and hemp, and post-consumer paper waste.

⚙	*Green Field Paper Co., USA*	316
♻	• *Renewable and recycled materials*	325
	• *Encourages forest resource conservation*	330

Treesaver range

The Treesaver range includes craft, packaging grade and printing papers, and uses 100% waste paper to create recycled papers such as MG Greentreesaver Kraft, MG Green Envelope and MG Treesaver Plus Kraft, used for envelopes.

⚙	*Smith Anderson & Co. Ltd, UK*	321
♻	• *Recycled content*	325

Continuum

Old denim, worn-out money notes and industrial cotton waste are recycled in these papers. Crane's has been recycling waste textiles and paper since 1801. The tradition continues with the Continuum line of tree-free papers with 50% cotton fibre and 50% hemp fibre.

⚙	*Crane & Co., USA*	314
♻	• *Renewable resources*	325
	• *Conservation of forest resources*	330

Vision® and Re-vision®

Kenaf fibre – long used as an alternative to tree pulp for paper manufacturing – is the principal raw material for this range of 100% tree-free and chlorine-free papers. All of the pulps and papers are manufactured in the US in compliance with EPA regulations.

⚙	*Vision Paper, KP Products Inc., USA*	322
♻	*• Renewable materials*	325
	• Conservation of forest resources	330

Alden & Ott inks

Alden & Ott manufacture a range of heat-set soy-based inks with about 20–25% soy content, and the colour pigments, avoid the use of heavy metals.

⚙	*Alden & Ott, USA*	312
♻	*• Renewable materials*	325
	• Cleaner production	326

Flint printing inks

A wide range of vegetable-based inks for offset and lithographic printing.

⚙	*Flint Ink, USA*	315
♻	*• Renewable materials*	325

EcoTech™

This range of bio-based inks is made of 61% renewable-resource content. The inks are low in VOCs, with only 5% or less.

⚙	*INX International, USA*	317
♻	*• Renewable materials*	325

Textiles

Green Cotton®

Well before 'organic' became a common word, companies such as Novotex were re-examining the sustainable features of their businesses. Sources of raw materials were analyzed and it was discovered that hand-picked cotton from pesticide-free South-American sources required less cleaning than intensively grown 'commercial' cotton. Long-fibre cottons were selected to provide a yarn that could be woven to facilitate dyeing with water-based dyes and reduce chemical additives throughout the production process. As a result Green Cotton is free of chlorine, benzidine and formaldehyde. Waste water generated in processing is chemically and biologically cleaned in situ. Supply-chain management, and cleaner and quieter production have also created a healthier environment for employees.

⚙	*Novotex A/S, Denmark*	319
♻	*• Clean production of organic textiles*	326
	• Reduced toxins	325

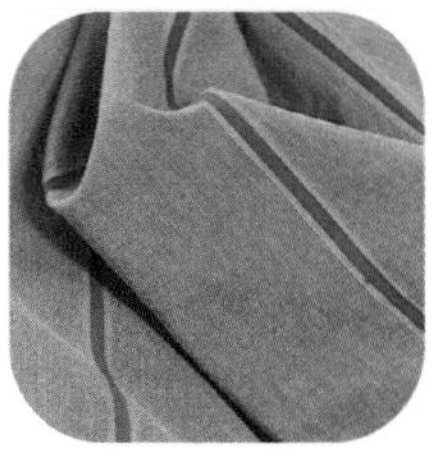

Bincho-Charcoal Border

Charcoal has long been known as an agent to filter and purify air and water, and also has good insulating characteristics. This predominantly wool/silk fabric has a border of charcoal fibres incorporated to take advantage of its positive properties.

⚙	*Reiko Sudo & Nuno Corporation, Japan*	319
♻	*• Natural purifying agents*	325
	• Some renewable fibre	325

Ingeo™ fibres

These high-performance synthetic fibers are made from a biopolymer derived from maize plants through the patented NatureWorks PLA process. Extruded as Ingeo™ fibre, it can be applied to the manufacture of clothing, furnishing fabrics, fibrefill, and non-wovens. The plant sources are 100% annually renewable, and at end-of-life Ingeo is completely biodegradable.

⚙	*NatureWorks LLC, USA*	318
♻	*• Renewable annual crop fibre*	330
	• Compostable	325

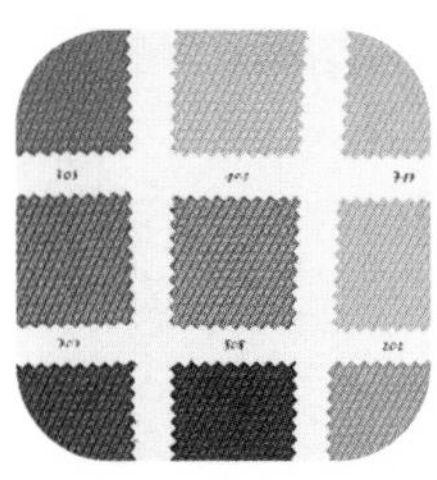

Climatex® LifeguardFR™

Gessner continues to set high standards in ecological textile design with this range of fire-retardant upholstery/furnishing fabrics. In collaboration with the independent German environmental institute (EPEA), Clariant, a leading chemical producer, and Lenzing AG, a fibre manufacturer, Rohner Textil undertook extensive laboratory trials to understand the full environmental impacts of the flame retardants. Climatex® LifeguardFR™ emerged as one of the most advanced ecological textiles with fire retardant certification meeting standards worldwide. It is made of wool and the cellulose fibre Redesigned LenzingFR™, extracted from beech trees. The Colors Kollektion uses environmentally sound chemicals and 16 dyes from Ciba developed with EPEA.

⚙	*Gessner AG, Switzerland*	316
♻	*• Avoidance of toxic and hazardous manufacturing*	326
	• Safe chemicals and dyes	325
	• Durable	325

Eco-yarns

Sourcing knitting, crochet and embroidery yarns can be a time consuming affair, so Annie Sherburne's eco-haberdashery shop is a great facility. Her eclectic yarns range from organic hemp, cotton and wool, to her 'Eco-Annie' brand, recycled white yarn suitable for natural dyeing. Yarns denoting the origin of the sheep include Wensleydale, Blue Faced Leicester, Welsh and Gotland, plus there are some rare breed wools, such as Alpaca from UK herds.

⚙	*Various, supplied by Annie Sherburne, UK*	321
♻	*• Recycled content* *• Virgin, local/ national and organic textile sources*	325 325

Terrazzo Felt 'Colour Chips'

This non-woven, needle-punched, blanket-type fabric fuses dye-chips into a 100%-natural-coloured alpaca-wool felt over a core of polyester organdy to produce unique pieces of material.

⚙	*Nuno Corporation, Japan*	319
♻	*• Renewable and recyclable materials*	325

Argyll Range

Designer Jasper Morrison has built on a long Scottish tradition of weaving woollen textiles by creating a new range of furnishing fabrics for Bute Fabrics in vibrant, contemporary colours, while retaining the durability and warm surface textures associated with traditional crafted products. Bute Fabrics sources much of their raw materials locally and adopts clean production, minimizing the use of harmful substances during processing, as an integral part of their environmental policy. These fabrics are suitable for restoration projects and for new furniture.

⚙	*Bute Fabrics Ltd, UK*	313
♻	*• Renewable materials* *• Clean production*	325 326

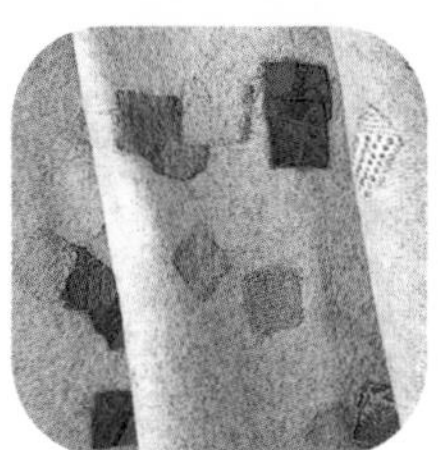

Terrazzo Felt 'Nuno'

Industrial waste snippets of various Nuno fabrics and outtakes in raw wool are combined in a needle-punched technique to create an interesting textured terrazzo effect. The constituents are 85% alpaca wool with 15% Nuno production waste.

⚙	*Nuno Corporation, Japan*	319
♻	*• Renewable and recyclable materials*	325

Organic flax linen

Harvested on the plains of Inner Mongolia, organic-certified flax is used to create fine pure linens varying in weight from 4.2 oz/sq. yd to 5.4 oz/sq. yd (142 g/sq. m to 183 g/sq m). Retting, to break down the fibres from the harvested stems, is done in artificial dew ponds rather than in watercourses, thereby eliminating waste-water pollution. This attention to detail is seen in the wide range of textiles the company produces as pure fibre or blends of organic cotton, hemp, soybean and bamboo.

⚙	*Pickering Intnl Inc., USA*	319
♻	*• Organic fibre production* *• Tough, durable fibres*	326 325

Hemp textiles

The hemp plant is said to have over 50,000 documented uses, and has been used as an essential plant fibre for many thousands of years due to its strength and durability. Industrial hemp is a fantastic cash crop that requires little or no fertilizer, suppresses weeds, leaves the soil in good condition, and produces 5–10 tonnes of fibre per hectare. It is a durable fibre suitable for rope making, webbing and canvas. In the fashion arena it has long been associated with 'hair shirts' and hippies, which has led to a significant failure to recognize the full potential of hemp textiles. Because it stretches less than other natural fibres, it retains its shape in garments. Hemp is also very porous, and holds dyes and colour well, and allows air to circulate easily. The fibre is often mixed with cotton, silk and wool. The range of textures, colours and drapeability of hemp textiles is diverse, as the catalogue of textiles at Hemp Traders shows.

⚙	*Hemp Traders, USA*	316
♻	*• Renewable resource*	325

Cantiva™

Hemp is a very strong natural fibre, naturally resistant to salt water, mould, mildew and UV light. Many different pure hemp or hemp/natural-fibre fabrics are designed using the Cantiva™ brand hemp fibre Hemptex®. Fabrics range from heavy-duty pure hemp canvas to lightweight hemp/silk or hemp/cotton mixtures weighing between 92 and 193 g/sq. m (2.7 and 5.7 oz /sq. yd). Bulk or wholesale orders are produced in contractual arrangements with a Chinese mill.

⚙	*Green China Textiles Co., USA*	316
♲	*• Renewable materials*	325

Romanian hemp fabrics

The international hemp textiles market is currently dominated by Chinese manufacturers, but there are examples of quality weaves from European sources, including Romania. EnviroTextiles is encouraging supply from both Asian and European sources with the philosophy that as hemp production increases the range of textiles will expand. Romanian hemp fabrics, R300, R301 and R302, provide a good choice of naturally coloured fibres of different weaves and weights from 6.5 oz to 10.4 oz/sq. yd (220 g/sq. m to 352.6 g/sq m), which have a range of applications in apparel and home furnishing/upholstery markets.

⚙	*EnviroTextiles, LLC, USA*	315
♲	*• Renewable fibres* *• Durable, long-life textiles*	325 325

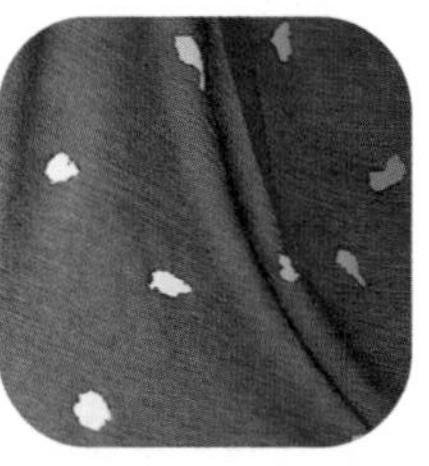

Green fabric (Eco-green fabric)

This is a fully bio-degradable, maize-starch fibre developed by Mitsui Chemical and Kanebo Synthetics in Japan. It is fully compostable by micro-organisms to release water and CO_2. Using a Dobby loom, threads of the fibre are 'overspun' to create a delicate crepe 800 mm (31½ in) wide.

⚙	*Reiko Sudo & Nuno Corporation, Japan*	319
♲	*• Biodegradable*	325

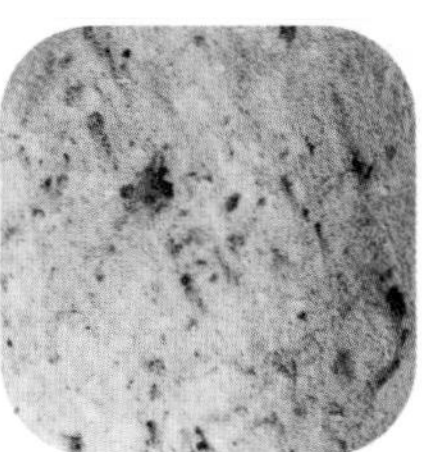

TENCEL®

TENCEL® is a fibre that uses natural raw materials in the form of lyocell cellulose derived from wood pulp harvested from managed forests. This fibre is processed through the unique TENCEL® 'closed loop' solvent spinning process, which is economical in its use of water and energy and uses a non-toxic solvent that is continuously recycled. The resultant TENCEL® fibre is soft, breathable, absorbent and fully biodegradable. A wide variety of luxurious surface finishes are achieved by abrading the wet fibres. TENCEL® is suitable for knitted and woven fabrics, is softer in feel yet stronger than cotton and provides a good surface for printing and dyeing.

⚙	*Lenzing Fibers, Austria*	317
♲	*• Renewable, compostable materials* *• Clean production* *• Low-energy production*	325 326 326

Miscellaneous

Bean-e-doo™

Franmar Chemical, Inc. manufacture a multipurpose, industrial-strength cleaner, Bean-e-doo™, derived from soy beans.

⚙	*Franmar Chemical Inc., USA*	315
♲	*• Renewable materials* *• Reduction in toxic chemicals and VOCs*	325 325

Bio T®

Bio T® is a general-purpose cleaner derived from terpene, which is suitable for use in the manufacturing industries and public-sector maintenance.

⚙	*BioChem Systems, USA*	313
♲	*• Derived from renewable materials*	325

SoySafe

SoySafe is a range of biodegradable, non-toxic cleaners and paint removers derived from soy beans.

⚙	*SoySafe, USA*	321
♲	*• Renewable materials* *• Non-toxic*	325 326

Fern vine (Yan lipao)

This abundant jungle fern is found in southern Thailand where it has supported the vibrant Yan lipao basketry industry for over 150 years. After removal of the outer layer, the pith is dried in the shade, then polished and smoothed, giving a characteristic black or brown colour. It is tough, durable and versatile as a weaving material. In recent years, Italian companies, such as Lino Codato, have created new furniture collections using the woven vine over hardwood frames thereby encouraging continuity of the skill and knowledge while offering European consumers contemporary furniture in natural materials. Any significant increase in world demand may require careful management of jungles to prevent depletion of stocks, so Yan lipao production should be monitored and sustainable harvesting adopted where possible.

⚙	*Tropical and sub-tropical countries*	
♻	• *Renewable material* • *Retention of craft skills*	325 324

SOY Gel™

Removal of paints, urethanes and enamels from timber surfaces is facilitated by application of SOY Gel™, made from 100% American-grown soy beans. This stripper is odourless and easy to use. Simply apply it to the surface to be treated, allow it time to work into the paint, then strip it off with hand-operated scrapers. Like most strippers it is a caustic and potentially harmful substance but it does not have the high VOC content of synthetic-based strippers and originates from annually grown soy beans.

⚙	*Franmar Chemical Inc., USA*	315
♻	• *Renewable material* • *Reduced off-gassing*	325 326

Sundeala and Celotex Sealcoat

Sundeala is a soft board manufactured from unbleached recycled newsprint available in a range of natural colours. Celotex Sealcoat Medium Board is also made from recycled newsprint and is coloured with natural mineral pigments. Both boards are suitable for interior applications, pinboards, noticeboards, exhibition displays and furniture.

⚙	*Sundeala Ltd, UK*	321
♻	• *Recycled materials* • *Clean production*	325 326

Water hyacinth (Eichornia crassipes)

Originating from tropical and sub-tropical regions of South America, the water hyacinth has invaded the southern United States and many other areas of the globe. This invasion results in the extinction of local flora and fauna and causes flooding by blockage of natural and man-made water systems. It is remarkably productive, one plant having the ability to produce 1,200 daughter plants in just four months. This biomass is now being harvested, prepared and dried to produce a tough, light but biodegradable weaving material. It is suitable for use as natural undyed material but can be dyed for different aesthetic effects. Italian companies, such as Lino Codato, have created new furniture collections using water hyacinth and, in doing so, encourage local conservation work and provide employment.

⚙	*Tropical and sub-tropical countries*	
♻	• *Renewable material*	325

ECOSTIX™

A whole new family of starch adhesive biopolymers called ECOSPHERE™ has been created by Ecosynthetix by redesigning starch molecules to form high solids dispersion in water without resorting to the use of heat and caustic chemicals. Tack and drying times are much faster than traditional starch adhesives. ECOSTIX™ are water-borne pressure-sensitive 'smart' adhesives (PSAs) whose adhesive properties turn off when subject to specific external conditions. Using patented ECOMER™ technology these smart adhesives solve the problem of recycling millions of discarded 'stickies'. The United States Postal Service has approved ECOSTIX™ as part of its Benign Stamp Program. The environmental footprint of using ECOSTIX™ with paper is reduced by improving recycling pulpability, lower VOCs and a biodegradable content.

⚙	*EcoSynthetix, USA*	314
♻	• *Renewable, compostable materials* • *Reduced VOCs* • *Improved recyclability*	325 325 327

Boards and Sheeting

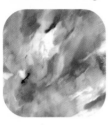

Bottle range

Commingled recycled plastic sheeting, of varying thicknesses, has become a familiar material. One is high-density polyethylene (HDPE) from post-consumer bottles. Typically the re-manufacturing process produces flow patterning and streaks towards the edges of the sheet. Colour palettes vary according to the supply of recycled bottles collected and the seasons.

⚙	*Smile Plastics, UK*	321
♻	*• 100% recycled content*	325

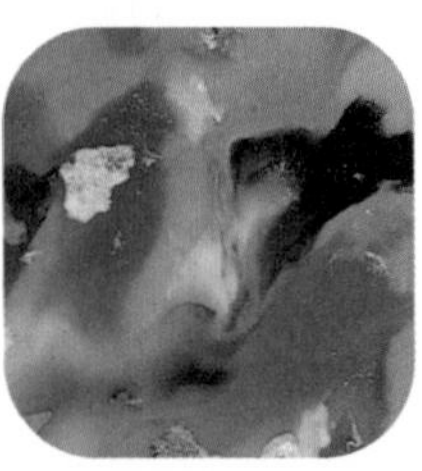

Origins Patterns

Recycled 100% post-consumer HDPE is used to manufacture new plastic boards suitable for a wide variety of applications.

⚙	*Yemm & Hart, USA*	323
♻	*• 100% recycled content*	325

ChoiceDek®

This product is manufactured from a mixture of recycled polyethylene (HDPE and LDPE) plastics and waste wood fibre. ChoiceDek are plank sections suitable for decking.

⚙	*Advanced Environmental Recycling Technologies, USA*	312
♻	*• Recycled materials*	325

Origins Solids

Commingled, recycled post-consumer polythene plastics are extruded into sheets suitable for a wide range of uses from packaging to laminates. Individual colours are carefully blended into specific colour formulas, and a huge range of colours are available.

⚙	*Yemm & Hart, USA*	323
♻	*• 100% recycled content*	325

FlexForm®

FlexForm® is a range of natural-fibre composites that serve the diverse needs of the automotive, agricultural and aeronautic industries, as well as other markets. It is a combination of natural fibres (bast fibres such as jute, kenaf, hemp and flax) and synthetic polymer fibres (polypropylene, polyester and/or maleic polypropylene) according to the required specification. The Material Safety Data Sheet indicates that it is biologically inert or degradable and has no VOC emissions. Available as rolls or profiled sheeting, it can be made into densities varying from 450 g/sq. m to 2400 g/sq. m (13.3 oz/sq. yd to 70.8 oz/sq. yd).

⚙	*FlexForm Technologies, USA*	315
♻	*• Renewable materials*	325
	• Partly biodegradable	325
	• Recyclable	325

Stokbord™ and Centriboard

Stokbord™ is a smooth or embossed low-density polyethylene (LDPE) sheet available in standard thicknesses of 6, 9, 12 and 14 mm (about ⅕ to ½ in). It is constituted from 40–50% post-consumer waste and 50–60% industrial/commercial waste. Centriboard is available in three grades: L – a smooth LDPE sheet, 1.5 mm to 18 mm thick (1/20 to 7/10 in); H – smooth HDPE sheet, 2 mm to 6 mm thick (about 1/12 to 1/4 in); and P – smooth polypropylene sheet, 2 mm to 6 mm thick.

⚙	*Centriforce Products Ltd, UK*	313
♻	*• Recycled materials*	325
	• Reduction in embodied energy (compared with virgin plastics)	326
	• Encouraging conservation of timber resources	330

Tectan

Used drinks cartons are mixed with industrial scrap from the carton-manufacturing plants under the Duales System Deutschland scheme to provide the ingredients for this tough, water-resistant board. The raw material is shredded, then compressed under heat and pressure, causing the polyethylene fraction to melt and bond the particles. It is suitable for building and furniture manufacturing.

⚙	*Entwicklungs-gesellschaft für Verbundmaterial Diez mbH, Germany*	315
♻	*• Recycled materials*	325

Wellies

There is a substantial market for children's wellington boots, yet many of these lead very short lives before being discarded. Smile Plastics brings thousands of these back into the recycling loop: old wellies are shredded and subjected to heat and up to 1,000 tonnes of pressure to produce flexible, soft, water-resistant sheets 2 mm thick and suitable for a wide variety of applications.

Smile Plastics, UK	321
• *Recycled material*	325

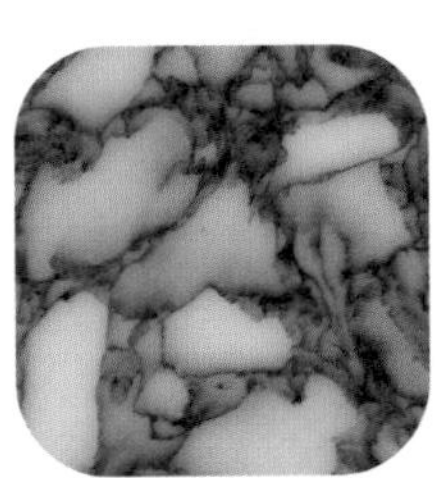

Dapple

Industrial foodstuff containers are shredded and mixed with Smile Plastics' factory waste. There is variation between batches but special mixes can be made to order.

Smile Plastics, UK	321
• *100% recycled content*	325

Plastic Profiles

Plastic profiles

Profiles and stakes in a variety of shapes are manufactured from recycled plastics.

WKR GmbH Germany	323
• *Recycled content*	325

DUROSAM®

Jute is an important natural fibre for agriculture and industry in India, Bangladesh and China, often used for sacking, packaging materials and twine. Like hemp, sisal and coir, it is durable and strong but, for a number of years, these fibres have been replaced by synthetic fibres. AB Composites set themselves the task of finding new applications for jute. The outcome is Natural Fibre Thermoset Composite DUROSAM®, a composite of jute and polymer and a unique system of PREPEG (a method of preparing and treating the polymer during manufacturing).

AB Composites Pvt Ltd, India	312
• *Encourages agricultural cash crop* • *Based on renewable material*	325 325

DPR Plaswood

Reclaimed polythene and polypropylene – 30% waste from supermarkets and 70% production factory waste – are re-blended into extruded profiles suitable for uses requiring tough, rot-free materials.

British Polythene Industries, UK	313
• *Recycled materials* • *Reduction in embodied energy (compared with virgin plastics)* • *Encouraging conservation of timber resources*	325 326 330

Plastic lumber

Commingled, recycled plastics are extruded into a variety of rectangular sections, making an alternative material for traditional uses such as decking and outdoor furniture.

Yemm & Hart, USA	323
• *Recycled content*	325

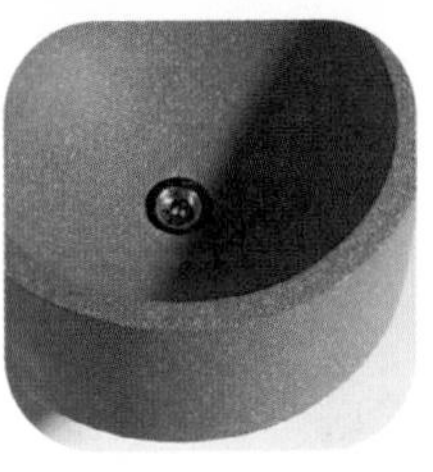

DURAT®

DURAT® is a smooth, silky surfaced, waterproof polyester-based plastic containing 50% recycled material. It is available in a standard range of 46 colours including pure bright hues and terrazzo- and granite-like abstract patterns. Tonester use DURAT® for a wide range of bathroom and kitchen fittings and fixtures. It is fully recyclable.

Tonester, Finland	322
• *Recycled content* • *Recyclable*	325 325

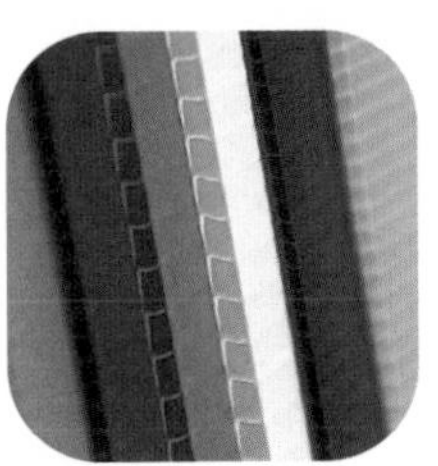

Correx Akylnx

This lightweight twin-walled PP sheet is made from 100% production and customers' returned waste. It is utilized for packaging, self-assembly storage systems and tree shelters.

Kaysersberg Plastics, France	317
• *Recycled materials*	325

Holloplas

Centriforce has supplied recycled finished products to construction, industrial, agricultural and recreational markets in over 30 countries. It offers an extensive range of hollow extruded profiles using a blend of recycled plastic from waste from retail distribution (40–50%) and industrial and commercial waste including film, pipe and packaging (50–60%). Standard sections are suitable for decking, tongue-and-groove flooring, fencing, railings and street furniture.

⚙	*Centriforce Products Ltd, UK*	313
♻	• *Recycled materials* • *Reduction in embodied energy (compared with virgin plastics)* • *Encouraging conservation of timber resources*	325 326 330

EcoClear®

EcoClear® is a resin and film made from recycled PET, which is suitable for beverage and food packaging.

⚙	*Wellman, Inc., USA*	323
♻	• *Recycled content*	325

Govaplast®

A range of square, round, rectangular and tongue-and-groove profiles is produced using recycled polyethylene and polypropylene plastics. Colours include charcoal grey, grey-green and mid-brown. The tongue-and-groove is used in everything from fabrication of equestrian buildings to outdoor planters.

⚙	*Govaerts Recycling NV, Belgium*	316
♻	• *Recycled and recyclable content*	325

Plastic profiles

A variety of round and square profiles and stakes are made from recycled plastics.

⚙	*Hahn Kunststoffe GmbH, Germany*	316
♻	• *Recycled content*	325

Mobiles

There are now about 45 million mobile phones in circulation in the UK and an estimated 15 million are discarded annually, many ending up in shredders and then landfill sites. A phone fashion industry has emerged, encouraging users to customize their phones with colourful new plastic covers. New regulations, in the form of the Waste Electrical and Electronic Equipment Directive from the European Union, will shortly require all manufacturers of mobile phones and a wide range of electronic equipment to take responsibility for recycling and safe disposal of these products. Recycling the plastic covers to create new materials is one solution. This process requires no additives or resins, just the old phone covers, and it produces an eclectic range of colours according to the nature of the waste streams.

⚙	*Smile Plastics, UK*	321
♻	• *Recycled materials*	325

Polymers

Acousticel

Recycled rubber fibre and granules are bonded with a special latex to bitumenized felt-cardboard to make rolls or panels for sound insulation on floors and walls. R10 is a 10 mm (½ in) thick roll and M20AD is supplied in 1 sq. m panels. Sound insulation of 43dB to 45dB is achievable when applying M20AD to single-thickness concrete block walls and fixing two layers of 12.5 mm plasterboard over the panels. Cork granules are added where better thermal insulation is needed.

⚙	*Sound Service, UK*	321
♻	• *Recycled and renewable materials*	325

SoyOyl® Biosynthetic polymers

USSC manufactures a range of specialist polyurethane foams using soy-bean oil. Like synthetic PU, the USSC foams are suitable for everything from loose-fill packaging to panels and shoe parts.

⚙	*Urethane Soy Systems Co., USA*	322
♻	• *Renewable content*	325

Recopol™

This is a co-mingled polymer that combines pre-consumer industrial waste and post-consumer waste from the automotive, electronic and appliance industries into a versatile engineering-grade plastic that can be moulded and routed. Production of Recopol™ generates 50% less CO_2 and requires 80% less energy than virgin ABS. Wharington makes a series of standard Recopol™ shell mouldings for different seating types. Colour diversity varies according to waste streams for each batch.

⚙	*Wharington International Pty Ltd, Australia*	323
♻	• *Recycled and renewable materials* • *Tough engineering grade plastic*	325 325

Duraplast® Biodegradable

This composite board is made up of an outer laminate of water-proofed polystyrene and an inner core of polystyrene using 15% recyclate, which are bound together with a bio-resin. The structure of the polystyrene is altered by the bio-resin, enabling microbial activity in the soil/landfill to break down the constituents into inert bio-mass/mass as tested according to ASTM testing standards, one of which is equivalent to ISO 14855. Degradation takes 1 to 5 years depending upon conditions.

⚙	*The Gilman Brothers Co., USA*	316
♻	• *Biodegrades to inert material* • *Some recycled content*	325 325

RecyTop

This CFC-free protection and drainage sheeting, 22–35 mm (1–1⅓ in) thick is used in civil engineering, for a range of applications requiring water and/or gas drainage, such as tunnel construction and underground parking. It is made of closed-cell polyethylene foam from recycling waste. Individual foam flakes are heat-bonded without further additives or adhesives.

⚙	*Schmitz Foam Products BV, the Netherlands*	320
♻	• *Recycled materials*	325

Rubber granulate

This company manufactures rubber granulate, 0.5 mm to 30 mm (up to 1 in) diameter particles, from 100%-reclaimed scrap tyres. The granulate can be bonded with virgin natural or synthetic rubber and elastomers and is ideal for play surfaces or other uses to reduce impact damage.

⚙	*Playtop, UK*	319
♻	• *Recycled materials*	325

FLO-PAK® Loosefill and SUPER 8 ®Loosefill

These figure-of-eight loosefill packaging chips are made from 100% recycled polystyrene. The form permits interlocking, improves packing ability and offers improved protection, as it won't chip or break. The SUPER 8 Loosefill is recyclable and biodegradable, decomposing completely within 9 to 60 months.

⚙	*FP Intnl, USA*	315
♻	• *100%-recycled* • *Recyclable* • *Biodegradable*	325 325 325

Paints

Falu Rödfärg Paint

Natural copper-based mineral pigments from the mines in Falun, Sweden are the main constituents of this range of water-based exterior wood paints. Available as a liquid or powder formulation, it can be applied by spraying or brushing to protect surfaces from the weather and microbial or insect attack. This is really a lithosphere product composed of a non-bio, non-organic association of earth pigments of silicon, iron, copper and/or zinc. Additionally, the paint does not contain synthetic solvents and is therefore able to return to nature.

⚙	*Stora Enso, Sweden*	321
♻	• *Bountiful raw mineral resources* • *Non-organic but natural pigments* • *No solvents*	325 325 325

Biora, Aqua range

Biora is a range of water-based acrylic resins suitable for application on walls, ceilings and other interior surfaces. Qualifying for the EU eco-label, these paints and varnishes have less environmental impact than conventional paints, have low VOCs and are solvent-free. Teknos are also certified to ISO 9001 and ISO 14001 and are working with the Swedish Paint Makers Organization to develop tools, such as lifecycle analysis, to make further improvements.

⚙	*Teknos Oy, Finland*	321
♻	*• Cleaner production* *• ISO 14001 certified* *• EU eco-label*	326 330 330

Ecos

This company claims to manufacture the only solvent-free odourless paints and varnishes in the world, with zero VOC content, independently tested by the US EPA and the Swedish National Testing & Research Institute. The Ecos range is, however, based upon synthetic resins, albeit non-allergenic, harmless resins, processed from crude oil, so it is not from a renewable source.

⚙	*Ecos Organic Paints, UK*	314
♻	*• Free of toxins (VOCs and vinyl chloride)*	325

Pinturas Proa

A range of water-borne, vinyl polymer interior paints containing less than 45% volatiles is certified with an EU eco-label. The company is also registered with the Spanish eco-label certification authority, AENOR, and participates in the Punto Verde (Green Dot) packaging disposal scheme.

⚙	*Pinturas Proa, Spain*	319
♻	*• EU eco-label* *• Reduction in toxic substances* *• Recycled and recyclable packaging*	330 325 329

YOLO Colorhouse® Outside

This is a zero VOC, 100% acrylic paint for outdoor applications that gives a durable finish because of a high percentage (40%) of solids. It is manufactured without solvents and other harmful chemicals.

⚙	*YOLO Colorhouse LLC, USA*	323
♻	*• VOC-free* *• Green Seal™ certified*	325 330

Safecoat Flat Zero VOC

Safecoat manufactures diverse paint formulations. Flat Zero VOC is virtually odourless and is a fast-drying paint that does not contain any formaldehyde, crystalline silica, ammonia or ethylene glycol. It qualifies for the US eco-building LEED standard. The Safecoat range also includes sealers, stains, primers and enamels.

⚙	*AMF Safecoat, USA*	312
♻	*• Formaldehyde- and VOC-free*	325

Innetak and Bindoplast

At the paint manufacturing plant at Malmö, Sweden, Akzo Nobel produce over 30 million litres (6.6 million gallons) of decorative coatings and 16 million litres (3.5 million gallons) of industrial coatings. Since 1995 the company has set itself a series of environmental targets, such as reducing the emissions of solvents to the air by 50% between 1995 and 1999 and reducing the total energy consumption per litre of paint manufactured by 5% between 1995 and 2000. Innetak and Bindoplast are decorative, water-based emulsion paints, which were the first brand in Europe to receive the EU eco-label.

⚙	*Nordsjö (Akzo Nobel Dekorativ), Sweden*	319
♻	*• EU eco-label ensuring low VOCs and general reduction in toxicity* *• Reduction in emissions to air*	330 328

iPaint interior clay paint

This interior-grade paint finish contains naturally coloured clays and sands, obtained from European sources. The finish is matte and the colour range is earthy, but tonal variation is easily achieved by diluting with water for brush or roller application. There are no added chemicals or pigmentation. The company also produces a range of clay plasters, waterproof lime finishes, renders and more.

⚙	*Tierrafino, the Netherlands*	322
♻	• *Natural, abundant lithosphere materials*	325
	• *Safe, non-toxic*	325

devinegreen Breathable Wall Finish®

This paint formulation is just one of the finishes of the ultra-low VOC, acrylic water-based latex paints from Devine Color. The devinegreen Breathable Wall Finish® provides an eggshell surface for interior walls, applicable by brush, roller or sprayer. Its yogurt-like consistency provides good, smooth coverage.

⚙	*Devine Color, USA*	314
♻	• *Ultra-low VOC content*	325

Keim paints

All the paints manufactured by Keim use inorganic materials that are abundant in the geosphere, including potassium silicate binders, mineral fillers and earth oxide colour pigments. Granital is an exterior paint with a range of 350 colours suitable for all mineral substrates, Concretal protects concrete against corrosion and Biosil is a water-borne, silicate-based paint suitable for interior applications. Ecosil is a recently introduced interior-quality paint, which is water-based, contains no chemical solvents and is VOC-free. Keim are certified to ISO 14001 and ISO 9001.

⚙	*Keim Mineral Paints Ltd, UK*	317
♻	• *Abundant geosphere materials*	325
	• *Non-toxic*	325

Textiles

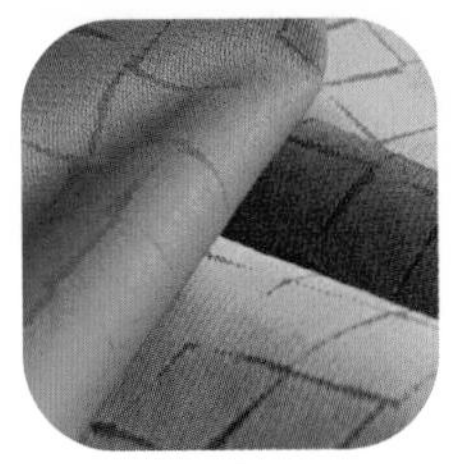

Eco Intelligent Polyester®

In their landmark book *Cradle to Cradle*, William McDonough and Michael Braungart proposed the idea of 'technical nutrients' that perform in closed-loop manufacturing systems: with Eco Intelligent Polyester® the Victor Group realizes this goal. This certainly looks like the 'next Industrial Revolution' mentioned by the authors, as it signals responsible eco-manufacturing. Eco Intelligent Polyester® is produced with hydroelectric renewable power, is free of toxins such as antimony, chlorine and PBTs, uses optimized chemicals and dyes, and is designed for perpetual recycling and re-manufacture. Suitable for office interiors, furniture and healthcare environments, Eco-intelligent® Polyester sets the benchmark for synthetic durable textiles.

⚙	*Victor Group Canada*	322
♻	• *Renewable energy and low waste manufacturing*	327
	• *ISO 14001 EMS certification*	330
	• *Cradle2Cradle gold certification*	330

EcoSpun®

Wellman is one of the world's leading manufacturers of yarn and textiles using PET from recycled drinks bottles. Wellman supply the furnishing and clothing industries, including Patagonia, the outdoor clothing company (p. 155). Ecospun® is a specialist fibre made using recycled plastics.

⚙	*Wellman, Inc., USA*	323
♻	• *Recycled content*	325

Elex Tex™

Conductive fibres are woven with traditional, natural yarns to create a flexible textile suitable for a variety of applications such as electronic clothing, roll-up keyboards and so on.

⚙	*Eleksen Ltd, UK*	314
♻	• *Dual-function material*	328

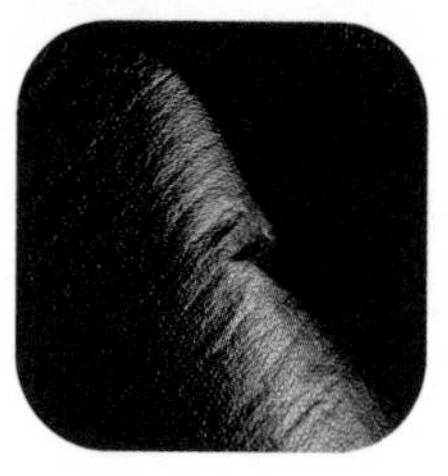

Otterskin

This 100%-polyester, non-woven, needle-punched fabric is made from recycled PET bottles. A surface coating of polyurethane provides wind- and waterproofing, yet the material is breathable and retains body heat.

⚙	*Nuno Corporation, Japan*	319
♻	*• Recycled materials*	325

Trevira NSK/Trevira CS

This is a recyclable fabric made from two types of polyester yarn, Trevira NSK, which gives strength, and Trevira CS, which is hypoallergenic and acts as a flame retardant. Being 100% polyester, it can be reworked by pleating, dyeing and printing and needs no flame-proof coating.

⚙	*Trevira GmbH, Germany*	322
♻	*• Recyclable • Cleaner technology*	325 326
🔍	*iF Design Award, 2000*	332

Take/Bamboo Hexagonal Pattern

Personal hygiene is given a boost if you wear a garment made with Take, as it uses copper sodium-chlorophyll as a catalyst during the manufacture of this silk fabric to generate strong anti-bacterial and odour-suppressing properties. It is made from a mixture of natural and synthetic fibres (rayon 45%, silk 38%, nylon 10% and polyurethane 7%). The copper sodium-chlorophyll is extracted from brush bamboo (*Kuma sasa*) and is sold as a commercial digestive medicine, while the residues provide natural fibre. In a full lifecycle analysis Take will offer considerable reductions during the usage phase of any garment, since washing frequency can be reduced, saving water and detergent consumption.

⚙	*Kazuhiro Ueno & Nuno Corporation, Japan*	319
♻	*• Natural anti-bacterial agents • Some renewable fibre*	 325

Stomatex

Stomatex is a breathable fabric made of neoprene and polyethylene, which mimics transpiration, the process of evaporation of moisture from leaves. Perspiration vapour generated by the wearer is collected in small depressions on the inside of the fabric where tiny pores allow the vapour to escape to the outside. Stomatex is activated only by sufficient body perspiration, so this is a responsive, 'smart' textile.

⚙	*Stomatex, UK*	321
♻	*• Improved personal health with breathable fabric*	328

Velcro®

Velcro® is a combination of two nylon fabrics, one woven with a surface of hooks and the other with a smooth surface with loops. When juxtaposed the two fabrics adhere, as the hooks take up in the loops, creating a strong 'adhesive' bond.

⚙	*Velcro, USA*	322
♻	*• Temporary bonding system allowing reuse of textiles*	327

Terratex

Made entirely of recycled PET recycled plastic bottles, Terratex is a tough, versatile, recyclable fabric for furnishing and similar applications.

⚙	*Interface, Inc., UK/USA*	317
♻	*• Recycled and recyclable materials*	325

Tyvek

With its durability and high chemical resistance, Tyvek was originally developed by DuPont for protective clothing but has since been used for haute-couture fashion and as a paper substitute for envelopes, stationery and various printed media. Tyvek is fully recyclable.

⚙	*DuPont, USA*	314
♻	*• Recyclable synthetic material*	325

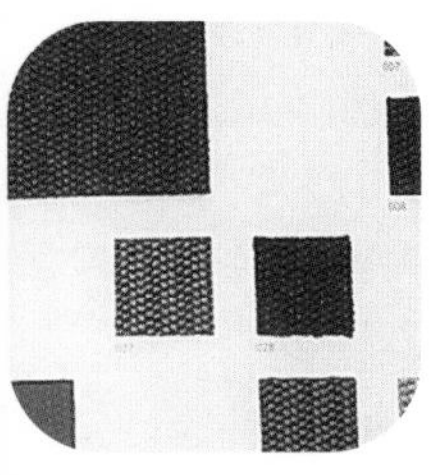

Kvadrat textiles

Kvadrat is a manufacturer of high-quality woven and printed furnishing and curtain textiles for the contract and retail markets. The company pays attention to reducing impact upon the environment and does not use AZO dyes or those that contain heavy metals, nor any chlorine-based chemicals. In 1997 Kvadrat received ISO 14001 certification for environmental management. The textiles have a long lifespan, and a number of them, including Molly, Hallingdal and Hacker, have been awarded the EU 'Flower' eco-label, as has Pure, which is made of 100% organic cotton. The vast majority of the curtains manufactured by Kvadrat are made of Trevira CS.

⚙	*Kvadrat A/S, Denmark*	317
♻	• *Recycled content* • *ISO 14001 certification* • *EU eco-label*	325 330 330

Miscellaneous

Syndecrete®

Syndecrete® is a chemically inert, zero out-gassing, concrete-like material composed of cement and up to 41% recycled or recovered materials from industrial or consumer waste. Typical wastes include HDPE, crushed recycled glass, wood chips and brass screw shavings. Pulverized fly ash (PFA), a waste residue from coal-fired power stations, is added to reduce the cement requirement by up to 15% and recovered polypropylene fibre scrap provides a 3D matrix to increase the tensile strength of this composite recyclate concrete. It is easily worked and polished to create a contemporary terrazzo look.

⚙	*Syndesis, Inc., USA*	321
♻	• *Recycled materials* • *Reduction in embodied energy of manufacture*	325 326

Tire Veneer

Yemm & Hart have expanded their range of recyclate polymers with the addition of the Tire Veneer range made of recycled tyres (the grindings from the re-treading process), coloured EPDM rubber granules and cork dust, all held together with a urethane binder. There is a range of colours available, with up to 20% colour specks or chips in primary or neutral colours; bespoke colour mixes are also possible. Tire Veneer is available as 18 in (457 mm) tiles or 36 in (914 mm) wide roll, 4 mm to 9 mm (⅙ in to ⅓ in) thick, and can be used as a flooring material or vibration dampeners, in addition to a variety of other interior and exterior applications.

⚙	*Yemm & Hart, UK/USA*	323
♻	• *Recycled materials* • *Recyclable*	325 325

4.0 Resources

Ábalos & Herreros
Victor Hugo, 1 3 Dcha
Madrid 28004, Spain
T +34 (0)91523 4404
W www.abalos-herreros.com

agnès b with Veja
c/o Veja (p. 322)

Aisslinger, Werner
Studio Aisslinger
Heidestrasse 46–52
10557 Berlin, Germany
T +49 (0)30 315 05 400
E studio@aisslinger.de
W www.aisslinger.de

Akeler Developments Ltd.
Parkway
Marlow, Buckinghamshire
SL7 1YL, UK
T +44 (0)16 2889 1685

Alessandro Zampieri Design
c/o [re]design (p. 309)

Allard, Helena and Cecilia Falk
c/o Iform (p. 316)

Allson, Gary
Hawthorne Cottage
Florence Terrace
Falmouth
Cornwall, TR11 3RR, UK
T +44(0) 7760401026
E garyallson@hotmail.com
W www.garyallson.co.uk

&made
Studio 214, 18–22
Creekside
London SE8 3DZ, UK
T +44(0) 79161 70293
E info@and-made.com
W www.and-made.com

Apotheloz, Christophe
c/o Brüggli Produktion & Dienstleistung (p. 313)

Arosio, Pietro
c/o Zanotta SpA (p. 323)

Arup
13 Fitzroy Street
London, W1T 4BQ, UK
T +44 (0)20 7636 1531
W www.arup.com

Atypyk
17 rue Lambert
75018 Paris, France
T + 33 (0)1 46 06 28 32
E contact@atypyk.com
W www.atypyk.com

Augustin, Stephan
Augustin
Produktentwicklung
Tengstrasse 45
80796 Munich, Germany
T +49 (0)89 27 306 90
E stephan@augustin.net
W www.augustin.net

Azumi, Shin and Tomoko
UK
E info@shinazumi.com
W www.shinazumi.com

The Baobab Tree
UK
T +44 (0)845 388 1475
E enquiries@thebaobabtree.co.uk
W www.thebaobabtree.co.uk

Baroli, Luigi
c/o Baleri Italia (p.312)

Bartsch Design GmbH
Alter Holzhafen 19
23966 Wismar, Germany
T +49 (0)3841 758 160
E mail@bartsch-design.de
W www.bartsch-design.de

Battaglia, Vanessa & Brendan Young
Studiomold
30 High Street
Fenstanton, PE28 9JZ, UK
T +44 (0)1480 390281
E design@studiomold.co.uk
W www.studiomold.co.uk

Behar, Yves
c/o fuseproject (p. 306)

Behnisch Architekten
163A Rotebühlstrasse
70197 Stuttgart, Germany
T +49 (0)711 60 77 20
E ba@behnisch.com
W www.behnisch.com

Bergne, Sebastian
Sebastian Berne Ltd.
2 Ingate Place
London SW8 3NS, UK
T +44 (0)20 7622 3333
E mail@sebastianbergne.com
W www.sebastianbergne.com

Berthier, Marc
Design Plan Studio
141 Boulevard St Michel
75005 Paris, France
T +33 (0)143 26 49 97
E dpstudio@wanadoo.fr

Bey, Jurgen
Studio Jurgenbey
Postbus 909
3000 AX Rotterdam
the Netherlands
T +31 (0)10 425 8792
E studio@jurgenbey.nl
W www.jurgenbey.nl

Bill Dunster Architects
ZEDfactory Ltd
21 Sandmartin Way
Wallington
Surrey SM6 7DF, UK
T +44 (0)20 8404 1380
E info@zedfactory.com
W www.zedfactory.com

BioRegional Development Group
BedZED Centre
24 Helios Road
Wallington
Surrey, SM6 7BZ, UK
T +44 (0)20 8404 4880
E info@bioregional.com
W www.bioregional.com

Blanca, Oscar Tusquets
Cavallers 50
08034 Barcelona, Spain
T +34 (0) 932065580
E info@tusquets.com
W www.tusquets.com

Blue Marmalade
36 Dalmeny Street
Edinburgh EH6 8RG, UK
T +44 (0)131 553 7766
W www.bluemarmalade.co.uk

Bocchietto, Luisa
c/o Serralunga (p. 321)

Boeri, Cini and Tomu Katayanagi
c/o Fiam Italia SpA (p. 315)

Borgersen, Tore and Espen Voll
c/o Iform (p. 316)

Boym, Constantin and Laurene with Rebecca Wijsbeek
c/o Moooi (p. 318)

Brown, Julian
Studio Brown
6 Princes Buildings
George Street
Bath BA1 2ED, UK
T +44 (0)1225 481 735
E julian@studiobrown.com

Brustad, Tore Vinje
c/o Permafrost (p. 309)

Burkhardt, Roland
c/o Sunways (p. 321)

Büro für Form
Birkenallee 15b
82049 Pullach, Germany
T +49 (0) 89 51 085 6943
E info@buerofuerform.de
W www.buerofuerform.de

Büro für Produktgestaltung
Brendstrasse 83
75179 Pforzheim, Germany
T +49 (0)7231 442 115
E f-neubert@s-direktnet.de

Cahen, Antoine
Les Ateliers du Nord
Place du Nord 2
1005 Lausanne, Switzerland
T +41 (0)21 320 58 07
E a.cahen@adn-design.ch
W www.adn-design.ch

Campana, Fernando and Humberto
São Paulo, Brazil
E campana@campanadesign.com.br
W www.campanas.com.br

Caruso, Jerome
c/o Herman Miller Inc. (p. 316)

Cattle, Christopher
Grown Furniture
UK
E info@grown-furniture.co.uk
W www.grown-furniture.co.uk

Cobonpue, Kenneth
Estrada Nacional 107
Nr.830 Aquas Santa
4445 Maia, Portugal
T +(351) 2297 41542
E europe@kennethcobonpue.com
W www.kennethcobonpue.com

Cohda Design Ltd.
Studio 6, Design Works
William Street
Felling, Gateshead
Tyne and Wear NE10 0JP
UK
T +44(0)191 423 6247
E info@cohda.com
W www.cohda.com

Colwell, David
David Colwell Design, UK
Trannon Studio
Llawr Y Glyn, Caersws
Powys SY17 5RH, UK
T +44(0)1686 430 434
E info@davidcolwell.com
W www.davidcolwell.com

Committee
c/o Moooi (p. 318)

CONBAM Advanced Bamboo Applications
An der Vogelstange 40
52511 Geilenkirchen
Germany
T +49 (0) 2451 4 82 45 45
E info@conbam.de
W www.conbam.de

Corchero, Elena (and Stefan Agamanolis)
UK
E elena@lostvalues.com
W www.lostvalues.com

Cornellini, Deanna
c/o G. T. Design (p. 315)

Cottrell & Vermeulen
1B Iliffe Street
London SE17 3LJ, UK
T +44 (0)20 7708 2567
E info@cv-arch.co.uk
W www.cottrellandvermeulen.co.uk

Cranmore, Louisa
c/o [re]design (p. 309)

Crasset, Matali
Matali Crasset Productions
26 rue du Buisson
75010 Paris, France
T +33 (0)1 42 40 99 89
E matali.crasset@wanadoo.fr
W www.matalicrasset.com

Crummey, Alan and Artein Hossein
Artal Designs, UK
c/o [re]design (p. 309)

Dahlström, Björn
c/o David Design (p. 314)

Day, Robin
c/o Magis SpA (p. 318)

Demakersvan
Marconistraat 52
3029 AK Rotterdam
the Netherlands
T +31 (0)10 2447474
E info@demakersvan.com
W www.demakersvan.com

Design Academy Eindhoven
Emmasingel 14
P.O. Box 2125
5600 CC Eindhoven
the Netherlands
T +31 (0)40 239 39 39
E info@designacademy.nl
W www.designacademy.nl

Design Tech
Zeppelinstrasse 53
72119 Ammerbuch, Germany
T +49 7073 918 90
E info@designtech.eu
W www.designtech.eu

Deuber, Christian
Bruchstrasse 35b
6003 Lucerne, Switzerland
T +41 (0)41 360 86 65
E info@christiandeuber.ch
W www.christiandeuber.ch

Dillon, Jane and Tom Grieves
Studio Dillon
16 DKH, 8 Dog Kennel Hill
London SE22 8AA, UK
T +44 (0)207 326 0804
E mail@studiodillon.com
W www.studiodillon.com

Disch, Rolf
SolarArchitektur
Projektentwicklung
Sonnenschiff
Merzhauser Strasse 177
79100 Freiburg, Germany
T + 49 761 459 44 0
E info@rolfdisch.de
W www.rolfdisch.de

Ditzel, Nanna
Nanna Ditzel Design A/S
Trepkesgade 2
2100 Copenhagen, Denmark
E dd@nanna-ditzel-design.dk
W www.nanna-ditzel-design.dk

DIY KYOTO
1 Temple Yard
Temple Street
London E2 6QD, UK
T +44 (0)207 729 7500
E info@diykyoto.com
W www.diykyoto.com

Draigo
UK
W www.draigo.com

Ecke: Design
Dernburgstrasse 7
14057 Berlin, Germany
T +49 (0)30 34770 90
E berlin@eckedesign.de
W www.eckedesign.de

Ecosistema Urbano
Estanislao Figueras 6
28008 Madrid, Spain
T +34 (0)915591601
E info@ecosistemaurbano.com
W www.ecosistemaurbano.org

Edwards, Alison
c/o [re]design (p. 309)

Eldøy, Olav
Stokke AS (p. 321)

Elegant Embellishments
Axel-Springer Strasse 39
10969 Berlin, Germany
T +49 (0)30 30345003
E info@elegantembellishments.net
W www.elegantembellishments.net

Emilio Ambasz & Associates, Inc.
200 W. 90th Street
Suite 11A
New York, NY 10024, USA
T +1 212 751 3517
E info@ambasz.com
W www.emilioambasz.com

Enthoven Associates Design Consultants
Lange Lozanastraat 254
2018 Antwerp, Belgium
T +32 (0)3 203 53 00
E eadc@ea-dc.com

Erik Krogh Design
Efterårsvej 1 B
2920 Charlottelund
Denmark
T +45 (0)3963 4333
E erkr@dk-designskole.dk

Faudet, Jochem
T +44 (0)7726456333
E info@jochemfaudet.com
W www.jochemfaudet.com

Feilden Clegg Bradley Studios LLP
Bath Brewery
Toll Bridge Road
Bath BA1 7DE, UK
T +44 (0)1225 852 545
E bath@fcbstudios.com
W www.fcbstudios.com

Feldmann + Schultchen
Design Studios
Himmelstrasse 10–16
22299 Hamburg, Germany
T +49 (0)4051 00 00
E mail@fsdesign.de
W www.fsdesign.de

Fergus, Lucy
Re-silicone
Studio 200
Cockpit Arts Deptford
18–22 Creekside
London SE8 3DZ, UK
T +44 (0)7815 089 121
E info@re-silicone.co.uk
W www.re-silicone.co.uk

FloSundK Architektur + Urbanistik
Bleichstrasse 24
66111 Saarbrücken, Germany
T +49 (0)681 3799710
E info@flosundk.de
W www.flosundk.de

Foersom, Johannes and Peter Hiort-Lorenzen
c/o Lammhults (p. 317)

Force4 and KHRAS
KHR Arkitekter
Kanonbådsvej 4
1437 KBHK, Denmark
T +45 (0)4121 7000
E khr@khr.dk
W www.khras.dk

Forcolini, Carlo and Giancarlo Fassina
c/o Luceplan (p. 318)

Formgestaltung
Alexander Schnell-Waltenberger
Sponheimstrasse 22
75177 Pforzheim, Germany
T +49 (0) 7231 3 33 64
E Schnell-Waltenberger@formgestaltung.de

Förster, Monica
Monica Förster Design Studio
Asögatan 194
116 32 Stockholm, Sweden
T +46 (0)8 611 22 09
E monica@monicaforster.se
W www.monicaforster.se

Fortunecookies
(Jacob Jurgensen Ravn)
Denmark
E jacob@fortunecookies.dk
W www.fortunecookies.dk

Frank, Ryan
Mentmore Studios
1 Mentmore Terrace
London E8 3PN, UK
T +44 (0)798 414 6383
E info@ryanfrank.net
W www.ryanfrank.net

Fredrikson, Patrick and Ian Stallard
Fredrikson Stallard
2 Glebe Road
London, E8 4BD, UK
T +44 (0)20 7254 9933
E info@fredriksonstallard.com
W www.fredriksonstallard.com

Frey, Patrick and Markus Boge
c/o Nils Holger Moormann GmbH (p. 308)

Front
Tegelviksgatan 20
116 41 Stockholm, Sweden
T +46 (0)8 710 01 70
E everyone@frontdesign.se
W www.frontdesign.se

fuseproject
528 Folsom Street
San Francisco, CA 94105
USA
T +1 415 908 1492
E info@fuseproject.com
W www.fuseproject.com

Futurefarmers
499 Alabama Street #114
San Francisco, CA 94110
USA
T +1 415 522 2124
E info@futurefarmers.com
W www.futurefarmers.com

GAAN GmbH
Gartenstrasse 4
8902 Urdorf-Zurich
Switzerland
T +41 (0)44 363 52 00
E info@gaan.ch
W www.gaan.ch

Gamper, Martino
E studio@gampermartino.com
W www.gampermartino

Gehry, Frank O.
Gehry Partners, LLP
12541 Beatrice Street
Los Angeles, CA 90066
USA
T +1 310 482 3000
W www.foga.com

gmp – Architekten
von Gerkan, Marg und Partner
Elbchaussee 139
22763 Hamburg, Germany
T +49 (0)40 88 151 0
W www.gmp-architekten.de

Goldsworthy, Kate
UK
W www.kategoldsworthy.co.uk

Goodman
Level 10, 60 Castlereagh Street
Sydney, NSW 2000
Australia
T www.goodman.com

Grcic, Konstantin
Konstantin Grcic Industrial Design
Schillerstrasse 40
80336 Munich, Germany
T +49 (0)89 5507 9995
E mail@konstantin-grcic.com
W www.konstantin-grcic.com

Green Map System
New York, USA
E info@greenmap.org
W www.greenmap.org

Gschwendtner, Gitta
Unit F1
2–4 Southgate Road
London N1 3JJ, UK
T +44 (0)2072492021
E mail@gittagschwendtner.com
W www.gittagschwendtner.com

Guixé, Martí
Spain
E info@guixe.com
W www.guixe.com

Guy Martin Furniture
Crown Studios
Old Crown Cottage
Greenham, Crewkerne
Somerset TA18 8QE, UK
T +44 (0)1308 868122
E guy@guy-martin.com
W www.guy-martin.com

Guynn, Mark D.
Systems Analysis Branch, NASA Langley Research Center
Hampton, VA 23681, USA
W rasc.larc.nasa.gov

Häberli, Alfredo & Christophe Marchand
Design Development
Seefeldstrasse 301a
8008 Zurich, Switzerland
T +41 (0)44 380 32 30
E studio@alfredo-haeberli.com
W www.alfredo-haeberli.com

Hanspeter Steiger Designstudio
Weineggstrasse 42
8008 Zurich, Switzerland
T +41 (0)43 499 00 44
E info@hanspetersteiger.com
W www.hanspetersteiger.com

Hargreaves, Dominic
Studio 3
Crewkerne Court
Bolingbroke Walk
London SW11, UK
E studio@dominichargreaves.com
W www.eyetohand.com

Haygarth, Stuart
33 Dunloe Street
London E2 8JR, UK
T +44(0)20 7503 4142
E info@stuarthaygarth.com
W www.stuarthaygarth.com

Hecht, Sam
c/o Droog (p. 314)

Heikkinen-Komonen Architects
Kristianinkatu 11–13
00170 Helsinki, Finland
T +35 8 9 751 02 111
E ark@heikkinen-komonen.fi
W www.heikkinen-komonen.fi

Hendrikse, Piet and J. P. S.
Q Drum (Pty) Ltd
P.O. Box 4099
Pietersburg 0700
South Africa
T +27 (0)15 297 6762
E info@qdrum.co.za
W www.qdrum.co.za

Henrichs, David
DH Product Design
Levetzowstrasse 23A
10555 Berlin, Germany
T +49 (0)30 39105830
E DavidHenrichs@hotmail.com

Hereford & Worcester County Council
Technical Services Department
County Hall,
Spetchley Road
Hereford and Worcester
WR5 2NP, UK
T +44 (0)1905 766 422

Herreros Arquitectos
Calle Princesa 25
El Hexágono
28008 Madrid, Spain
T +34 (0) 91 522 77 69
E estudio@herrerosarquitectos.com
W www.herrerosarquitectos.com

Heufler, Gerhard
Heufler Design
Körösistrasse 5
80101 Graz, Austria
E gerhard.heufler@utanet.at
W www.fh-joanneum.at/ide

Holroyd, Amy Twigger
Keep & Share
Lugwardine Court
Lugwardine, HR1 4AE, UK
T +44 (0)1432 851 162
E amy@keepandshare.co.uk
W www.keepandshare.co.uk

Holz Box ZT GmbH
Colingasse 3
6020 Innsbruck, Austria
T +43 (0)512 561478
E mailbox@holzbox.at
W www.holzbox.at

Hort, Michael and Danny Gasser
c/o Sycamore Technology (p. 310)

Höser, Christoper
c/o Glas Platz (p. 316)

IDRA International Design Resource Awards
Design Resource Institute
347 NW 105th Street
Seattle, WA 98177, USA
T +1 206 289 0949
E Designwithmemory@cs.com
W www.designresource.org

Indoorlandscaping
25 Seidlstrasse, 3rd floor
80335 Munich, Germany
T +49 (171) 480-4804
E munich@indoorlandscaping.com
W www.indoorlandscaping.com

Ito, Setsu
Italy
W www.studioito.com

Jakobsen, Hans Sandgren
Færgevej 3
8500 Grenaa, Denmark
T +45 (0)86 32 00 48
E mail@hans-sandgren-jakobsen.com
W www.hans-sandgren-jakobsen.com

Jakus, Josh
1025 Carleton Street #4
Berkeley, CA 94710, USA
T +1 510 868 8475
E info@joshjakus.com
W www.joshjakus.com

Jansen, Willem
c/o Design Academy Eindhoven (p. 305)

Joachim, Mitchell, Lara Greden and Javier Arbona
c/o Terreform 1 (p. 310)

John Patrick ORGANIC
535 W. 24th Street
5th floor
New York, NY 10011, USA
T +1 212 206 7179
E johnpatrickorganic@gmail.com
W www.johnpatrickorganic.com

Jongerius, Hella
JongeriusLab
Baan 74
3011 CD Rotterdam
the Netherlands
T +31 (0)10 477 0253
E info@jongeriuslab.com
W www.jongeriuslab.com

Jørgensen, Carsten
c/o Bodum (p. 313)

Kämpfen, Beat
Kämpfen für Architektur
Badenerstrasse 571
8049 Zurich, Switzerland
T +41 (0)44 344 46 20
E info@kaempfen.com
W www.kaempfen.com

Karrer, Beat
Studio Beat Karrer
Zimmerlistrasse 6
8004 Zurich, Switzerland
T +41 (0)1 400 55 00
E karrer@beatkarrer.net
W www.beatkarrer.net

Karpf, Peter
c/o Iform (p. 316)

Kaufmann, Johannes and Oskar Leo
Johannes Kaufmann Architektur
Sagerstrasse 4
6850 Dornbirn, Austria
T +43 (0)5572 23690
E office @jkarch.at
W www.jkarch.at

Kekeritz, Timm
Traumkrieger GmbH
Einfelder Schanze 101a
24536 Neumünster
Germany
E timm@traumkrieger.de
W www.traumkrieger.de/virtualwater

Kelly, Rachel
Interactive Wallpaper
Unit 4 & 5 Barn Farm
Bonningate nr Kendal
Cumbria LA8 8JX, UK
T +44 (0)1539 822511
E studio@interactivewallpaper.co.uk
W www.interactivewallpaper.co.uk

Klug, Ubald
33 rue Croulebarbe
75013 Paris, France
T +33 (0)1 44 33 13 882
F +33 (0)1 45 35 31 54

Komplot Design
Amager Strandvej 50
2300 Copenhagen
Denmark
T +45 32 96 32 55
E komplot@komplot.dk
W www.komplot.dk

Koponen, Olavi
Apollonkatu 23 B 39
00100 Helsinki, Finland
T +358 9 441 096
E olavi.koponen@kolumbus.fi
W www.kolumbus.fi/olavi.koponen

Krier, Sophie
c/o Moooi (p. 318)

Le Laboratoire
4 rue de Bouloi
75001 Paris, France
W www.lelaboratoire.org

Lakic, Sacha
Sacha Lakic Design
W www.sachalakic.com

de Larratea, Lili
Rethink Games
The Hub, 5 Torrens Street
London EC1V 1NQ, UK
T +44 (0) 78 0075 5269
E info@rethinkgames.com
W www.playrethink.com

Lawton, Tom
The Firewinder Company Ltd
The Old Library
44 High Street
Malmesbury, Wiltshire
SN16 9AT, UK
T +44 (0)845 680 1590
E sales@firewinder.com
W www.firewinder.com

Lee, Marcus
FLACQ Architects Ltd
4 John Prince's Street
London W1G 0JL, UK
T +44 (0)20 7495 5755
E info@flacq.com
W www.flacq.com

de Leede, Annelies
c/o Goods (p. 316)

Lehanneur, Mathieu
14 rue de Jeuneurs
75002 Paris, France
T +33 (0)9 71 49 88 77
E m@mathieulehanneur.com
W www.mathieulehanneur.com

Lewis, David
c/o Vestfrost A/S (p. 322)

Liddle, Richard
c/o Cohda Design Ltd.
(p. 305)

Lindsay Johnston Architecture
16 Milson's Passage
Hawkesbury River via
Brooklyn
NSW 2083, Australia
T +61 2 9985 1262
E lindsay@rivertime.org
W www.rivertime.org/lindsay

Lo, K. C.
31 Finsbury Park Road
London N4 2JY, UK
T +44 (0)20 7689 1582
E email@kclo.co.uk
W www.kclo.co.uk

Lovegrove, Ross
Lovegrove Studio
21 Powis Mews
London W11 1JN, UK
T+44 (0)207 229 7104
W www.rosslovegrove.com

Marczynski, Mike
c/o Business Lines (p. 313)

Mario Bellini Architects S.r.l.
Piazza Arcole 4
20 143 Milan, Italy
T +39 (0)2 5815 191
E info@mariobellini.com
W www.mariobellini.com

Marquina, Nani (and Adriana Miquel and Care&Fair)
c/o Nani Marquina (p. 318)

Meda, Alberto and Francisco Gomez Paz
Alberto Meda Industrial
Design
via Savona, 97
20144 Milan, Italy
T +39 (0)2 42290157
E info@albertomeda.com
W www.albertomeda.com

Meller Marcovicz, Gioia
Gioia Limited
P.O. Box 381
30100 Venice, Italy
T +39 (0)41 795213
E info@gioia.org.uk
W www.gioia.org.uk

Metcalfe, Christopher
New Zealand
E info@
christophermetcalfe.com
W www.christopher
metcalfe.com

Millar, Andrew
c/o [re]design (p. 309)

Miller, Edward Douglas
c/o Remarkable Pencils Ltd
(p. 320)

Miller, Giles
8 Fairfield Road
London E3 2QB, UK
T +44 (0)7780 924689
E studio@gilesmiller.com
W www.gilesmiller.com

Minakawa, Gerard
c/o IDRA (p. 307)

MIO
446 North 12th Street
Philadelphia, PA 19123
USA
T +1 215 925 9359
E info@mioculture.com
W www.mioculture.com

Mithun Architects
Pier 56, 1201 Alaskan Way,
Suite 200
Seattle, WA 98101, USA
T +1 206 623 3344
E mithun@mithun.com
W www.mithun.com

Le Moigne, Nicolas
17 Avenue de Jurlgoz
1006 Lausanne
Switzerland
E nicolas_lm@hotmail.com
W www.nicolaslemoigne.com

Molo Design
1470 Venables Street
Vancouver, V5L 2G7
Canada
T +1 604 696 2501
E info@molodesign.com
W www.molodesign.com

Morrison, Jasper
Jasper Morrison Ltd
Paris, France and London
E mail@jaspermorrison.com
W www.jaspermorrison.com

Morsbags
UK
E admin@morsbags.com
W www.morsbags.com

N Fornitore
c/o Purves & Purves
(p. 320)

Navone, Paola
c/o Gervasoni SpA (p. 316)

Neggers, Jan
c/o Goods (p. 316)

Negroponte, Nicolas with Yves Behar, Martin Schnitzer, Bret Recor and Design Continuum
c/o fuseproject (p. 306)

Newson, Marc, Jens Martin Skibsted and Beatrice Santiccioli
c/o Biomega (p. 313)

NEXT Architects
P.van Vlissingenstr 2a
1096 BK Amsterdam
the Netherlands
T +31 (0)20 4630463
W www.nextarchitects.com

Nils Holger Moormann GmbH
An der Festhalle 2
83229 Aschau im
Chiemgau, Germany
T +49 (0) 8052 90450
E info@moormann.de
W www.moormann.de

Nishimura, Rentaro
UK
E mail@rentaro.co.uk
W www.rentaro.co.uk

Nydahl, Andreas
Näsbydalsvägen 2, 14 tr
183 31 Täby, Sweden
T +46 (0)73 587 63 96
E contact@andreasnydahl.com
W www.andreasnydahl.com

O'Leary, John D.
Bluegreen&co
Unit 2 Belleknowes
Industrial Estate
Inverkeithing, Fife
KY11 1HY, Scotland
T +44 (0) 7927189250
E bluegreenandco@aol.com
W www.bluegreenandco.com

O'Neill, Cj
UK
T +44 (0)771 253 7000
E cj@cjoneill.co.uk
W www.cjoneill.co.uk

Opsvik, Peter
c/o Stokke AS (p. 321)

Pakhalé, Satyendra
Atelier Satyendra Pakhalé
R. J. H. Fortuynplein 70
1019 WL Amsterdam
the Netherlands
T +31 (0)20-4197230
E info@satyendra-pakhale.com
W www.satyendra-pakhale.com

Paul Morgan Architects
Level 10, 221 Queen Street
Melbourne Victoria 3000
Australia
W +61 3 9605 4100
E office@
paulmorganarchitects.com
W www.paulmorgan
architects.com

Peebles, Patrick
FanWing
T +44 (0)785 537 4006
E peebles@fanwing.com
W www.fanwing.com

Permafrost
Design Studio
Borggata 1
0650 Oslo, Norway
T +47 (0)23 24 04 50
E mail@permafrost.no
W www.permafrost.no

Philips Design
Emmasingel 24
Building HWD
5611 AZ Eindhoven
the Netherlands
T +31 (0)40 27 59000
W www.design.philips.com

Piercy Conner Architects
Studio A, Jacks Place
6 Corbet Place
London E1 6NH, UK
T +44 (0)20 7426 1280
E info@piercyconner.co.uk
W www.piercyconner.co.uk

Pillet, Christophe
c/o Ceccotti Collezioni Srl
(p. 313)

Prause, Philipp
Beckmanngasse 9A/19
1140 Vienna, Austria
T +43 (0)1 876 4039
E office@prausedesign.com
W www.prausedesign.com

ProduktEntwicklung Roericht
c/o Wilkhahn (p. 323)

PROFORM Design
Lener + Rossler +
Hohannismeier Gbr
Seehalde 16
71364 Winnenden
Germany
T +49 (0)7195 919100
E proform_design@t-online.de
W www.industriedesign.com

Raddisshme Design House
Room B5
60 Xian Lie Dong Hen Lu
Tianhe District
Guangzhou, 510500 China
T +86 20 28296215
E info@raddisshme.com
W www.raddisshme.com

Raffield, Tom
The Old Grammar School
West Park, Redruth
Cornwall TR15 3AJ, UK
T +44(0)7968 621955
E contact@tomraffield.com
W www.tomraffield.com

rawstudio
T +44(0)845 680 8443
E jane@rawstudio.co.uk
W www.rawstudio.co.uk

[re]design
1 Summit Way
Crystal Palace
London SE19 2PU, UK
T +44 (0)20 840 641 60
E info@redesigndesign.org
W www.redesigndesign.org

Refsum, Bjørn
c/o Stokke AS
(p. 321)

Reln
c/o Wiggly Wigglers(p. 323)

Rizzatto, Paolo
c/o Serralunga (p. 321)

Garth, Roberts
c/o Zanotta (p. 323)

Roije, Frederik
c/o Design Academy
Eindhoven (p. 305)

Rossi, Diego and Raffaele Tedesco
c/o Luceplan (p. 318)

Roth, Antonia
FH Hannover
Postfach 92 02 51
30441 Hannover, Germany
E antonia@gmx.de
W www.www.fh-hannover.de

Roy Tam Design
Sherborne, Dorset, UK
T +44 (0)780 853 5863
E roytamdesign@
btinternet.com
W www.eco-furniture.co.uk

Salm, Jaime (and Roger Allen)
c/o MIO (p. 308)

Sanders, Mark
MAS Design
Axis House
77A Imperial Road
Windsor SL4 3RU, UK
E mark@MAS-Design.com
W www.mas-design.com

Scarff, Leo
Rear 91 Ballybough Road
Dublin 3, Ireland
T/F +353 (0)1 836 3135
E leo@leoscarffdesign.com
W www.leoscarffdesign.com

Schneider, Wulf and Partners
Schellbergstrasse 62
70188 Stuttgart, Germany
T +49 (0)711 286 49 00
E bfg@profwulfschneider.de
W www.profwulfschneider.de

Sempé, Inga
c/o edra (p. 314)

Shigeru Ban Architects
5-2-4 Matsubara, Setagaya
Tokyo, Japan
T +81-(0)3-3324-6760
E Tokyo@
ShigeruBanArchitects.com
W www.shigerubanarchitects.co

Shine, Benjamin
Benjamin Shine Studio
UK
W www.benjaminshine.com

Simmering, Michaele and Johannes Pauwen
Kalon Studios
Los Angles, USA and Berlin,
Germany
E studio@kalonstudios.com
W www.kalonstudios.com

Simon White Design
UK
T +44(0)7817 979 967
E simon@
simonwhitedesign.co.uk
W www.simonwhitedesign.co.uk

Smart Cities at MIT Media Lab
The Media Laboratory
Massachusetts Institute
of Technology
Building E15
77 Massachusetts Avenue
Cambridge, MA 02139, USA
W cities.media.mit.edu

Sodeau, Michael
c/o Gervasoni SpA
(p. 316)

Spiegelhalter, Thomas
WAH 204, 850 W 37 Street
Los Angeles, CA 90089
USA
T +1 323 377 3685
E spiegelhalterstudio@
googlemail.com
W www.sustainabilitythomas
spiegelhalter.com

Sprout Design Ltd.
One Bermondsey Square
London SE1 3UN, UK
T +44(0)20 7645 3790
E info@sproutdesign.co.uk
W www.sproutdesign.co.uk

Starck, Philippe
Ubik
18/20, rue du Faubourg
du Temple
75011 Paris, France
T + 33(0)1 48 07 54 54
W www.starck.com

Startup, Jasper
Startup Design
No 4, 126A Albion Road
London N16 9PA, UK
T +44 (0) 207 923 1223
W www.startupdesign.co.uk

Steinmann, Peter and Herbert Schmid
c/o Atelier Alinea AG
(p. 312)

Stiletto DESIGN VERTRieB
Auguststrasse 2
10117 Berlin, Germany
T +49 (0)30 280 94 614
E hallo@siltetto .de
W www.stiletto.de

Studio 7.5
Nithackstrasse 7
10585 Berlin, Germany
T +49 (0)30 341 70 34
E contact@seven5.com
W www.seven5.com

Studio Aisslinger
Heidestrasse 46–52
10557 Berlin, Germany
T +49 (0)30 315 05 400
E studio@aisslinger.de
W www.aisslinger.de

Studio Jacob de Baan (and Frank de Ruwe)
Nieuwevaart 128
1018 ZM Amsterdam
the Netherlands
E info@jacobdebaan.nl
W www.jacobdebaan.com

Stumpf, Bill and Don Chadwick
c/o Herman Miller, Inc.
(p.316)

Sugasawa, Mitsumasa
c/o Atelier Alinea AG
(p. 312)

Suppanen, Ilkka and Pasi Kohhonen
Studio Suppanen
Sturenkatu 13
00510 Helsinki, Finland
T +358 9 622 787 37
E info@suppanen.com
W www.suppanen.com

Sutton Vane Associates
Dimes Place
106–108 King Street
London W6 0QP, UK
T +44 (0)20 8563 9370
W www.sva.co.uk

Sycamore Technology
P.O. Box 621
Milsons Point, Sydney
NSW 1565, Australia
T +61 (2) 9410 1847
E info@sycamorefan.com
W www.sycamorefan.com

Tabu, Ryuichi
R-design Studio
E ryuichi@r-designstudio.com
W www.r-designstudio.com

Takahashi, Koji
c/o IDRA (p. 307)

Tato, Belinda and Jose Luis Vallejo
c/o Ecosistema Urbana
(p. 305)

Tea Un Kim
South Korea
T +44 (0)7525 032 323
E greenteaun@gmail.com
W www.greenteaun.com

Terreform 1
33 Flatbush Avenue
7th Floor
Brooklyn, NY 11217, USA
T +1 617 285 0901
E info@terreform.org
W www.terreform.org

Tesnière, François and Anne-Charlotte Goût
3bornes Architectes
70 rue Jean Pierre Timbaud
75011 Paris, France
T + 33 (0)1 47 00 78 27
E 3bornes@magic.fr

Thirlwell, Sarah
UK
T +44 (0)794 731 6288
E me@sarahthirlwell.com
W www.sarahthirlwell.com

Timpe, Jakob
Prinzenstrasse 39
10969 Berlin, Germany
E studio@jakobtimpe.com
W www.jakobtimpe.com

Todiwala, Jamsheed
c/o [re]design (p. 309)

Tolstrup, Nina
Studiomama
21-23 Voss Street
London E2 6JE, UK
T +44(0)207 0330 408
E tolstrup@studiomama.com
W www.studiomama.com

Tonges, Christoph
c/o CONBAM (p. 305)

Tortel Design
6, Villa du Clos de Malevart
75011 Paris, France
T +33 (0)1 43550370
E tortel.design@infonie.fr
W www.tortel.design.fr

Trubridge, David
44 Margaret Avenue
Havelock North
New Zealand
T/F +64 (0)6 877 46 84
E trubridge@clear.net.nz
W www.davidtrubridge.com

Turner, Lucy
Higher Market Studios
19 Higher Market Street
Penryn
Cornwall TR10 8ED, UK
T +44 (0)1326 374191
E lucy@
highermarketstudio.co.uk
W www.highermarketstudio
.co.uk

Uhuru Design
160 Van Brunt St., 3rd Fl
Brooklyn, NY 11231, USA
T +1 718 855 6519
E info@uhurudesign.com
W www.uhurudesign.com

University of Technology Eindhoven
P.O. Box 513
5600 MB Eindhoven
the Netherlands
T +31 40 247 91 11
W w3.tue.nl/en

van der Meulen, Jos
c/o Goods (p. 316)

van der Poll, Marijn
c/o Droog (p. 314)

VarioPac Disc Systems GmbH
Hangbaumstrasse 13
32257 Bünde, Germany
T +49 (0)5221 7684 17
W www.variopac.com

Veloland Schweiz
Stiftung Veloland Schweiz
Postfach 8275
3001 Bern, Switzerland
T +41 (0)31 307 47 40
E info@veloland.ch
W www.veloland.ch

Vestergaard Frandsen
Chemin de Messidor 5–7
1006 Lausanne
Switzerland
T +41 (0) 21 310 7333
W www.lifestraw.com
www.vestergaard-frandsen.com

Vuarnesson, Bernard
Sculptures-Jeux
18 Rue Domat
75005 Paris, France
T +33 (0)1 43 54 20 39
W www.sculpturesjeux.fr

Wanders, Marcel
Marcel Wanders Studio
P.O. Box 11332
1001 GH Amsterdam
the Netherlands
T +31 (0)20 4221339
E joy@marcelwanders.nl
W www.marcelwanders.com

Weatherhead, David
Top Floor,2a
Alexandra Grove
London N4 2LG, UK
T +44 (0) 78 1315 1842
E david@
davidweatherhead.com
W www.davidweatherhead.com

WEmake
1 Summit Way
Crystal Palace
London SE19 2PU, UK
T +44 (0)79 62 10 87 82
E sarah@wemake.co.uk
W www.wemake.co.uk

Wettstein, Robert A.
Josefstrasse. 188
8005 Zurich, Switzerland
T +41 44 271 01 16
E robert.wettstein@gmx.ch
W www.wettstein.ws

Willat, Boyd
c/o Willat Writing
Instruments (p. 323)

Worn Again
Rich Mix, Unit CO2
35–47 Bethnal Green Road
London E1 6LA, UK
T +44 (0) 207 739 0189
E info@wornagain.co.uk
W www.wornagain.co.uk

Wortmann, Constantin
c/o Büro für Form(p. 304)

Yoshioka, Tokujin
c/o Driade (p. 314)

Young, Carol
Undesigned
1953 1/2 Hillhurst Avenue
Los Angeles CA 90027
USA
T +1 323 663 0088
E contact@undesigned.com
W www.undesigned.com

Young, Michael
c/o Magis SpA (p. 318)

21st Century Health Ltd
2 Fitzhardinge Street
London W1H 6EE, UK
T +44 (0)20 7935 5440
E info@21stcenturyhealth.co.uk
W www.21centuryhealth.co.uk

3M Deutschland GmbH
Carl-Schurz-Strasse 1
41453 Neuss, Germany
T +49 (0)2131 14 0
W www.3m.com

A. Winther A/S
Rygesmindevej 2
8653 Them, Denmark
T +45 86 84 72 88
E win@a-winther.dk
W www.a-winther.com

AB Composites Pvt. Limited
1/1b/18 Ramkrishna
Naskar Lane
Calcutta 700 010, India
T +91 (0)2370 5982/6348
E anukul@cal2.vsnl.net.in
W www.abcomposites.net

ABG Ltd
Unit E7,
Meltham Mills Road
Meltham,West Yorkshire
HD7 4DS, UK
T +44 (0)1484 852 096
W www.abg-geosynthetics.com

Advanced Environmental Recycling Technologies, Inc.
P.O. Box 1237
Springdale, AR 72765, USA
T +1 800 951 5117
E sales@choicedek.com
W www.choicedek.com

Advanced Vehicle Design
The Barn
Warrener Street
Sale Moor
Sale, Cheshire
M33 3GE, UK
T +44 (0)161 969 9692
E bob@windcheetah.co.uk
W www.windcheetah.co.uk

AEG Electrolux Plc
W www.aeg-electrolux.com
www.electrolux.com

aerodyn Energiesysteme GmbH
Provianthausstrasse 9
24768 Rendsburg
Germany
T +49 (0)4331 12750
E info@aerodyn.de
W www.aerodyn.de

AeroVironment, Inc.
Corporate HQ
825 S. Myrtle Drive
Monrovia, CA 91016, USA
T +1 626 357 9983
W www.aerovironment.com

Airbus
W www.airbus.com

Airnimal Europe Ltd
T +44 (0)1954 782020
E info@airnimal.eu
W www.airnimal.eu

Alden & Ott
616 E. Brook Drive
Arlington Heights
IL 60005, USA
T +1 847 956 6509
W www.aldenottink.com

Allison Transmission, Inc.
W www.allison
transmission.com

Amasec Airfil Ltd.
Unit 1, Colliery Lane
Exhall, Coventry
Warwickshire CV7 9NW, UK
T +44 (0)24 7636 7994
W www.airfil.com

American Excelsior Company
850 Avenue H East
Arlington, TX 76011, USA
T +1 800 777 7645
E sales@americanexcelsior.com
W www.amerexcel.com

AMF Safecoat
T +1 800 239 0321
E info@afmsafecoat.com
W www.afmsafecoat.com

Ampair Ltd
Park Farm, West End Lane
Warfield
Berkshire, RG42 5RH, UK
T +44 (0)1344 303 311
E info@ampair.com
W www.ampair.com

Amtico International
Kingfield Road
Coventry
Warwickshire CV6 5AA, UK
T +44 (0)121 745 0800
W www.amtico.com

Anglepoise Ltd
Unit A10 Railway Triangle
Walton Road
Farlington, PO6 1TN, UK
T +44 (0) 2392 224450
E info@anglepoise.co.uk
W www.anglepoise.com

Anthologie Quartett
Schloss Hünnefeld
49152 Bad Essen, Germany
T +49 (0)5472 94090
E info@anthologiequartett.de
W www.anthologiequartett.de

Apple
1 Infinite Loop
Cupertino, CA 95014, USA
T +1 408 996 1010
W www.apple.com

Architectural Systems, Inc.
150 W. 25th Street, 8th floor
New York, NY 10001, USA
T +1 212 206 1730
E sales@archsystems.com
W www.archsystems.com

Artemide SpA
Via Bergamo, 18
20010 Pregnana Milanese
(MI), Italy
T +39 (0)2 935 181
E info@artemide.com
W www.artemide.com

Atelier Alinea AG
Bernstrasse 229
3627 Heimberg, Switzerland
T +41 (0)33 438 32 72
E info@atelieralinea.ch
W www.atelieralinea.ch

Atelier Satyendra Pakhalé
Zeeburgerpad 50
1019 AB Amsterdam
the Netherlands
T +31 (0)20 419 72 30
E info@satyendra-pakhale.com
W www.satyendra-pakhale.com

Athletic Polymer Systems
708 S Temescal Street
Suite 101
P.O. Box 788
Corona, CA 92878, USA
T +1 951 273 7984
W www.tartan-aps.com

Auro Pflanzenchemie AG
Alte Frankfurter Strasse 211
38122 Braunschweig
Germany
T +49 (0)531 281 41 0
E info@auro.de
W www.auro.de

Baccarne Design
Gentbruggekouter 5
9050 Gent, Belgium
T +32 (0)9 232 44 21
E info@baccarne.be
W www.baccarne.be

Bags of Change
25 Kew Gardens Road
Richmond
Surrey TW9 3HD, UK
E info@bagsofchange.co.uk
W www.bagsofchange.co.uk

Baleri Italia SpA
8, Via Felice Cavallotti
20122 Milan, Italy
T +39 (0)2 760 239 54
E info@baleri-italia.com
W www.baleri-italia.com

Bamboo Hardwoods
4100 4th Ave. South
Seattle, WA 98134, USA
T +1 800 607 2414
E info@
bamboohardwoods.com
W www.bamboohardwoods.com

Beacon Press
Bellbrook Park
Uckfield, East Sussex
TN22 1PL, UK
T +44 (0)1825 768611
E print@beaconpress.co.uk
W www.beaconpress.co.uk

Benza Design
413 W. 14 Street, #301
New York, NY 10014, USA
T +1 718 383 1334
E ingo@benzadesign.com
W www.benzadesign.com

Better Energy Systems Ltd.
85–87 Bayham Street
London NW1 0AG, UK
T +44 (0) 207 424 7999
E media@solio.com
W www.solio.com

Big Agnes
735 Oak Street
Steamboat Springs, CO
80487, USA
T +1 970 871 1480
E info@bigagnes.com
W www.bigagnes.com

BioChem Systems
P.O. Box 132196
The Woodlands, TX 77393
USA
T +1 800 777 7870
W www.biochemsys.com

Biolan Oy
P.O. Box 2
27501 Kauttua, Finland
T +358 (0)2 549 1600
E export@biolan.fi
W www.naturum.fi

Biomega
Skoubogade 1,1. MF
1158 Copenhagen
Denmark
T +45 (0)70 20 49 19
E info@biomega.dk
W www.biomega.dk

BioShield Paint Company
3005 S. St. Francis Suite 2A
Santa Fe, NM 87505, USA
T +1 800 621 2591
E info@bioshieldpaint.com
W www.bioshieldpaint.com

Biotec
Werner-Heisenberg
Strasse 32
46446 Emmerich, Germany
T +49 (0)2822 92510
E info@biotec.de
W www.biotec.de

BJ Green LLC
1701 Chamberlain Drive
NW
Wilson NC, 27896, USA
T +1 252 243 1534
E jjones@nc.rr.com
W www.bjgreeninsulation.com

Blackwall Ltd
UK
T +44 (0)113 245 2244
E info@straight.co.uk
W www.blackwall-ltd.com

Bodum AG
Kantonstrasse 100
6234 Triengen, Switzerland
T +41 (0)41 935 4500
W www.bodum.com

The Body Shop International plc
Care Centre
Building 4, Hawthorn Road
Wick, Littlehampton
West Sussex, BN17 7LT, UK
T +44 (0)800 0929090
W www.bodyshop.com

Boeing
100 North Riverside
Chicago, IL 60606 USA
T +1 312 544 2000
W www.boeing.com

Bourgeois SC
364 rue des Epinettes
BP71
74210 Faverges, France
T +33 (0)4 50 32 58 58
E contact@bourgeois.coop
W www.bourgeois.coop

BP plc
International Headquarters
1 St James's Square
London SW1Y 4PD, UK
T +44 (0)20 7496 4000
E careline@bp.com
W www.bp.com

Brammo Inc.
550 Clover Lane
Ashland, OR 97520, USA
T +1 541 482 9555
E info@brammo.com
W www.brammo.com

British Polythene Industries
96 Port Glasgow Road
Greenock PA15 TGL, UK
T +44 (0)1475 501 000
W www.bpipoly.com

Brompton Bicycle Ltd
Lionel Road South
Brentford
Middlesex TW8 9QR, UK
T +44 (0)20 8232 8484
W www.brompton.co.uk

Brook Crompton
St Thomas Road
Huddersfield
West Yorkshire, HD1 3LJ
UK
T +44 (0)1484 557 200
E csc@brookcrompton.com
W www.brookcrompton.com

Brüggli Produktion & Dienstleistung
Hofstrasse 5
8590 Romanshorn
Switzerland
T +41 (0)71 466 94 94
E info@brueggli.ch
W www.brueggli.ch

Buderus Heiztechnik GmbH
Sophienstrasse 30–32
35576 Wetzlar, Germany
T +49 (0)6441 418 0
E info@buderus.de
W www.buderus.de

Building Research Establishment (BRE)
Bucknalls Lane
Watford WD25 9XX, UK
T +44 (0)1923 664000
E enquiries@bre.co.uk
W www.bre.co.uk

Business Lines
The Old Motor House
Underley, Kearstwick
Kirby Lonsdale
Cumbria LA6 2DY, UK
T +44 (0)15242 71200
E sales@businesslinesltd.com
W www.checkpoint-safety.com

Bute Fabrics Ltd
Barone Road
Rothesay, Isle of Bute
PA20 0DP, UK
T +44 (0)1700 50 37 34
W www.butefabrics.com

Candidus Prugger SaS
Via Johan Kravogl 10
39040 Bressanone BZ, Italy
T +39 (0)472 834530
E prugger.candidus@dnet.it
W www.bendywood.com

Canon Deutschland GmbH
Europark Fichtenhain A10
47807 Krefeld, Germany
W www.canon.de

Cappellini SpA
Via Marconi 35
22060 Arosio, Italy
T +39 (0)31 759 111
E cappellini@cappellini.it
W www.cappellini.it

Cargill Inc.
P.O. Box 9300
Minneapolis, MN 55440, USA
T +1 952 742 7575
W www.cargill.com

Ceccotti Collezioni
P.O. Box 138
Viale Sicilia 4/a
56021 Cascina, Pisa, Italy
T +39 (0)50 701 955
E info@ceccotti.it
W www.ceccotti.it

Centriforce Products Ltd
14–16 Derby Road
Liverpool L20 8EE, UK
T +44 (0)151 207 8109
E sales@centriforce.co.uk
W www.centriforce.com

ClassiCon
Sigmund-Riefler-Bogen 5
81829 Munich, Germany
T +49 (0)89 74 81 330
E info@classicon.com
W www.classicon.com

Clivus Multrum
15 Union Street
Lawrence, MA 01840, USA
T +1 800 425 4887
E forinfo@clivusmultrum.com
W www.clivusmultrum.com

Comatelec Schréder Group GiE
T +33 (0)148 161788
W www.schreder.com

Crane & Company
30 South Street
Dalton, MA 01226, USA
T +1 800 268 2281
E customerservice@crane.com
W www.crane.com

CuteCircuit LLC
1240 North Shore Drive
Stockholm, ME 04783, USA
E cute@cutecircuit.com
W www.cutecircuit.com

Daimler AG
W www.daimler.com
www.smart.com

Dalsouple
Showground Road
Bridgwater
Somerset TA6 6AJ, UK
T +44 (0)1278 727777
E info@dalsouple.com
W www.dalsouple.com

David design
Skeppsbron 3
211 21 Malmö, Sweden
T +46 (0)40 300 000
E info@daviddesign.se
W www.daviddesign.se

De Vecchi
Via Lombardini 20
20143 Milan, Italy
E info@devecchi.com
W www.devecchi.com

Devine Color
668 McVey #81
Lake Oswego, OR 97034
USA
T +1 503 387 5840
E contact@devinecolor.com
W www.devinecolor.com

Deutsche Bahn AG
W www.deutschebahn.com
www.bahn.de/international/view/en

DeWeNe Ltd.
c/o Humphries Kirk
40 High West Street
Dorchester DT1 1UR, UK
T +44 (0)845 6344 088
E info@DeWeNe.com
W www.dewene.com

Disc-O-Bed GmbH
2150 Northmont Parkway
Suite A
Duluth, GA 30096, USA
T +1 770 295 2292
E usa@discobed.com
W www.discobedusa.com

Domeau & Pérès
21 rue Voltaire BP 68
92250 La Garenne
Colombes cedex, France
T +33 (0)1 47 60 93 86
E info@domeauperes.com
W www.domeauperes.com

Dr Bronnner's Magic Soaps
P.O. Box 28
Escondido, CA 92033, USA
T +1 760 743 2211
E customers@drbronner.com
W www.drbronner.com

Driade SpA
Via Padana inferiore 12,
29012 Fossadello di Caorso
(PC), Italy
E export@driade.com
W www.driade.com

Droog
Staalstraat 7a/b
1011 JJ Amsterdam
the Netherlands
T +31 (0)20 523 5050
E info@droog.com
W www.droog.com

DuPont
T +1 302 774 1000
E info@dupont.com
W www.dupont.com

Duralay
Interfloor Ltd
Broadway
Haslingden, Rossendale
BB4 4LS, UK
T +44 (0)1706 213 131
E sales@interfloor.com
W www.interfloor.com/duralay.html

Durex
W www.durex.com

Dyson Ltd
Tetbury Hill
Malmesbury, Wiltshire
SN16 0RP, UK
T +44 (0)800 298 0298
E askdyson@dyson.co.uk
W www.dyson.co.uk

Earthshell Corporation
1107 Springfield Road
Lebanon, MO 65536, USA
T +1 866 387 3233
E support@earthshell.com
W www.earthshell.com

Eastman Chemical Company
P.O. Box 431
Kingsport, TN 37662, USA
T +1 423 229 2000
W www.eastman.com

Eco-Boudoir
P.O. Box 57539
London NW6 9FL, UK
T +44(0)207 209 4818
E info@eco-boudoir.com
W www.eco-boudoir.com

EcoDomo LLC
T +1 301 424 7717
E info@ecodomocom
W www.ecodomo.com

Eco Kettle
Product Creation Ltd.
Jasmine House
High Street, Henfield
West Sussex BN5 9HN, UK
T +44 (0)1273 495888
E ecohelpline@productceation.co.uk
W www.ecokettle.com

Eco Timber International
1611 Fourth Street
San Rafael, CA 94901, USA
T +1 415 258 8454
W www.ecotimber.com

ECORE International
715 Fountain Avenue
Lancaster, PA 17601, USA
T +1 866 883 7780
W www.ecoreintl.com

Ecos Organic Paints
Unit 34
Heysham Business Park
Middleton Road
Heysham LA3 3PP, UK
T +44 (0)1524 852 371
E mail@ecospaints.com
W www.ecospaints.com

EcoSynthetix
3900 Collins Road,
Lansing MI 48910, USA
T +1 866 326 7849
E info@ecosythetix.com
W www.ecosynthetix.com

Ecovative Design LLC
1223 Peoples Avenue
Troy, NY 12180, USA
T +1 518 690 0399
E info@evocativedesign.com
W www.ecovativedesign.com

Ecover
W www.ecover.com

Ecozone (UK) Ltd
P.O. Box 59165
London NW2 9HG, UK
T +44 (0) 845 230 4200
E customer@ecozone.co.uk
W www.ecozone.co.uk

edra SpA
P.O. Box 28
Perignano (PI), Italy
T +39 (0)587 616660
E edra@edra.com
W www.edra.com

Ejector Systems GmbH
Hellerweg 180
32052 Herford, Germany
T +49 (0)5221 76840
E info@ejector.de
W www.ejector.de

Eleksen Ltd
Old Repeater Station
Brompton-on-Swale
North Yorkshire DL10 7JH
UK
T +44 (0)870 0727 272
E incoming@eleksen.com
W www.electrotextiles.com

Elephant Dung Paper
11 Caroline Close
Kingskerswell
Newton Abbot
Devon TQ12 5, UKJ
E sales@greenelephant.org.uk
W www.greenelephant.org.uk
www.rainbowgifts-usa.com
www.elephantdungpaper.com

El Naturalista
Camino de Labiano 30
31192 Mutilva Alta, Spain
T +34 948 85 27 67
E info@elnaturalista.com
W www.elnaturalista.com

Emeco
805 Elm Avenue
Hanover, PA 17331, USA
T +1 800 366 5951
E info@emeco.net
W www.emeco.net

Entwicklungsgesellschaft für Verbundmaterial Diez mbH
Industriestrasse 17–25
65582 Diez, Germany
T +49 (0)64 32 10 61
E info@evd-diez.de
W www.tectan.de

EnviroGLAS
7704 San Jacinto Place
Suite 200
Plano, TX 75024, USA
T +1 972 608 3790
E communications@enviroglasproducts.com
W www.enviroglasproducts.com

Environ Biocomposites
221 Mohr Drive
Mankato, MN 56001, USA
T +1 507 388 3434
E sales@environbiocomposites.com
W www.environbiocomposites.com

EnviroTextiles LLC
T +1 970 945 5986
E info@envirotextile.com
W www.envirotextile.com

Ernst Schausberger & Co. GmbH
Heidestrasse 19
4623 Gunskirchen, Austria
T + 43 (0)7246 6493 0
E office@schausberger.com
W www.schausberger.com
www.disklev.com

Escofet 1886 SA
Ronda Universitat 20
08007 Barcelona, Spain
T +34 (0)93 318 5050
E informacion@escofet.com
W www.escofet.com

Excel Industries Limited
Maerdy Industrial Estate
Rhymney
Gwent NP22 5PY, UK
T +44 (0) 1685 845 200
E sales@excelfibre.com
W www.excelfibre.com

Feetz
Apolloweg 339
8239 DC Lelystad
the Netherlands
T +31 (0)320-257315
E info@feetz.nl
W www.feetz.nl

Festo & Co.
Corporate Design
KC-C1, Rechbergstrasse 3
73770 Denkendorf
Germany
T +49 (0)711 347 3886
W www.festo.com

Fiam Italia SpA
Via Ancona 1/b
61010 Tavullia
Pesoro, Italy
T +39 (0)721 200 51
E info@fiamitalia.it
W www.fiamitalia.it

Fiat
W www.fiat.com

Filsol Solar Ltd
15 Ponthenri Industrial Estate
Ponthenri
Llanelli, Carmarthenshire
SA15 5RA, UK
T +44 (0)1269 860 229
E info@filsolsolar.com
W www.filsolsolar.com

Fingermax GmbH
Lindwurmstrasse 99
80337 Munich, Germany
T/F +49 (0)89 267417
E klaus-peter-frank@gmx.de
W www.fingermax.de

Fisher Space Pen Co.
711 Yucca Street
Boulder City, NV 89005
USA
T +1 800 634 3494
E fisher@spacepen.com
W www.spacepen.com

Fiskars Corporation
P.O.Box 235
(Mannerheimintie 14 A)
00101 Helsinki, Finland
T +358 (0)19 277721
F +358 (0)19 230986
W www.fiskars.fi

Fiskeby Board AB
601 02 Norrköping
Sweden
T +46 11 15 57 00
E info@fiskeby.comT
W www.fiskeby.com

FlexForm Technologies
4955 Beck Drive
Elkhart, IN 46516, USA
T +1 586 598 7880
W www.flexformtech.com

Flint Ink
26b, Boulevard Royal
2449 Luxembourg
E info@flintgrp.com
W www.flintgrp.com

Flow Control Water Conservation Ltd
Conservation House
Brighton Street
Wallasey
Merseyside CH44 6QJ, UK

Ford Motor Company
W www.ford.com

FP International
1090 Mills Way
Redwood City, CA 94063
USA
T +1 650 261 5300
W www.fpintl.com

Franmar Chemical, Inc.
P.O. Box 5565
Bloomington, IL 61702
USA
T +1 800 538 5069
E franmar@franmar.com
W www.franmar.com

Fredericia Furniture A/S
Treldevej 183
7000 Fredericia, Denmark
T +45 (0)75 92 33 44
E sales@fredericia.com
W www.fredericia.com

Freeplay Energy
71 Gloucester Place
London W1U 8JW, UK
T +44 (0)20 7935 5226
E info@freeplayenergy.com
W www.freeplayenergy.com

Fritz Becker KG
Am Königsfeld 15
33034 Brakel, Germany
T +49 (0)5272 6009 0
E bernd.schulte@becker-kg.de
W www.becker-kg.de

Fritz Hansen A/S
Allerødvej 8
3450 Allerød, Denmark
T +45 48 17 23 00
W www.fritzhansen.com

Front Corporation
3-13-1 Takadanokata
Shinjuku-ku 169
Tokyo, Japan
T +81 (0)3 3360 3391

Fujitsu Technology Solutions GmbH
Germany
T +49 (0)89 62060-0
W ts.fujitsu.com

G. T. Design
Via del Barroccio 14/A
40138 Bologna, Italy
T +39 (0)51 535 951
E info@gtdesign.it
W www.gtdesign.it

General Electric
W www.ge.com

General Motors
W www.gm.com

Gervasoni SpA
Viale del Lavoro, 88
33050 Pavia di Udine (UD)
Italy
T +39 (0)432 656 611
E ino@gervasoni1882.com
W www.gervasoni1882.com

Gessner AG
Florhofstrasse 13
8820 Wädenswil, Switzerland
T +41 (0)44 789 86 00
E fabrics@gessner.ch
W www.climatex.com

Gibson Guitars
309 Plus Park Boulevard
Nashville, TN 37217, USA
T +1 800 444 2766
E service@gibson.com
W www.gibson.com

The Gilman Brothers Company
Gilman Road
P.O. Box 38
Gilman, CT 06336
T +1 800 852 4220
E sales@gilmanbrothers.com
W www.gilmanbrothers.com

Glas Platz
Auf den Pühlen 5
51674 Wiehl-Bomig
Germany
T +49 (0)2261 7890 0
E info@glas-platz.de
W www.glas-platz.de

GLASSX AG
Technoparkstrasse 1
8005 Zurich, Switzerland
T +41 (0)44 445 17 40
E info@glassx.ch
W www.glassx.ch

Glindower Ziegelei GmbH
Alpenstrasse 47
14542 Glindow, Germany
T +49 (0)3327 66490
E info@ziegelmanufaktur.com
W www.ziegelmanufaktur.com

Goods
218 Prinsengracht
1016 HD Amsterdam
the Netherlands
T +31 (0)20 625 8405
E goods@goods.nl
W www.goods.nl

Goodwin Heart Pine Company
106 SW 109th Place
Micanopy, FL 32667, USA
T +1 800 336 3118
E Goodwin@heartpine.com
W www.heartpine.com

Govaerts Recycling NV
Industriepark Kolmen
Kolmenstraat 1324
3570 Alken, Belgium
T +32 (0)11 59 01 60
E info@govaertsrecycling.com
W www.govaertsrecycling.com

Green China Textiles
Hemp Textiles
International
Bellingham, WA 98225
USA
T +1 360 650 1684
E yitzac@comcast.net
W www.greenchinagroup.com

Green Field Paper Company
7196 Clairemont Mesa Blvd
San Diego, CA 92111, USA
T +1 888 402 9979
W www.greenfieldpaper.com

Greenlight Surfboard Supply
2060 Bennett Rd
Philadelphia, PA 19116
T +1 215 805-1506
E info@
greenlightsurfsupply.com
W www.greenlightsurfsupply.com

GUBI A/S
Frihavnen
Klubiensvej 7-9, Pakhus 53
2100 Copenhagen
Denmark
T +45 (0)33 32 63 68
E gubi@gubi.dk
W www.gubi.com

Hahn Kunststoffe GmbH
Gebäude 1027
55483 Hahn-Flughafen
Germany
T +49 (0)6543 9886 0
E info@hahnkunststoffe.de
W www.hahnkunststoffe.de

Hans Grohe AG
Auestrasse 5–9
77761 Schiltach, Germany
T +49 (0)7836 51 0
W www.hansgrohe.com

Hanson Building Systems
Birchwood Way
Cotes Park Industrial Estate
Somercotes, Derby
DE55 4NH, UK
T +44 (0)1773 602432
E mmc@hanson.biz
W www.heidelbergcement.com/uk/en/hanson

Haworth, Inc.
One Haworth Center
Holland, MI 49423, USA
T +1 616 393 3000
W www.haworth.com

Heller S.r.l.
Corso Matteotti, 1
20121 Milan, Italy
T +39 02 76316553
E info@helleronline.it
W www.helleronline.com

Hemp Traders
11301 West Olympic Blvd.
Suite 121-514
Los Angeles, CA 90064
USA
T +1 310 637 3333
E contact@hemptraders.com
W www.hemptraders.com

Heradesign
A Business Unit of Knauf
Insulation GmbH
Ferndorf 29
9702 Ferndorf, Austria
E office@heradesign.at
W www.heradesign.at

Herman Miller Inc.
855 East Main Ave.
P.O. Box 302
Zeeland, MI 49464, USA
T +1 616 654 3000
W www.hermanmiller.com

Hock & Co. KG
Industriestrasse 2
86720 Nördlingen
Germany
T +49 (0)9081 80500-0
E info@thermo-hanf.de
W www.thermo-hanf.de

Honda
W www.honda.com

Howies Ltd
Bath House Road
Cardigan SA43 1JY, UK
T +44(0)1239 61 41 22
E info@howies.co.uk
W www.howies.co.uk

Hülsta GmbH & Co. KG
Karl-Hüls-Strasse 1
48703 Stadtlohn, Germany
T +49 (0)2563 86-0
E huelsta@huelsta.de
W www.huelsta.de

Hunton Fiber AS
PB 235
1372 Asker, Norway
T +44 (0)1933 68 26 83
E hunton@hunton.no
W hunton.sircon.net

Huopaliike Lahtinen Ay
Partalantie 267
42100 Jämsä, Finland
T +358 14 768 017
W www.huopaliikelahtinen.fi

Husqvarna AB
Box 30224
104 25 Stockholm, Sweden
T +46(0)36 14 65 00
W www.husqvarna.com

Icebreaker New Zealand Ltd
Level 2, Hope Gibbons
Building
7–11 Dixon Street
Wellington 6011
New Zealand
T +64 4 385 9113
E cs.nz@icebreaker.com
W www.icebreaker.com

Ifö Sanitär AB
Box 140,
295 22 Bromölla, Sweden
T +46 (0)456 480 00
E info@ifo.se
W www.ifo.se

Iform
Sundspromenaden 27
Box 5055
200 71 Malmö, Sweden
T +46 (0)40 30 36 10
E info@iform.net
W www.iform.net

IKEA
W www.ikea.com

Indigenous Designs
2250 Apollo Way, Suite 400
Santa Rosa, CA, 95407
USA
T +1 707 571 7811
E info@indigenousdesigns.com
W www.indigenousdesigns.com

Induced Energy Ltd
Westminster Road,
Brackley, NN13 7EB, UK
T +44 (0)1280 705 900
W www.inducedenergy.com

Inka Paletten GmbH
Bahnhofstrasse 21
85635 Siegertsbrunn / Munich, Germany
T +49 (0)81 02 77 42 0
E info@inka-paletten.de
W www.inka-paletten.de

Innovia Films Inc.
290 Interstate North Cir SE
Suite 100
Atlanta, GA 30339, USA
T +1 877 822 3456
E kristin.cromie@innoviafilms.com
W www.innoviafilms.com

Interface, Inc.
2859 Paces Ferry Road
Suite 2000
Atlanta, GA 30339, USA
T +1 770 437 6800
W www.interfaceglobal.com

Interlübke Gebr. Lübke GmbH & Co. KG
Ringstrasse 145
33373 Rheda-Wiedenbrück
Germany
T +49 (0)5242 121
E info@interluebke.de
W www.interluebke.com

International Food Container Organization (IFCO) Systems GmbH
Zugspitzstrasse 7
82049 Pullach, Germany
T +49 (0)89 744 91 0
E info@ifcosystems.de
W www.ifcosystems.de

Interstuhl Büromöbel GmbH & Co. KG
Brühlstrasse 21
72469 Messstetten-Tieringen, Germany
T +49 (0) 7436 871
E info@interstuhl.de
W www.interstuhl.de

INX International Ink Co.
150 North Martingale Road
Suite 700
Schaumburg, IL 60173, USA
T +1 800 631 7956
E info@inxintl.com
W www.inxinternational.com

Kartell
Via delle Industrie 1
20082 Noviglio (MI), Italy
T +39 (0)2 90012 1
E kartell@kartell.it
W www.kartell.com

Kaysersberg Plastics
Madleaze Industrial Estate
Bristol Road
Gloucester, GL1 5SG, UK
T +44 (0)1452 316 500
W www.kaysersberg-plastics.com

Keim Mineral Paints Ltd
Muckley Cross, Morville
Near Bridgnorth
Shropshire WV16 4RR, UK
T +44 (0)1746 714 543
W www.keimpaints.co.uk

Kelly Kettle Company
Newtown Cloghans
Knockmore PO
Ballina
County Mayo, Ireland
T +353 87 2864321
W www.kellykettle.com

Kirei
412 N. Cedros Ave
Solana Beach, CA 92075
USA
T +1 619 236 9924
E info@kireiusa.com
W www.kireiusa.com

Kopf Solarschiff GmbH
Meboldstrasse 12
72172 Sulz-Kastell, Germany
T +49 (0)7454 94490-12
E info@kopf-solarschiff.de
W www.kopf-solarschiff.de

Kronospan Schweiz AG
Dekorative Holzwerkstoffe
Willisauerstrasse 37
6122 Menznau, Switzerland
T +41 (0)41 494 94 94
E kronospan@kronospan.ch
W www.kronospan.ch

KTM Industries, Inc.
3327 Ranger Road
Lansing, MI 48906, USA
T +1 877 938 6738
E info@ktmindustries.com
W www.ktmindustries.com

Kvadrat A/S
Lundbergsvej 10
8400 Ebeltoft, Denmark
T +45 (0) 8953 1866
E kvadrat@kvadrat.dk
W www.kvadrat.dk

Lammhults Möbel AB
Box 26
Växjövägen 41
360 30 Lammhult, Sweden
T +46 (0)472269500
E info@lammhults.se
W www.lammhults.se

LCDA (La Chanvrière de l'Aube)
Rue Général de Gaulle
B.P. 602
10208 Bar sur Aube Cedex
France
T +33 (0)3 25 92 31 92
W www.chanvre.oxatis.com

Leclanché SA
Avenue des Sports 42
1401 Yverdon-les-Bains
Switzerland
T +41 (0)24 424 6500
E contact@leclanche.ch
W www.leclanche.ch

LEDtronics
23105 Kashiwa Court
Torrance, CA 90505, USA
T +1 800 579 4875
E info@ledtronics.com
W www.ledtronics.com

Lenzing Fibers AG
Werkstrasse 2
4860 Lenzing, Austria
E office@lenzing.com
W www.lenzing.com

Lexon Design Concepts
6, rue Louis Pasteur
92100 Boulogne
Billancourt France
T +33 (0)1 41 10 20 00
E contact@lexon-design.com
W www.lexon-design.com

Light Corporation
14800 172nd Avenue
Grand Haven, MI 49417
USA
T +1 800 544 4899
E info@lightcorp.com
W www.lightcorp.com

Light Projects Ltd
23 Jacob Street
London SE1 2BG, UK
T +44 (0)20 7231 8282
E info@lightprojects.co.uk
W www.lightprojects.co.uk

LINPAC Environmental
Leafield Way
Leafield Industrial Estate
Corsham
Wiltshire SN13 9UD, UK
T +44 (0)1225 816 500
E envinfo@linpac.com
W www.linpac-environmental.com

Lino Codato Collection
Via Tiziano, 19
31021 Mogliano Veneto (TV), Italy
T +39 (0)41 5970067
E hc@hc-homecollection.com
W www.lccitalia.it

Living Tree Paper Company
1430 Willamette Street
Suite 367
Eugene, OR 97401, USA
T +1 800 309 2974
E info@livingtreepaper.com
W www.livingtreepaper.com

Livos Pflanzenchemie GmbH & Co. KG
Auengrund 10
29568 Wieren, Germany
T +49 (0)58 25 88 0
E info@livos.de
W www.livos.de

Lloyd Loom of Spalding
Wardentree Lane
Pinchbeck
Spalding PE11 3SY, UK
T +44 (0)1775 712 111
E info@lloydloom.com
W www.lloydloom.com

Loftcube GmbH
Maximilianstrasse 35a
80539 Munich, Germany
T +49 (0)89 242 18 111
E info@loftcube.net
W www.loftcube.net

Lotus Cars USA Inc.
2236 Northmont Parkway
Duluth, GA 30096, USA
T +1 800 245 6887
W www.lotuscars.com

LSK Industries Pty Ltd
92 Woodfield Boulevarde
Caringbah
NSW 2229, Australia
T +61 (0)2 9525 8544

Luceplan SpA
Via E.T. Moneta, 40
20161 Milan, Italy
T +39(0) 02 662421
E info@luceplan.com
W www.luceplan.com

Lusty's Lloyd Loom
Geoffery Lusty
Hoo Lane
Chipping Campden
GL55 6AU, UK
T +44 (0)1386 840379
E GeoffreyLusty@aol.com
W www.lloyd-loom.co.uk

Magis SpA
Via Magnadola 15
31045 Motta di Livenza
(TV), Italy
T +39 (0)422 862600
E info@magisdesign.com
W www.magisdesign.com

MAN Nutzfahrzeuge AG
Dachauer Strasse 667
80995 Munich, Germany
T +49 (0)89 15 80 01
E info@man-mn.com
W www.man-mn.com

Marine Current Turbines Ltd
The Court
The Green
Stoke Gifford
Bristol BS34 8PD, UK
T +44 (0)117 979 1888
E sylvie.head@
marineturbines.com
W www.marineturbines.com

Marlec Engineering Company Ltd
Rutland House
Trevithick Road
Corby NN17 5XY, UK
T +44 (0)1536 201 588
E sales@marlec.co.uk
W www.marlec.co.uk

Mathmos
96 Kingsland Road
London E2 8DP, UK
T +44(0)20 7549 2700
E mathmos@mathmos.co.uk
W www.mathmos.com

Meadowood Industries, Inc.
P.O. Box 257
Belmont, CA 94002, USA
E info@
meadowoodindustries.com
W www.meadowood
industries.com

Meld USA
3001-103 Spring Forest
Road
Raleigh, NC 27616, USA
T +1 919 790 1749
E info@meldusa.com
W www.meldusa.com

Mercedes-Benz
W www3.mercedes-benz.com

Metabo AG
Metabo-Allee 1
76622 Nürtingen, Germany
T +49 (0)7022 72 0
E metabo@metabo.de
W www.metabo.de

Metabolix, Inc.
21 Erie Street
Cambridge, MA 02139
USA
T +1 617 492 0505
E info@metabolix.com
W www.metabolix.com

Ming Cycle Industrial Co., Ltd.
No. 50, Lane 462
Guang Shing Road
Taiping City, Taichung
Hsien, Taiwan, R.O.C.
T + 886 4 22713395
E contact@strida.com
W www.strida.com

Monodraught Ltd
Halifax House
Cressex Business Park
High Wycombe, HP12 3SE
UK
T +44 (0)1494 897 700
E info@monodraught.com
W www.monodraught.com

Montis
Postbus 153
5100 AD Dongen
the Netherlands
T +31 (0)162 377 777
E info@montis.nl
W www.montis.nl

Moonlight Aussenleuchten
Via Gabbio 8
6612 Ascona, Switzerland
T +41 (0)91 792 1760
E info@moonlight-swiss.ch
W www.moonlight.info

Moooi
Minervum 7003
P.O. Box 5703
4801 EC Bredav
the Netherlands
T +31 (0)76 578 4444
E info@moooi.com
W www.moooi.com

Moroso SpA
Via Nazionale 60
33010 Cavalicco
Udine, Italy
T +39 (0)432 577111
E www.moroso.it

Moixa Energy Ltd
AP09 10–11 Archer Street
London W1D 7AZ, UK
T +44 (0)207 734 1511
E info@moixaenergy.com
W www.moixaenergy.com

MUJI (Ryohin Keikaku Co., Ltd)
W www.muji.com

Nani Marquina
Església 10, 3er D
08024 Barcelona, Spain
T +34 932 376 465
E info@nanimarquina.com
W www.nanimarquina.com

Natural Collection
Dept 7306
Sunderland SR9 9XZ, UK
T +44 (0)845 3677 001
E info@naturalcollection.com
W www.naturalcollection.com

NatureWorks LLC
15305 Minnetonka
Boulevard
Minnetonka, MN 55345
USA
T +1 952 742 057
E mary_rosenthal
@natureworksllc.com
W www.natureworksllc.com

NEC Corporation
7-1, Shiba 5-chome
Minato-ku, Tokyo 108-8001
Japan
W www.nec.com

New Leaf Paper
116 New Montgomery
Street
Suite 830
San Francisco, CA 94105
USA
T +1 415 291 9210
E info@newleafpaper.com
W www.newleafpaper.com

Next Home Collection e.K.
Amsterdamer Strasse
145–147
50735 Cologne, Germany
T +49-(0)221 71505 0
E info@next.de
W www.next.de

Nigel's Eco Store
T +44 (0)800 288 8970
W www.nigelsecostore.com

Nighteye GmbH
c/o PROFORM Design
(p. 309)

Nokia
W www.nokia.com

Nora Systems GmbH
Höhnerweg 2–4
69469 Weinheim, Germany
T +49 6201 80 6633
E birgitta.knoch@nora.com
W www.nora.com

Nordsjö (Akzo Nobel Dekorativ)
205 17 Malmö, Sweden
T +46 (0)66 81 94 00
E norge@akzonobel.com
W www.nordsjo.no

Novotex A/S
Ellehammervej 8
7430 Ikast, Denmark
T +45 (0)70 70 25 02
E info@green-cotton.dk
W www.green-cotton.dk

Nuno Corporation
B1F Axis Building
5-17-1 Roppongi
Minato-ku, Tokyo
106-0032, Japan
T +81 (0)3 3582 7997
E info@nuno.com
W www.nuno.com

Nutshell Natural Paints Ltd.
Unit 3, Leigham Units
Silverton Road
Matford Park
Exeter EX2 8HY, UK
T +44 (0)1392 823760
E enquiries@nutshellpaints.co.uk
W www.nutshellpaints.co.uk

OFFECCT AB
Box 100
543 21 Tibro, Sweden
T +46 (0)504 415 00
E support@offecct.se
W www.offecct.se

The Old Fashioned Milk Paint Co., Inc
436 Main Street
Groton, MA 01450, USA
T +1 978 448 6336
W www.milkpaint.com

Omlet Ltd
Tuthill Park
Wardington OX17 1RR, UK
T +44(0)845 450 2056
E info@omlet.co.uk
W www.omlet.co.uk

One Laptop Per Child, OLPC
P.O. Box 425087
Cambridge, MA 02142
USA
T +1 617 452 5663
E information@laptop.org
W laptop.org/en

Optare Group Ltd
Manston Lane
Leeds, LS15 8SU, UK
T +44 (0)113 264 5182
E info@optare.com
W www.optare.com

OSMO
Unit 24 Anglo Business Park
Smeaton Close
Aylesbury HP19 8UP, UK
T +44 (0)1296 481220
E info@osmouk.com
W www.osmouk.com

Osprey Packs
115 Progress Circle
Cortez, CO 81321, USA
T +1 866 284 7830
E info@ospreypacks.com
W www.ospreypacks.com

Oxfam
W www.oxfam.org.uk

Pace Industries, Inc.
1400 Industrial Street
Reedsburg, WI 53959, USA
T +1 800 524 6777
E service@pace-industries-inc.com
W www.pace-industries-inc.com

Pactiv Corporation
1900 West Field Court
Lake Forest, IL 60045, USA
T +1 888 828 2850
W www.pactiv.com

Pallucco Italia SpA
Via Azzi 36
31040 Castagnole de Paese (TV), Italy
T +39 (0)422 438800
E infopallucco@palluccobellato.it
W www.palluccobellato.it

Panda Flooring Co.
1 Grange Park
Thurnby LE7 9QQ, UK
T +44 (0)116 241 4816
E david@pandaflooring.co.uk
W www.pandaflooring.co.uk

Paneltech International LLC
2999 John Stevens Way
Hoquiam, WA 98550, USA
T +1 360 538 9815
E joan@paperstoneproducts.net
W www.paperstoneproducts.com

Paperback
Unit 2
Bow Triangle Business Centre
Eleanor Street
London E3 4NP, UK
T +44 (0)20 8980 5580
E sales@paperback.coop
W www.paperback.coop

Paradigma Deutschland GmbH
Ettlinger Strasse 30
76307 Karlsbad, Germany
T +49 (0)7202 922 0
E info@paradigma.de
W www.paradigma.de

Patagonia
T +1 775 747 1992
W www.patagonia.com

PCD Maltron Ltd
Castlefields
Stafford ST16 1BU, UK
T +44 (0)845 230 3265
E sales@maltron.co.uk
W www.maltron.com

The Peabody Group
45 Westminster Bridge Road
London, SE1 7JB, UK
T +44 (0)20 7021 4000
W www.peabody.org.uk

Pedalite International Ltd.
Hamilton House
111 Marlowes
Hemel Hempstead
HP1 1BB, UK
T +44 (0)1442 450 483
E customerhelp@pedalite.com
W www.pedalite.com

Philips Lighting
Royal Philips Electronics
Amstelplein 2
Breitner Center
P.O. Box 77900
1070 MX Amsterdam
the Netherlands
T +31 (0)20 59 77777
W www.lighting.philips.com

Pickering International
888 Post Street
San Francisco, CA 94109
USA
T +1 415 474 2288
E contact@picknatural.com
W www.pickhemp.com

Pinturas Proa
San Salvador de Budiño
Gánderas de Prado
36475 Porriño, Spain
T +34 (0)986 34 6525
E admonproa@pinturasproa.com
W www.pinturasproa.com

Plantic Technologies Limited
51 Burns Road
Altona
3018, Victoria, Australia
T +61 3 9353 7900
E info@plantic.com.au
W www.plantic.com.au

Playtop Ltd
Wesley House, Brunel Drive
Newark, Nottinghamshire
NG24 2EG, UK
E sales@playtop.co.uk
W www.playtop.co.uk

Plexwood
Bijleveldstraat 6
3521 EL Utrecht
the Netherlands
T +31 (0)30 252 10 14
E info@plexwood.com
W www.plexwood.nl

Pli Design Ltd.
Unit D 15
62 Tritton Road
London SE21 8DE, UK
T +44(0) 20 8670 6857
W www.plidesign.co.uk

Polti SpA
Via Ferloni 83
22070 Bulgarograsso (CO)
Italy
T +39 (0)31 939 111
E polti@polti.com
W www.polti.it

Porous Pipe Ltd
Standroyd Mill
Cottontree, Colne
Lancashire BB8 7BW, UK
T +64 (0)3 572 8977
W www.porouspipe.co.uk

Potatopak NZ Ltd
P O Box 746
Blenheim 7240
New Zealand
T +44 (0)1963 362744
E richard@potatoplates.com
W www.potatopak.com

Powabyke Ltd
3 Wood Street
Queen Square
Bath BA1 2JQ, UK
T +44 (0)1225 443 737
E sales@powabyke.com
W www.powabyke.com

Prairie Forest Products
P.O. Box 279
Neepawa
Manitoba, Canada
T +1 204 476 7700
E info@prairieforest.com
W www.prairieforest.com

Preserve
681 Main Street
Waltham, MA 02451
USA
T +1 888 354 7296
E info@preserveproducts.com
W www.preserveproducts.com

Prismo Travel Products
5 Drumhead Road
Chorley North Industrial Park
Chorley PR6 7BX, UK
T +44 (0)845 121 4455
W prismo.co.uk

Purves & Purves
Unit 7 Mill Farm Business Park, Millfield Road
Hounslow, Middlesex
TW4 5PY, UK
T +44 (0)20 8893 4000
E furniture@purves.co.uk
W www.purves.co.uk

Q Drum (Pty) Ltd
P.O. Box 4099
Pietersburg 0700
South Africa
T +27 (0)15 297 6762
E info@qdrum.co.za
W www.qdrum.co.za

QK Honeycomb Products Ltd
Creeting Road
Stowmarket
Suffolk IP14 5AS, UK
T +44 (0)1449 612 145
E sales@qkhoneycombproducts.co.uk
W www.quintonkaines.co.uk

Quanta Computer Inc.
T www.quanta.com.tw

Radius GmbH
Hamburger Strasse 8a
50321 Brühl, Germany
T +49(0)2232 7636 0
E info@radius-design.de
W www.radius-design.de

RainTube Technologies
215 S 4th Street
Jacksonville, OR 97530, USA
T +1 866 724 6356
E info@raintube.com
W www.raintube.com

Recovery Insulation Ltd
3 Mowbray Street
Unit 12
Sheffield S3 8EN, UK
T +44 (0) 114 2499459
E info@recovery-insulation.co.uk
W www.recovery-insulation.co.uk

Remarkable (Pencils) Ltd
The Remarkable Factory
Midland Road
Worcester, WR5 1DS, UK
T +44 (0)20 8741 1234
E info@remarkable.co.uk
W www.remarkable.co.uk

Re-New Wood
104 N. W. 8th
P.O. Box 1093
Wagoner, OK 74467, USA
T +1 800 420 7576
W www.renewwood.com

Renfe
Avenida Pio XXI, 110
28036 Madrid, Spain
T+34 (0)91 300 62 58
E prensa@renfre.es
W www.renfe.es

Rent-a-Green Box
3505 Cadillac Avenue
Building F-9
Costa Mesa, CA 92626
USA
T +1 888 900 7225
W www.rentagreenbox.com

Rethink Games
The Hub
5 Torrens Street
Islington
London EC1V 1NQ, UK
T +44 (0)78 0075 5269
E info@rethinkgames.com
W www.playrethink.com

Rexite SpA
Via Edison 7
20090 Cusago (Milan)
Italy
T +39 (0)29 039 0013
W www.rexite.it

riese und müller GmbH
Haasstrasse 6
64293 Darmstadt
Germany
T +49 (0)6151 366 86 0
E team@r-m.de
W www.r-m.de

Robert Cullen & Sons
10 Dawsholm Avenue
Glasgow G20 0TS, UK
T +44 (0)141 945 2222
E sales@cullen.co.uk
W www.cullen.co.uk

Rodman Industries
P.O. Box 88
Oconomowoc, WI 53066
USA
T +1 262 569 5820
E info@rodmanindustries.com
W www.rodmanindustries.com

Rolls-Royce International Limited
65 Buckingham Gate
London SW1E 6AT, UK
T +44 (0)20 7222 9020
W www.rolls-royce.com

Röthlisberger
Sägeweg 11
3073 Gümligen, Switzerland
T +41 (0)31 950 21 40
E kollektion@roethlisberger.ch
W www.roethlisberger.ch

Sanford
2707 Butterfield Road
Oak Brook, IL 60523, USA
T +1 800 323 0749
E mail@sanford.com
W www.sanford.com

Save A Cup Recycling Company
Suite 2, Bridge House
Bridge Street
High Wycombe, HP11 2EL
UK
T +44 (0)1494 510167
E info@save-a-cup.co.uk
W www.save-a-cup.co.uk

Schäfer Werke GmbH
Pfannenbergstrasse 1
57290 Neunkirchen
Germany
T +49 (0)2735 787 01
E info@schaefer-werke.de
W www.schaefer-werke.de

Scherzer & Co. AG
Friesenstrasse 50
50670 Cologne, Germany
E info@scherzer-ag.de
W www.scherzer-ag.de

Schiebel Elektronische Geräte GmbH
Margaretenstrasse 112
1050 Vienna, Austria
T +43 (0)1 546 26 0
E info@schiebel.net
W www.schiebel.net

Schmitz Foam Products BV
P.O. Box 1277
(Prodktieweg 6)
6040 KG Roermond
the Netherlands
T +31 (0)475 370 270
E sales@schmitzfoam.com
W www.schmitzfoam.com

SEAT
W www.seat.com

Second Nature UK Ltd
Soulands Gate
Dacre, Penrith
Cumbria CA11 0JF, UK
T +44 (0)17684 86285
E info@secondnatureuk.com
W www.secondnatureuk.com

Segway Inc.
14 Technology Drive
Bedford, NH 03110, USA
T +1 603 222 6000
W www.segway.com

Serralunga
Via Serralunga,9
13900 Biella (BI), Italy
T +39 (0)15 2435711
E info@serralunga.com
W www.serralunga.it

ShelterWorks Ltd.
P.O. Box 1311
Philomath, OR 97370, USA
T +1 541 929 8010
W www.faswall.com

ShetkaWorks L.L.C.
435 West Industrial Street
P.O. Box 38
Le Center, MN 56057, USA
T +1 507 357 4177
W www.shetkastone.com

Sherburne, Annie
179A Goldhurst Terrace
London NW6 3ER, UK
T +44 (0) 207 328 2182
E info@anniesherburne.co.uk
W www.anniesherburne.co.uk

Smile Plastics
Mansion House
Ford, Shrewsbury, SY5 9LZ
UK
T +44 (0)1704 509888
E sales@smile-plastics.co.uk
W www.smile-plastics.co.uk

Smith Anderson Group Ltd.
Well Brae
Falkland, Fife KY15 7AY, UK
T +44 (0)1337 855555
W www.smithanderson
packaging.co.uk

Smith & Fong Company
475 Sixth Street
San Francisco, CA 94103
USA
T +1 866 835 9859
E info@plyboo.com
W www.plyboo.com

Solanyl Biopolymers Inc.
Box 1119
Carberry, Manitoba
ROK OHO, Canada
T +1 204 834 3500
E info@Solanyl.ca
W www.solanyl.ca

Solarcentury
91–94 Lower Marsh
London SE1 7AB , UK
T +44 (0)20 7803 0100
E enquiries@solarcentury.com
W www.solarcentury.co.uk

Solar Sailor Holdings Ltd
The Bentleigh, Suite 206
1 Katherine Street
Chatswood NSW 2067
Australia
T +61 2 9924 6400
E admin@solarsailor.com.au
W www.solarsailor.com

Sound Service (Oxford)
Crawley Mill
Witney OX29 9TJ, UK
T +44 (0)845 363 7131
E tech@soundservice.co.uk
W www.soundservice.co.uk

SoySafe Products, Inc.
20 N. Lincoln Street
Batavia, IL 60510, USA
T +1 866 359 9401
E info@soysafe.com
W www.soysafe.com

Soy Works Corporation
T +1 630 853 4327
E soyworks@msn.com
W www.soyworks
corporation.com

SRAM Corporation
1333 N. Kingsbury
4th Floor
Chicago, IL 60622, USA
T +1 312 664 8800
W www.sram.com

Staber Industries, Inc.
4800 Homer Ohio Lane
Groveport, OH 43125, USA
T +1 614 836 5995
E sales@staber.com
W www.staber.com

StarchTech, Inc.
720 Florida Avenue
Mineapolis, MN 55426
USA
T +1 800 597 7225
E sti@startech.com
W www.starchtech.com

Steelcase, Inc.
T +44 (0)207 421 9000
E info_uk@steelcase.com
W www.steelcase.com

Stokke AS
Håhjem
6260 Skodje, Norway
T +47 (0)70 24 49 00
E info@stokke.com
W www.stokke.com

Stomatex Ltd
Moorfoot
Bathpool
Cornwall PL15 7NW, UK
T +44 (0)1579 380173
E info@stomatex.com
W www.stomatex.com

Stora Enso International Office
Level 2 – West Wing
1 Sheldon Square
London W2 6TT, UK
T +44 20 7121 0880
W www.storaenso.com

Suck UK
31 Regent Studios
8 Andrews Rd
London E8 4QN, UK
T +44 (0)20 7923 0011
E Joanna@suck.uk.com
W www.suck.uk.com

Sundeala
Middle Mill
Cam
Dursley GL11 5LQ, UK
T +44 (0)1453 540 900
E sales@sundeala.co.uk
W www.sundeala.co.uk

Sun-Mar Corporation
600 Main Street
Tonawanda, NY 14150 USA
T +1 905 332 1314
E compost@sun-mar.com
W www.sun-mar.com

Sunways AG
Macairestrasse 3–5
78467 Konstanz, Germany
T +49 (0)7531 99677 0
E info@sunways.eu
W www.sunways.eu

SympaTex Technologies GmbH
Feringastrasse 7A
85744 Unterföhring
Germany
T +49(0)89 940 058 0
E info@sympatex.com
W www.sympatex.com

Syndesis, Inc.
2908 Colorado Avenue
Santa Monica, CA 90403
USA
T +1 310 829 9704
E inquiries@syndecrete.com
W www.syndecrete.com

Tata Motors
W www.tatamotors.com

Teisen Products Ltd
Bradley Green
Redditch B96 6RP, UK
T +44 (0)1527 821621
E heat@farm2000.co.uk
W www.farm2000.co.uk

Teknos Oy
Takkatie 3
P.O. Box 107
00371 Helsinki, Finland
T +358 (0)9 506 091
E sales@teknos.fi
W www.teknos.fi

Tendo Co. Ltd
1-3-10 Midarekawa
Tendo
Yamagata 994-8601
Japan
T +81 (0)23 653 3121
E suga@tendo-mokko.co.jp
W www.tendo-mokko.co.jp

Tensar International
Cunningham Court
Shadsworth Business Park
Blackburn BB1 2QX, UK
T +44 (0)1254 262 431
E info@tensar.co.uk
W www.tensar.co.uk

Teragren LLC
12715 Miller Road NE
Suite 301
Bainbridge Island
WA 98110, USA
T +1 206 842 9477
E info@teragren.com
W www.teragren.com

Terra Plana
124 Bermondsey Street
London Bridge
London SE1 3TX, UK
T +44 (0)1458 449 081
E info@terraplana.com
W www.terraplana.com

Tesla Motors, Inc.
1050 Bing Street
San Carlos, CA 94070, USA
T +1 650 413 4000
W www.teslamotors.com

Thermo Technologies
9009 Mendenhall Court
Suite E
Columbia, MD 21045, USA
T +1 410 997 0778
E solar@thermotechs.com
W www.thermomax.com

Thonet GmbH
Michael Thonet Strasse 1
35066 Frankenberg
Germany
T +49 (0)6451 508 0
E info@thonet.de
W www.thonet.de

Tierrafino B.V.
Archangelkade 23
1013 BE Amsterdam
the Netherlands
T +31 (0)20 689 25 15
E info@tierrafino.com
W www.tierrafino.com

Timber Holdings Intl.
T +1 414 445 8989
E info@ironwoods.com
W www.ironwoods.com

Titan Wood Ltd
T +44 20 8150 8835c
E marketing@accoya.info
W www.titanwood.com

Tonester
Huhdantie 4
21140 Rymättylä, Finland
T +358 (0)2 252 1000
E tonester@durat.com
W www.durat.com

Tonwerk Lausen AG
Hauptstrasse 74
4415 Lausen, Switzerland
T +41 (0)61 927 95 55
E info@twlag.ch
W www.twlag.ch

Toyota Motor Corporation
W www.toyota.com

Trevira GmbH
Werk Bobingen
Max-Fischer-Strasse 11
86399 Bobingen Germany
T +49 (0)8234 82 0
E trevira.info@trevira.com
W www.trevira.de

Trevor Baylis Brands plc
The Enterprise Centre
West Wing
Spelthorne Civic Offices
Knowle Green
Staines TW18 1XB, UK
T +44(0)5601 290240
E business@
trevorbaylisbrands.com
W www.trevorbaylisbrands.com

TSA Inox
8, rue Jules Py
88210 Moussey, France
T + 33 (0)3 29 42 50 00
E commercial-tsainox
@tsainox.fr
W www.tsa-industries.com

UPM
Eteläesplanadi 2
P.O. Box 380
00101 Helsinki, Finland
T +358 (0)20 415 111
E info@upm-kymmene.com
W www.upm-kymmene.com

Urethane Soy Systems Company
P.O. Box 590
Volga, SD 57071, USA
T +1 888 514 9096
W www.soyoyl.com

U. S. Plastic Lumber
2300 Glades Road
Boca Raton, FL 33431, USA
T 888-733-2546
W www.usplasticlumber.com

Vaccari Ltd
52 Greenway
Crediton
Devon EX17 3LP, UK
T +44 (0)1363 777746
E info@vaccari.co.uk
W www.vaccari.co.uk

Vectrix Corporation
Tech Plaza III
76 Hammarlund Way
Middletown, RI 02842
USA
T +1 401 848 9993
E info@vectrixusa.com
W www.vectrix.com

Veja
E contact@veja.fr
W www.veja.fr

Velcro
406 Brown Avenue
Manchester, NH 03103
USA
T +1 800 225 0180
E marketing@velcro.com
W www.velcro.com

Venturi Automobiles
Gildo Pastor Center
7 rue du Gabian
98000 Monaco
T +377 99 99 52 00
E info@venturi.fr
W www.venturi.fr

Vestfrost A/S
Spangsbjerg Møllevej 100
6705 Esbjerg Ø, Denmark
T +45 (0)79 14 22 22
E info@vestfrost.dk
W www.vestfrost.com

Viceversa
Via Dello Stelli 3
Vallina Bagno a Ripoli
50010 Florence, Italy
T +39 (0)55 692041
E export@viceversa.com.hk
W www.viceversa.com

Victor Group
2805 90th Street
Saint-Georges, QC
G6A 1K1, Canada
T +1 418 227 9897
E info@victor.qc.ca
W www.victor-innovatex.com

Vision Paper
KP Products Inc.
P.O. Box 20399
Albuquerque, NM 87154
USA
T +1 505 294 0293
E info@visionpaper.com
W www.visionpaper.com

Vitra (International) AG
Headquarters
Klünenfeldstrasse 22
4127 Birsfelden
Switzerland
T +41 61 377 00 00
E info@vitra.com
W www.vitra.com

Volkswagen
W www.volkswagen.com

Volvo Articulated Haulers
405 08 Göteborg, Sweden
T +46 (0)31 660 000
W www.volvo.com

Waldemeister Bikes
Industriestrasse 26
79194 Freiburg, Germany
T +49 761 458 7821 0
E info@waldmeister-bikes.de
W www.waldmeister-bikes.de

WaterFilm Energy Inc
P.O. Box 128
Medford, NY 11763, USA
T +1 631 758 6271
E info@gfxtechnology.com
W www.gfxtechnology.com

Well Ausstellungssystem Gmbh
Schwarzer bär 2
30449 Hannover, Germany
T +49 (0)511 92881 10
E info@well.de
W www.well.de/en

Wellman, Inc.
3303 Port and Harbor Drive
Port Bienville Industrial
Park Bay St. Louis, MS
39520, USA
E info@wellmaninc.com
W www.wellmaninc.com

Werth Forsttechnik
Seelbach 5
66687 Wadern, Germany
T +49 (0)68 71 90 90 0
E info@weihnachtsbaum.de
W www.weihnachtsbaum.de

Weyerhaeuser Company
P.O. Box 9777
Federal Way, WA 98063, USA
T +1 800 525 5440
E iLevel@Weyerhaeuser.com
W www.weyerhaeuser.com

Wharington International Pty Ltd
48-50 Hargreaves Street
Huntingdale
3166 Victoria, Australia
T +61 3 9544 5533
E sales@wharington.com.au
W www.wharington.com.au

Wiggly Wigglers (UK)
Lower Blakemere Farm
Blakemere, HR2 9PX, UK
T +44 (0)1981 500 391
E wiggly@wigglywigglers.co.uk
W www.wigglywigglers.co.uk

Wilde & Spieth Designmöbel GmbH & Co.
Röntgenstrasse 1/1
73730 Esslingen, Germany
T +49 (0)711 351 303 0
E info@wilde-spieth.com
W www.wilde-spieth.com

Wilkhahn
Wilkening + Hahne GmbH & Co. KG
Fritz-Hahn-Strasse 8
31844 Bad Münder
Germany
T +49 (0)5042 9990
E info@wilkhahn.de
W www.wilkhahn.com

Willat Writing Instruments
W www.sensa.com

WISA Wood Products
P.O. Box 203
(Niemenkatu 16)
15141 Lahti, Finland
T +35 (8 0)204 15 113
E wood@upm-kymmene.com
W www.wisa.fi

WKR GmbH
Industriegebiet I/6
Entenpfuhl 10
67547 Worms, Germany
T +49 (0)6241 434 51
E info@wkr-gmbh.de
W www.wkr-gmbh.de

Xella International GmbH
Franz-Haniel Platz 6-8
47119 Duisburg, Germany
T +44 (0)800 523 5665
E info@xella.com
W www.xella.de

Xerox Corporation
T +1 800 275 9376
W www.xerox.com

XO
RN 19
77170 Servon, France
T +33 (0)1 60 62 60 60
E info@xo-design.com
W www.xo-design.com

Xootr LLC
2001 Rosanna Avenue
Scranton, PA 18509, USA
T +1 800 816 2724
E info@xootr.com
W www.xootr.com

Yemm & Hart
1417 Madison 308
Marquand, MO 63655, USA
T +1 573 783 5434
E info@yemmhart.com
W www.yemmhart.com

YOLO Colorhouse LLC
3909 NE Martin Luther King Jr Blvd
Suite 201
Portland, OR 97212, USA
T +1 877 493 8276
E info@yolocolorhouse.com
W www.yolocolorhouse.com

Zanotta SpA
Via Vittorio Veneto 57
20054 Nova Milanese, Italy
T +39 (0)0362 4981
E zanottaspa@zanotta.it
W www.zanotta.it

ZEM Europe GmbH
Wohllebgasse 11
Postfach 1252
8034 Zurich, Switzerland
T +41 (0)1 210 4774
E zem@zem.ch
W www.zem.ch

The design strategies described with each product in Objects for Living, Objects for Working and Materials are listed below. They are grouped according to one of five lifecycle phases, although some strategies are applicable to more than one phase. The Lifecycle Phases are: I Pre-production, including materials selection; II Manufacturing/Making/ Fabrication; III Distribution/ Transportation; IV Functionality and use; and V Disposal/ End-of-life/ New life. Other strategies that do not easily fit into this product lifecycle are described under the heading Miscellaneous. Reference should also be made to the Glossary of eco-design terms (p. 338).

I PRE-PRODUCTION PHASE

Adaptable designs – artefacts/spaces that have been designed to ensure they can be easily adapted as future conditions change.

Active citizen participation – designs that encourage citizens to become proactive in dealing with challenging contemporary social, environmental, economic and/or political issues.

Anti-fashion – a design that avoids temporary, fashionable styles.

Anti-obsolescence – a design that is easily repaired, maintained and upgraded so it is not made obsolete with changes in technology or taste.

Appropriate/intermediate technology – designs that apply appropriate levels of technology for the local economic, political, environmental and socio-cultural conditions.

Bespoke or one-off unique designs – imbues a distinct character to the design that differentiates it from those produced by mass manufacturing and so creates a distinct 'value' profile.

Carbon neutral or zero carbon designs – those designs that have a zero net output of the greenhouse gas carbon dioxide throughout their lifecycle.

Classic design – creating a design that will have socio-cultural durability.

Dematerialization – the process of converting products into services. A good example of dematerialization through timeshare of a product is a local community sharing a car 'pool' in which all individuals have the opportunity to use or hire a car when needed rather than own a car that stands idle for a large part of its life. Other examples include digital cameras where silver halide film is replaced by CCD chips, dematerializing part of the consumables cycle. Designing products used in the context of a dematerialized service may place unusual constraints on the design such as concentration on maintenance and longevity of parts.

Durability – design for easily maintained or repaired products to extend useful/ functional life.

Eco-awareness – designs that raise the users' awareness of ecological and/or sociological issues.

Energy efficient and/or energy self-sufficient – products and services to minimize energy consumption during the life of the product/service.

Ergonomic improvements – to encourage better utility, functionality and universal access.

Extending or expanding craft and/or vernacular traditions – design that builds on local narratives and traditions that tread lightly on ecological and social conditions.

Lifecycle Analysis (LCA) – the calculation of the key areas of environmental impact of a product or building and subsequent effort to minimize such impacts by design.

Local economy/employment focused – creating products or buildings that generate local employment and/or nurture the local economy using local materials and resources.

'Halfway' product – a product whose design and making/manufacturing is incomplete until the user has made his/her contribution. The aim is to deepen bonds between the user and the product, as the product is completed by the user.

Design to encourage or mimic natural systems – design that embraces biomimicry/biomimetics, passive house design, natural design, ecological design and more.

Open source design – available for the 'commons' and/or public domain. Made available to ensure that design can be improved through input from the wider community, encouraging innovation.

Personalization – enabling consumers to personalize a product in order to create a longer-lasting emotional bond and extend product life.

Product-service-system (PSS) – creating a service (based upon infrastructure, network and ICT provision) whose products have less environmental impact than individually owned and consumed products.

Product take-back – a system under which manufacturers agree to take back a product when it has reached the end of its useful life so that components and/or materials can be reused or recycled (see also Producer responsibility). This can fundamentally change the essence of the design and engage the designer in examining design for assembly (DfA), disassembly (DfD) and remanufacture.

Raising awareness – of our ecological and sociological conditions.

Reduced reliance on fossil fuels (oil, coal, gas) – because of diminishing world reserves and the phenomenon of 'peak oil'.

Re-humanizing technology – ensuring that technologies are embedded in a way to revitalize the humanistic, emotional and experiential dimensions of life.

Retention of craft skills/hand-making – designs that encourage a diverse skill base.

Reusable product – a product that can be reused at the end of its initial lifespan for an identical, similar or new use.

Service design – design to replace products with services such as 'on demand' usage rather than individuals owning their own products, or Product-Service-Systems

Social design – to increase social conviviality, awareness, interaction, cohesion and regeneration of individuals and communities.

Social networking – designs that encourage the social exchange of ideas, techniques and objects to increase the social capital and capacity of individuals and communities.

Synergistic and symbiotic designs – those that find a synergy or symbiosis with nature and help regenerate local ecology and ecosystems.

Universal design – the application of widely accepted practices, components, fixtures, materials and technologies suitable for a wide range of end-uses.

Use of existing manufacturing capacity – using existing factories and plant with a proven low environmental impact to produce other goods.

PRE-PRODUCTION: MATERIALS SELECTION

100% recycled materials – designs that utilize 100% recycled materials and do not use virgin materials, to conserve (finite) resources and minimize resource depletion.

Abundant materials from the lithosphere/geosphere – inorganic materials, such as stone, clay, minerals and metals from the earth's crust.

Biodegradable – decomposed by the action of microbes such as bacteria and fungi.

Biomass crops– carbon-based biological materials derived from living or recently living organisms

Biopolymers – plastics made from plants. Biopolymers can be composted and returned to nature.

Bio-regional materials – sourced from within the local bio-region to reduce transportation distance and use of other peoples' bio-regional ecological capacity; may increase cash crops for local communities.

Certified sources – materials that are independently certified and labelled as originating from sustainably managed resources, from recycled materials or conforming to a national or international eco-label.

Chemical-free materials – those free of chemicals or chemical compounds such as formaldehyde or VOCs.

Compostable – can be decomposed by microbes such as bacteria and fungi to release nutrients and organic matter.

Durable/extremely durable – tough, strong materials that do not break or wear and survive the life of the product or well beyond.

Eco-labelled materials – raw and processed materials certified to a nationally or internationally recognized eco-label certification scheme.

Fairtrade materials – raw materials generated under production schemes/ facilities endorsed by the Fairtrade Foundation or affiliates, with the intention of supporting local production and communities.

Lightweight – materials with a high strength-to-weight ratio.

Locally sourced materials – those originating from close proximity to the point of manufacturing or production.

Low-embodied-energy materials – materials that require relatively little energy to extract and manufacture.

Material Safety Data Sheets (MSDS) – material specifications are defined in internationally recognized MSDS.

Mono or single material – use of one material for an entire product or component, facilitating end-of-life recycling.

Non-toxic/Non-hazardous – not likely to cause loss of life or ill health to people or degradation of living ecosystems.

Precious materials – materials that can serve to ensure the socio-cultural longevity of a product or building.

Reclaimed – materials saved for reuse on demolition of the built environment.

Recyclable materials – components of products that can be used in a new product.

Recyclate – material that has been made into a new material comprising wholly or partially recycled materials. An alternative term is 'recycled feedstock'.

Recycled – materials that have been processed (such as cleaned, graded, shredded, blended), then remanufactured.

Recycled content – materials that include some recycled and some virgin content. If a material has 100% recycled content, it is a recycled material.

Reduction in materials used – reducing the materials required to deliver the required design functions.

Renewable – a material that can be extracted from resources which absorb energy from the sun to synthesize or create matter. These resources include primary producers, such as plants and bacteria, and secondary producers, such as fish and mammals.

Simplicity and truth in materials – where the materials 'speak' to the user and reveal their origins, connections and essence, thereby adding value to the object.

Stewardship sourcing – materials from certified sources and supply chain management.

Supply-chain management (green procurement) – is the process of specifying that the goods or materials of suppliers meet minimum environmental standards. The specification may be that the goods will come from certified sources (e.g., the Forest Stewardship Council, national or international eco-labels), carry recognized accreditation (e.g. ISO 14001, EMAS) or meet trade association standards (e.g. National Association of Paper Manufacturers' recycled-paper logo in the UK).

Sustainable/from sustainable sources – materials that originate from managed resources which are forecast to last for a very long time and/ or are well managed and/ or certified renewable resources (see above).

Waste materials – materials fabricated from production (factory), agricultural or consumer waste.

II MANUFACTURING/ MAKING/ FABRICATION PHASE

Production processes

Avoidance of toxic/ hazardous substances – avoiding substances liable to damage human health and living ecosystems.

Bespoke and one-off production tailored to the needs of an individual – creating better utility, function and pleasure for the user and likely to extend product life.

Bio-manufacturing – using nature to help fabricate products in situ. For example, 'manufacturing' natural gourds by training them in special shapes for later use as packaging; growing plants to produce biopolymers (natural plastics).

Clean production – systems are put in place to reduce the impact of manufacturing goods by minimizing the production of waste and emissions to land, air and water. Closed-loop recycling (see below) technologies are often incorporated into clean production.

Certified Environmental Management Systems (EMS) – such as ISO 14001 and EMAS, ensure specific standards aimed at minimizing environmental impacts of production systems.

Closed-loop recycling/ production – the process of introducing waste streams back into the manufacturing process in a continuous cycle without loss of waste from that cycle. The textile and chemical industries often recycle chemical compounds used in processing their end-products, resulting in cleaner production.

Cold construction/ manufacturing – methods that require no heat or pressure and hence reduce energy consumption and facilitate disassembly.

Design for assembly (DfA) – is a method of rationalizing and standardizing parts to facilitate the fixing together of components during production or manufacture.

Design for disassembly (DfD) – is a method of designing products to facilitate cost-effective, non-destructive breakdown of the component parts of a product at the end of its life so that they can be recycled or reused.

DIY – Do-it-yourself and design-it-yourself techniques especially suited to downloadable designs distributed via the Internet (see below).

'Downloadable' designs – designs that can be downloaded from the Internet – such as cutting plans, instructional videos – so that users can DIY and assemble their own designs.

Efficient use of raw and manufactured materials – reducing materials used and minimizing waste production.

Existing manufactured components – use of stocks of existing but unused components rather than making new components.

Innovation and revitalization of traditional (low impact) technologies – using inherently low environmental impact, traditional or craft, technologies in an innovative way, e.g. weaving plant fibres for furniture or boat construction.

Lightweight construction – reducing materials used but maintaining strength.

Low-energy manufacturing/ production/construction techniques/assembly – reducing the energy required to make components or products.

Non-toxic – use of natural, bio-degradable or non-toxic materials and finishes.

Reduced resource consumption – reducing materials used, especially raw materials extracted from the environment.

Reduction in embodied energy of materials and construction – considering the production process as an energy flow and trying to reduce the total energy used.

Reduction in use of consumables – reducing consumables used during the manufacturing process.

Reduction of production waste – achieved by more efficient designs and/or manufacturing processes, to reduce costs, production times and resource depletion.

Regulation and standards compliant – manufacturing compliant with latest sector specific regulations and legislation. For example, within the electrical and electronic sector: Restriction of Hazardous Substances (RoHS), Waste Electrical and Electronic Equipment (WEEE) and Energy Use Products (EuP) Directives in the European Union Standards for sustainable building include LEED in the USA and BREAM in the UK.

Reusable buildings – demountable, modular buildings, which can be transported and reassembled in new locations.

Self-assembly – the final assembly is done by the consumer, thereby saving energy in the fabrication process.

Simple, low-cost construction – manufacturing with simple, inexpensive tooling and low-energy processes.

Support grass-roots initiatives and communities – though the choice of manufacturing location and processes, e.g. social enterprises, community-based enterprises.

Zero waste production – the elimination of waste from the production process.

Recycling and reuse

Design for recyclability (DfR) – is a design philosophy that tries to maximize positive environmental attributes of a product, such as ease of disassembly, recyclability, maintenance, reuse or refurbishment, without compromising the product's functionality and performance.

Design for recycling (DfR) – considers the best methods to improve recycling of raw materials or components by facilitating assembly and disassembly, ensuring that materials are not mixed and appropriately labelling materials and components.

Materials labelling – assists with improved identification of materials for recycling.

Materials recycled at source – use of office, factory or domestic waste to make new products in situ.

Reuse of end-of-life components (remanufacturing) – taking back worn-out or old components/ products and refurbishing them to an 'as-new' standard for resale.

Reusing materials – reusing materials with out changing their original state. By comparison, recycling involves some reorganization or partial destruction of the material followed by reconstitution.

Reuse of redundant components – components formally manufactured for another use are re-employed in a new product.

Reused or second life objects – any complete object reused in a new product.

Single material components – components made of one material (a mono-material component).

Use of ready-mades/ready made components – components made for one product reapplied to a new or different type of product.

III DISTRIBUTION/ TRANSPORTATION PHASE

Flat-pack products –products that can be stored flat to maximize use of transport and storage space.

Lightweight products – products that have been designed to be lightweight, yet retain full functionality, and as a result require less energy to transport.

Reduced energy use during transport/reduction in transport energy – this can be achieved by careful design of products to maximize packing per unit area and minimize weight per product.

Reusable packaging – packaging that can provide protection on more than one trip.

Self-assembly – designs that are assembled by the consumer, therefore saving valuable space in transport and storage.

IV FUNCTIONALITY AND USE PHASE

Socially beneficial designs

Affordable and key worker housing – high-quality living spaces for sectors of society that need support.

Alternative modes of transport for improved choice of mobility – reduces dependency on high-environmental-impact products such as the car and affords improved mobility options for minority groups, such as the disabled.

An aid to reduce population growth – helps keep the balance between population and resource availability and so slows environmental degradation, social exclusion and other problems.

Community ownership – encourages group rather than individual ownership and so improves the efficiency of product usage.

Dependable designs – those that give the required functionality and service over many years, encouraging respect for the product, manufacturer and designer.

Design for need – A concept that emerged in the 1970s and was promoted by exponents such as the design academic Victor Papanek and by a landmark exhibition at the Royal College of Art, London, in 1976. Design for need concentrates on design for social needs rather than for creating 'lifestyle' products.

Emergency provision/ distribution of clean, safe water – products designed to reduce human mortality and disease.

Encourages local shopping – designs that encourage the user to shop locally.

Encourages recycling – products designed to facilitate recycling.

Encourages self-suffiency – products designed to facilitate self-suffiency

Equal access to information resources – products to enable minority groups, such as the disabled, to gain access to information resources.

Equal access for public services – products to enable minority groups, such as the disabled, full access to public services, such as transport.

Health improvements – products that improve the quality of local or personal environments through removing toxic substances, air purification, etc.

Hire rather than ownership – products designed for hire rather than for personal ownership, receiving more efficient and economical use.

Interaction – designs that invite the user to interact, modify and/or re-form the original design, engaging the user more deeply with the product/experience.

Personalization – designs that can be modified easily by the user to appeal to the user's own creative instincts and/or personality.

Reduced noise/noise pollution – products designed to minimize distress and disturbance caused by excessive noise.

Self-help design, design democracy – designs that encourage people to examine and implement their own design skills.

Tools for education, communication – designs that are, in themselves, tools to expand educational and communication possibilities.

Designs to reduce emissions/pollution/toxins

Free of CFCs and HCFCs – products, generally associated with the use of refrigerants, that do not use either chlorofluorocarbons (CFCs), which are greenhouse gases, or hydrochlorofluorocarbons (HCFCs), which are greenhouse gases and ozone-depleting gases.

Hybrid power – at least part of the energy used by a product/service comes from renewable sources.

Reduction in /avoidance of emissions (to water) – products whose production and use avoids or minimizes emissions of hazardous and toxic substances to water.

Reduction in /avoidance of emissions/pollution (to air) – products whose production and use avoids or minimizes emissions of hazardous and toxic substances to air, including greenhouse gases, hydrocarbons, particulate matter and cancer-causing substances (carcinogens).

Reduction in /avoidance of hazardous/toxic substances – products that are safe for human use because they contain little or no hazardous or toxic substances. There are international and national lists of banned substances, including chemicals and pesticides. Some companies produce their own lists, in addition to those that they legally have to comply with. Safe for human use does not necessarily mean safe for plant and other wildlife.

Zero emissions – refers to vehicles powered with electric motors or with hydrogen fuel-cell power systems that do not produce exhaust emissions of greenhouse gases (such as carbon dioxide, carbon monoxide, methane or oxides of nitrogen) or particulate matter (such as PM10s). A true zero-emission electric vehicle (EV) is one that uses electricity generated from renewable power rather than fossil fuel or nuclear sources.

Designs for improved functionality

Customizable – describes a product that the consumer can alter to his or her own specification or configuration.

Dual function – one product with two functions.

Easy to clean – saves time and consumables, and can extend the useful life of the product.

Emotionally durable product – where the designer has enabled customization, personalization, interaction or modification that ensures a deeper bond between user and product.

Improved ergonomics – products that are easier and more comfortable to use.

Improved health and safety – products that don't endanger health or safety or that promote better health.

Improved social well-being – products that encourage social interaction or deeper, more meaningful experience or where users contribute to making the experience.

Improved user-friendliness – products that are easier to understand and more fun to use.

Improved user functionality – products that serve their purpose better than previous designs.

Interactivity/ user involvement – engaging the user's abilities and skills in the product or building to improve the experience.

Modular design/modularity – products that can be configured in many ways to suit the user by changing the arrangement of individual modules. Modular design also offers the user the possibility of adding modules as needs require.

Multifunctional – a product capable of more than two functions.

Multi-use space – a space capable of being used for different types of functions.

Portable – a product that is easily transported for use in different locations.

Safe, i.e. non-toxic and non hazardous – a product without adverse effects on human health.

Universal/inclusive design – design that encourages use by a wide range of people with varying abilities; design that enables rather than disables.

Upgradable/upgradability – a product that is easy to upgrade by replacing old components/elements with new. This is especially important for technological products.

Designs to increase product lifespan/longevity

Design for ease of maintenance/maintainability – products with good instructions and easy access to maintain or service parts that wear.

Durable – products that are tough, owing to strong materials and high-quality manufacturing, and so resistant to use and wear.

Ease of repair/repairability – products easy to assemble/disassemble to repair worn or broken parts.

Designs to reduce energy consumption

Energy conservation – products designed to prevent loss of energy.

Energy efficient – products/ buildings designed to use energy efficiently.

Energy label standards – Products/buildings certified to independent energy consumption and conservation standards, e.g. EU Energy Label, Energy Star rating, Blue Angel Eco-label, BREAM/ Eco-home rating by the Building Research Establishment (UK).

Energy neutral – products/ buildings that generate as much energy as they consume.

Fuel economy – products that use less fossil fuel energy than an earlier generation of products and so cause reduced emissions to air over their lifetime.

Human-powered products – products that use energy supplied by humans.

Hybrid power – products that combine two or more power sources, for example, hybrid electric/petrol or fuel cell/electric cars.

Improved energy efficiency – products with improved usage or output per unit of energy expended.

Integrated energy control systems – conservation, generation, reclamation of energy in a product/ building using an integrated ICT-based system.

Integrated or intelligent transport systems – transport systems that permit a range of mobility products to be used to offer a choice of mobility paths for the user.

Low voltage – products capable of operating on 12-volt or 24-volt electricity supply rather than higher voltages.

Natural lighting – products that encourage the use of natural lighting (rather than consuming electricity).

Rechargeable (batteries) – products that encourage repeat battery use by recharging from a mains or renewable power supply, and so reduce waste production.

Renewable power – electricity generated from products that convert the energy of the sun, wind, water or geothermal heat from the earth's crust.

Solar power (passive) – products that produce light or heat by absorbing the energy of the sun.

Solar power (generation) – products that generate electricity by absorbing the energy of the sun. These typically include products equipped with photovoltaic panels.

Recycling and reduction of waste production

Recyclable packaging/ containers – packaging and containers made of materials that can be recycled.

Reduction in use of consumables – products that reduce the use of consumables such as paper, inks, batteries, oils and detergents.

Reusable packaging/ containers – packaging and containers that can be reused for repeat trips.

Designs to improve water usage

Water conservation – products that reduce water usage, and/or facilitate water collection.

Water generation (freshwater) – products that generate fresh water from contaminated surface or ground water, seawater or water-saturated air.

V DISPOSAL/END-OF-LIFE/ NEW LIFE PHASE

Conservation of landfill space – products that decompose to release landfill space or products that can be recycled, reused or remanufactured to avoid being sent to landfill.

Cradle2Cradle – the protocol established by William McDonough Braungart Chemistry (WMBC), USA, that certifies the extent of lifecycle and end-of-life strategies for clean production, recycling, and resource depletion.

Extending product life – giving an existing product/object a new lease of life by refurbishing, adding new design elements, repairing, re-configuring or other design approaches that keep the product from being disposed of.

Encouraging local composting/local biodegradation of waste – products that can be locally decomposed by the owner, so saving on the transport energy of waste collection and landfill space.

Producer responsibility/ product take-back – a system under which manufacturers agree to take back a product when it has reached the end of its useful life so that it can be disassembled and components or materials can be reused or recycled. (See also Pre-production phase.)

Recycling – products that are designed to be easily recyclable by being made of single materials or by being easily disassembled into materials or components that can be recycled.

Remanufacture – products that are easily disassembled for refurbishment or to remanufacture new products.

Reuse – products that are easily reused for the same or a new purpose or are easily disassembled for the components and/or materials to be reused.

MISCELLANEOUS STRATEGIES

Certification of Products – (see also Green Organizations, p. 332)

Eco-labels – labels attached to products which confirm that the manufacturers conform to independently certified standards in terms of reduced environmental impacts.

Independently certified labels – a variety of labels applied to products which signify that the products meet specific criteria for reduced environmental impacts, inclusion of recycled materials, and/or materials/ products from sustainable, renewable, and/or responsibly/ ethically managed sources.

Environmental management and business systems

Corporate environmental policy – a written statement defining a company's position on the environment with an on-going audit of progress over time. Existence of a corporate environmental policy *usually* indicates inclusion of environmental management systems and/or the use of basic eco-design strategies in everyday business.

Eco Management and Audit Scheme (EMAS) – an independently certified environmental management system, which operates in the European Union. Certification is awarded by national bodies in individual EU countries verified by the EMAS organization.

ISO 14001 – an international standard for environmental management schemes maintained by the International Standards Organization (ISO) in Geneva, Switzerland. New standards are emerging for lifecycle assessment (ISO 14040) and eco-labelling and environmental labels (draft ISO 14021).

ISO 9001 – an international standard for quality assurance maintained by the International Standards Organization (ISO) in Geneva, Switzerland. Certification is granted by independent national organizations accredited by the ISO.

Biodiversity

Animal-friendly products – products that are manufactured without harm to animals.

Conservation of land resources – using 'brownfield' or under-used urban sites/locations rather than building on virgin, agricultural or rural land.

Encouragement of food production – products that promote production of food in the home or garden.

Encouragement of conservation and biodiversity – products that assist in promoting conservation and diversity as a result of a corporate environmental or supply-chain management policy or by sourcing materials from habitats managed to maintain diversity.

Low ecological footprint – minimizing inputs and outputs of materials and energy per capita for a product/building.

Non-GM materials – non-Genetically Modified crops and their derivatives.

Protection against soil erosion – products used to avoid or reduce soil erosion by water or wind.

Spatial economy – minimizing spatial requirements per capita.

ACADEMIC AND RESEARCH

Attainable Utopias Network
UK
E info@attainable-utopias.org
W attainable-utopias.org
An ongoing and expanding network with diverse projects examining how design is evolving and responding to contemporary (sustainability) issues, with a focus on the metadesign approach encouraging participation in design.

Centre for Design at Royal Melbourne Institute of Technology (RMIT)
GPO Box 2476V
Melbourne
Victoria 3001, Australia
W www.cfd.rmit.edu.au

Centre for Environmental Assessment of Product and Material Systems (CPM)
Chalmers University of Technology
412 96 Göteborg, Sweden
T +46 (0)31 772 56 40
E contact@chalmers.se
W www.cpm.chalmers.se

Centre for Sustainable Design
University for the Creative Arts at Farnham
Falkner Road
Farnham
Surrey GU9 7DS, UK
T +44 (0)1252 892 772
E cfsd@ucreative.ac.uk
W www.cfsd.org.uk

Consortium on Green Design and Manufacturing (CGDM)
University of California
1115 Etcheverry Hall
Berkeley, CA 94720, USA
T +1 510 642 8657
E cgdm@newton.berkeley.edu
W cgdm.berkeley.edu

DEMI – Design for the Environment Multimedia Implementation Project
E info@demi.org.uk
W www.demi.org.uk
DEMI is a a multimedia design and environmental teaching and learning resource for higher education created by a consortium of universities in 2000/2001 that enables holistic thinking about the sustainability challenge.

Design Academy Eindhoven
Emmasingel 14
PO Box 2125
5600 CC Eindhoven
the Netherlands
T +31 (0)40 239 3939
E info@designacademy.nl
W www.designacademy.nl

Design for Sustainability Program
TU Delft Subfaculty of Industrial Design Engineering
the Netherlands
T +31 (0)15 278 9111
E info@tudelft.nl
W www.io.tudelft.nl/research

Imagination at Lancaster
UK
W www.imagination.lancaster.ac.uk
A creative research lab that focuses on democratizing innovation, Design for Sustainability, service design and social technologies.

Institute for Engineering Design – Austrian Ecodesign Information Point
Austrian Ministry of Transport, Innovation and Technology with Vienna University of Technology
Austria
W www.ecodesign.at

Sustainable Design Network
Loughborough University
Leicestershire LE11 3TU
UK
T +44 (0)1509 263171
W www.sustainabledesignnet.org.uk

Sustainable Everyday Project
Italy
W www.sustainable-everyday.net
The SEP is an open web platform to stimulate social conversation on possible sustainable futures, by exploring design scenarios, cases and events.

ARCHITECTURE

American Institute of Architects
1735 New York Avenue NW
Washington, DC 20006
USA
T +1 202 626 7300
E infocentral@aia.org
W www.aia.org
www.e-architect.com
The AIA has an extensive list of sustainability resources for practicing architects, including the 2030 Toolkit, initiatives, case studies and knowledge centres.

Architecture for Humanity and the Open Architecture Network
848 Folsom, Suite 201
San Francisco, CA 94107
USA
T+1 415 963 3511
W www.architectureforhumanity.org
www.openarchitecture.org
Established in 1999 by Cameron Sinclair and Kate Stohr, Architecture for Humanity has attracted thousands of designers and architects who donate their time and services to creating affordable, sustainable buildings for those in need. The Open Architecture Network was created to provide architects, designers, builders and their clients with an online forum to share architectural plans and drawings.

Association for Environment Conscious Building (AECB)
PO Box 32
Llandysul, Carmarthenshire
SA44 5ZA, UK
E sally@aecb.net
W www.aecb.net

BRE Trust
Bucknalls Lane
Garston
Watford WD25 9XX, UK
T +44 (0)1923 664 000
E secretary@bretrust.org.uk
W www.bretrust.org.uk
The BRE is a charitable organization whose objectives are to advance knowledge, innovation and communication in all matters concerning the built environment for public benefit.

Centre for Alternative Technology
Machynlleth
Powys SY20 9AZ, UK
T +44 (0)1654 705 950
E info@cat.org.uk
W www.cat.org.uk
Established in 1975 as a resource centre to encourage a more ecological way of living, CAT now demonstrates physical ways in which buildings, renewable-energy technology and waste-water treatment can reduce environmental impact. CAT has also published extensive DIY and professional guides on all aspects of low-impact technology.

Energy Efficiency and Renewable Energy
Mail Stop EE-1
Department of Energy
Washington, DC 20585
USA
T +1 202 586-9220
W www.eere.energy.gov

Royal Institute of British Architects (RIBA)
66 Portland Place
London W1B 1AD, UK
T +44 (0)20 7580 5533
E info@inst.riba.org
W www.architecture.com

Sponge
UK
W www.spongenet.org
Sponge is a network that provides fresh ideas for building, demonstrating how sustainable development can improve the quality of our built and natural environment.

SUST
UK
W www.sust.org
A collaborative resource portal between The Lighthouse, Glasgow and the Scottish Ecological Design Association, UK, the website has extensive practical guidance and a directory of materials.

U.S. Green Building Council
2101 L Street, NW
Suite 500
Washington, DC 20037, USA
T +1 202-742-3792
E info@usgbc.org
W www.usgbc.org

ORGANIZATIONS – SUSTAINABLE DESIGN, ECO-DESIGN, GREEN DESIGN, SOCIAL DESIGN AND DESIGN ACTIVISM

Ecodesign Foundation
Australia
W www.changedesign.org
This is the archived site for the EcoDesign Foundation and Change Design (now merged with the Society for Responsible Design).

Environmental Design Research Association (EDRA)
PO Box 7146
Edmond, OK 73083, USA
T +1 405 330 4863
E edra@edra.org
W www.edra.org
EDRA was founded in 1968 for the advancement of the art and science of environmental design and research with the goal of improving the understanding of interrelationships between people and their built and natural surroundings.

O2 Network
W www.o2.org
O2 Network coordinates participating O2 groups in 16 countries sharing information and promoting discussion about eco-design and sustainable design in order to integrate sustainability into the design process.

Scottish Ecological Design Association (SEDA)
35/1 Granton Crescent
Edinburgh EH5 1BN, UK
T +44 (0)1875 614 105
E info@seda2.org
W www.seda2.org

SlowLab
USA and the Netherlands
E info@slowlab.net
W www.slowlab.net
SlowLab's mission is to promote 'slowness' and 'slow design' as a positive catalyst for individual, socio-cultural and environmental well-being. By engaging individuals and leveraging the collaborative potential of communities, networks of cooperation are created and lead to new thinking and approaches.

Social Design Site
Germany
T +61 (0)500 589 500
E srd@green.net.au
W www.socialdesignsite.com
An organization dedicated to exploring the positive contribution of design to social development and contemporary issues.

Society for Responsible Design (SRD)
PO Box 326 Church Point
NSW 2105 Sydney
Australia
E srd@green.net.au
W www.green.net.au/srd

AWARDS

Design Preis Schweiz
D'S Design Center AG
P.O. Box 852
4901 Langenthal
Switzerland
T +41 (0)62 923 03 33
E designcenter@designnet.ch
W www.designpreis.ch

Good Design Award – Ecology Design Prize
Japan Industrial Design Promotion Organization (JIDPO)
G-Mark Division
5th floor, Midtown Tower
9-7-1 Akasaka, Minato-ku
Tokyo 107-6205 , Japan
T +81 (0)3 3435 5633
E info-e@g-mark.org
W www.g-mark.org/english/
The Good Design Selection System, with its G-Mark logo for winning products, was launched in 1957 and became the Good Design Award in 1998. Product categories are wide-ranging and attract thousands of entrants from Japan and the international design community. Special award categories include the 'Long Life Award' and the 'Sustainable Design Award'.

Industrie Forum Design Hannover (iF) GmbH
Messegelände
30521 Hannover, Germany
T +49 (0)511 89 31 125
E info@ifdesign.de
W www.ifdesign.de
This is one of the most prestigious annual design awards in Germany. Categories include a special Ecology Design Award and Universal Design Award, as well as Product Design Awards for office, business, communications, home, household, lighting, consumer electronics, lifestyle, public design, packaging design, textile design, building technology, industry, transport, medical and leisure. Winners of the Ecology Design Award are selected from any of the subcategories in the competition.

BUSINESS AND THE ENVIRONMENT

The Future 500
335 Powell Street
14th Floor
San Francisco, CA 94102
USA
T +1 415 294 7775
E info@future500.org
W www.future500.org

International Institute of Sustainable Development (IISD)
Head Office
161 Portage Avenue East, 6th floor
Winnipeg
Manitoba R3B 0Y4, Canada
T +1 204 958 7700
W www.iisd.org
www.iisd.ca

World Business Council for Sustainable Development (WBCSD)
4, chemin de Conches
Conches
1231 Geneva, Switzerland
T +41 (0)22 839 3100
E info@wbcsd.ch
W www.wbcsd.ch
Established in 1999, the WBCSD brings together a powerful group of global companies that are focused on embedding sustainability into their business models. New projects include the Eco-commons, a means of sharing intellectual property around green innovations, and on Ecosystem Resources.

CERTIFICATION, ECO-LABELS AND ENERGY LABELS

British Standards Institution (BSI)
389 Chiswick High Road
London W4 4AL, UK
T +44 (0)20 8996 9001
E cservices@bsigroup.com
W www.bsigroup.com

Duales System Deutschland GmbH – The Green Dot (Der Grüne Punkt)
Germany
E info@gruener-punkt.de
W www.gruener-punkt.de
Founded in 1990, this not-for-profit organization administers Der Grüne Punkt (the Green Dot) packaging recycling scheme to comply with the 1991 German Packaging Ordinance. Any packaging marked with the Green Dot is acceptable for recycling. All types of packaging are accepted including glass, wood, ceramics, ferrous and non-ferrous metals, plastics and paper. This scheme is now licensed to a number of organizations in other EU countries – the ARA System (Austria), Ecoembalajes España (Spain), FOST Plus (Belgium), Repak Ltd (UK), Sociedade Ponto Verde SA (Portugal) and VALORLUX asbl (Luxembourg).

EMAS (Eco-Management and Audit Scheme)
Belgium
E emas@iema.net
W www.iema.net/ems/emas

ENERGY STAR
US EPA
ENERGY STAR Hotline (6202J)
1200 Pennsylvania Ave NW
Washington, DC 20460
USA
T +1 888 782 7937
E energystar@optimuscorp.com
W www.energystar.gov
ENERGY STAR is a joint programme of the U.S. Environmental Protection Agency (EPA) and the U.S. Department of Energy. The ENERGY STAR label certifies eco-efficient office equipment, buildings and more.

EU Energy Portal
Belgium
E info@energy.eu
W www.energy.eu
A large portal site giving access to the latest thinking and practices on all energy matters, including energy generation, efficiency and conservation, CO_2 emissions and the EU Energy Label.

Forest Certification Resource Center (FCRC)
USA
E fcrc@metafore.org
W www.certifiedwood.org
Since 2001 the FCRC has been providing a comprehensive resource that helps businesses and consumers make informed decisions by providing accurate, objective information about forest certification.

Forest Stewardship Council (FSC)
In the UK:
FSC UK
11–13 Great Oak Street
Llanidloes
Powys SY18 6EB, UK
T +44 (0)1686 413 916
E info@fsc-uk.org
W www.fsc-uk.org

In the USA:
FSC USA
212 Third Avenue North
Suite 280
Minneapolis, MN 55401
USA
T +1 612 353 4511
E info@fscus.org
W www.fscus.org

Worldwide:
FSC Worldwide
Charles de Gaulle Strasse 5
53113 Bonn, Germany
T +49 (0) 228 367 66 0
W www.fsc.org/worldwide_locations.html
Founded in 1993, the FSC is an independent, not-for-profit, non-governmental organization, which is responsible for standard setting, trademark assurance and accreditation services for companies and organizations interested in responsible forestry. It is an international programme implemented by independent organizations that are evaluated, accredited and monitored by the FSC. In the UK the Soil Association Woodmark Scheme and the SGS Forestry QUALIFOR Programme are both accredited. Other accredited organizations include the Rainforest Alliance SmartWood Program and Scientific Certification Systems Forest Conservation Program (USA), Silva Forest Foundation (Canada), Skal (Netherlands) and the Institut für Marktökologie (Switzerland). In the Directory of FSC Endorsed Forests Worldwide, millions of hectares are certified in temperate, subtropical and tropical regions spanning 46 countries. The chain of custody is also inspected by the FSC, ensuring that endorsement with the FSC logo is not abused by agents, distributors, wholesalers or retailers.

Global Ecolabelling Network (GEN)
Canada
E gensecretariat@terrachoice.ca
W www.globalecolabelling.net
GEN is not accredited to issue eco-labels but keeps the most up-to-date list of 26 eco-labelling organizations worldwide and details of the type of products and materials currently covered. GEN links directly with most eco-labelling organizations' websites.

Group for Efficient Appliances (GEA)
c/o RAMBØLL
Teknikerbyen 31
2830 Virum, Denmark
T +45 (0)45 98 87 92
E kimj@ramboll.dk
W www.gealabel.org/
An association of energy-labelling authorities in European countries (includes Austria, Denmark, Finland, France, Germany, Sweden, the Netherlands and Switzerland), the European Energy Network and the European Association of Consumer Electronics Manufacturers (EACEM). Labels are available for a range of electronic equipment from PCs to TVs.

International Organization for Standardization (ISO)
1 ch. de la Voie-Creuse
Case postale 56
1211 Geneva 20,
Switzerland
T +41 (0)22 749 01 11
E central@iso.ch
W www.iso.ch

National Association of Paper Merchants (NAPM)
PO Box 2850
Nottingham NG5 2WW, UK
T +44 (0)115 841 2129
E info@napm.org.uk
W www.napm.org.uk
NAPM-approved recycled paper and boards are guaranteed to contain a minimum of 50%, 75% or 100% recycled fibre content from genuine paper and board waste, not mill waste.

Programme for Endorsement of Forest Certification Schemes (PEFC)
W www.pefc.org
The PEFC is an independent, non-profit, non-governmental organization, founded in 1999 which promotes sustainably managed forests through independent third-party certification. Originally a European organization it now has member organizations from 26 countries, which embrace 35 national certification schemes.

ReSy GmbH
Organisation für Wertstoffentsorgung GmbH
Hilpertstrasse 22
64295 Darmstadt
Germany
T +49 (0)6151 92 94 22
W www.resy.de/ind-eng.htm
This company certifies that the content of paper and corrugated board packaging is suitable for recycling in the German paper industry. The ReSy logo is used with the international recycling logo of the Mobius loop.

SmartWood Program
Rainforest Alliance
665 Broadway, Suite 500
New York, NY 10012, USA
T +1 212 677 1900
E info@ra.org
W www.rainforest-alliance.org
SmartWood is a program of the Rainforest Alliance, which encourages environmentally and socially responsible forestry management. SmartWood has certified more than 100 operations worldwide, which produce a wide range of certified lumber and products. The FSC has accredited SmartWood for its certification of forestry operations processes that now cover 53,557,709 million hectares (132,343,981 million acres) in 2,910 forestry operations in 64 countries.

ECO-MATERIALS

ATHENA Sustainable Materials Institute
629 St. Lawrence Street
PO Box 189
Merrickville, Ontario
K0G 1N0, Canada
T +1 613 269 3795
W www.athenasmi.org

BC (formerly The BioComposites Centre)
Bangor University
Deiniol Road
Bangor
Gwynedd LL57 2UW, UK
T +44 (0)1248 370 588
E bc@bangor.ac.uk
W www.bc.bangor.ac.uk
Since 1989 BC has examined the potential of bio-based alternatives to replace synthetic materials.

BRE Trust Companies and Green Book Live
Bucknalls Lane
Garston
Watford WD25 9XX, UK
T +44 (0)1923 664000
E enquiries@bre.co.uk
W www.bre.org.uk
www.greenbooklive.com
The BRE group includes a diverse resource base with access to environmentally sound products and services, BREAM building guidelines and much more. The Green Book Live is a free online database designed to help specifiers and end users identify products and services that can help to reduce their impact on the environment.

Ecolect
19 Rausch Street, Suite C
San Francisco, CA 94103
USA
T +1 877 326 5328
E sf@ecolect.net
W www.ecolect.net
Ecolect is an eco-material database for designers, architects and builders that provides the tools and resources to make it easier to design responsibly with state-of-the-art sustainable materials.

Institute for Local Self Reliance (USA)
927 15th St NW, 4th floor
Washington, DC 20005
USA
T +1 202 898 1610
E info@ilsr.org
W www.ilsr.org
The ILSR maintains an online database of materials called The Carbohydrate Economy Clearinghouse, which lists state by state the companies in the US that are manufacturing materials from biological sources. This includes biofuels, biocomposites, biopolymers, paints, finishes and cleaners, with examples of the use of waste or recycled raw materials.

Material ConneXion
60 Madison Avenue,
2nd Floor
New York, New York 10010
USA
T +1 212 842 2050
E info@materialconnexion.com
W www.materialconnexion.com
Material ConneXion maintains a database of over 4,500 materials – including materials derived from or containing recycled content – and 100 new sustainable materials recently added to the database and displayed at Milan Salone 2009.

National Non-Food Crops Centre (NNFCC)
Biocentre
York Science Park
Innovation Way
Heslington
York YO10 5DG, UK
T +44 (0)1904 435182
E enquiries@nnfcc.co.uk
W www.nnfcc.co.uk
NNFCC provides independent advice and information about renewable materials and technologies to industry, government and the general public in the UK.

New Uses Council
c/o BioDimensions
20 South Dudley, Suite 802
Memphis, TN 38103, USA
E info@newuses.org
W www.newuses.org
The New Uses Council is dedicated to developing and commercializing new industrial, energy and non-food consumer uses of bio-based renewable agricultural, forestry, livestock and marine products.

Programme for Endorsement of Forest Certification Schemes (PEFC)
W www.pefc.org
See above

Salvo (UK)
W www.salvo.co.uk
Established in 1992, Salvo is Europe's only association coordinating the activities of architectural salvage companies and reclaimed building materials suppliers. Although members are predominantly from the UK, listings include companies in Australia, Belgium, Canada, France, Ireland and the USA.

ECO SHOPS

Centre for Alternative Technology
Mail Order
Machynlleth
Powys SY20 9AZ, UK
T +44 (0)1654 705959
E mail.order@cat.org.uk
W www2.cat.org.uk/shopping

EcoMall
USA
E ecomall@ecomall.com
W www.ecomall.com

EthicalSuperstore.Com
16 Princes Park
Team Valley Trading Estate
Gateshead, Tyne and Wear
NE11 0NF, UK
T +44 (0) 800 999 2134
E enquiries@ethicalsuperstore.com
W www.ethicalsuperstore.com

The Green Stationery Company
Studio One
114 Walcot Street
Bath BA1 5BG, UK
T +44 (0)1225 480 556
E sales@greenstat.co.uk
W www.greenstat.co.uk

Natural Collection
Dept 7306
Sunderland SR9 9XZ, UK
T +44 (0)845 3677 001
E enquiries@naturalcollection.com
W www.naturalcollection.com

Nigel's Eco Store
55 Coleridge Street
Hove BN3 5AB, UK
T +44 (0)800 288 8970
E helpme@nigelsecostore.com
W www.nigelsecostore.com

Real Goods Solar
California and Colorado
USA
T +1 888 507 2561
E solar@realgoods.com
W www.realgoodssolar.com

Sustainability Store
USA
E john@sustainabilitystore.com
W www.sustainabilitystore.com

ENERGY

The British Wind Energy Association (BWEA)
Greencoat House
Francis Street
London SW1P 1DH, UK
T +44 (0)20 7901 3000
E info@bwea.com
W www.bwea.com
Promotes the use of renewable wind power and has an extensive list of publications for commercial and domestic generation, plus a list of members and suppliers.

Centre for Sustainable Energy, CSE
3 St Peter's Court
Bedminster Parade
Bristol BS3 4AQ, UK
T +44 (0)117 934 1400
E info@cse.org.uk
W www.cse.org.uk
CSE provides research, consultancy, education and training in sustainable energy technology and systems. It also has experience of delivering local and regional initiatives and lobbying to assist development of appropriate energy policies.

Energy Efficiency and Renewable Energy Network (EREN)
USA
W www.eere.energy.gov

The Environment Directory – Alternative Energy
California, USA
W www.webdirectory.com/Science/Energy/Alternative_Energy
This is the largest Internet directory of environmental organizations, and includes sites from over 100 countries.

GENERAL

International Network for Environment Management (INEM)
Ludwig Karg
Chairman
Osterstrasse 58
20259 Hamburg, Germany
T +49 (0)40 4907 1600
E l.karg@inem.org
W www.inem.org

INTERNATIONAL, NATIONAL AND FEDERAL AGENCIES

Department for Environment, Food and Rural Affairs
Eastbury House
30–34 Albert Embankment
London SE1 7TL, UK
T +44 (0)20 7238 6951
E helpline@defra.gsi.gov.uk
W www.defra.gov.uk

Envirowise
Envirowise Advice Line
Building 329
Harwell SIC
Didcot OX11 0QJ, UK
T +44 (0)800 585794
W www.envirowise.gov.uk
A government programme run by the Department of the Environment offering practical environmental advice for business.

European Environment Agency (EEA)
Kongens Nytorv 6
1050 Copenhagen K
Denmark
T +45 (0)3336 7100
W www.eea.eu.int

US Environmental Protection Agency
Ariel Rios Building
1200 Pennsylvania Avenue NW
Washington, DC 20460
USA
T +1 202 272 0167
W www.epa.gov

LIFECYCLE ANALYSIS (LCA) ORGANIZATIONS & TOOLS

ATHENA Sustainable Materials Institute
See p. 334

Boustead Consulting
Black Cottage
East Grinstead
Horsham
West Sussex RH13 8GH
UK
T +44 (0)1403 864 561
E sales@boustead-consulting.co.uk
W www.boustead-consulting.co.uk
Boustead Model version 5.0 is a lifecycle inventory and LCA tool drawing upon decades of experience to define inputs and outputs for thousands of raw and manufactured materials and processes.

Cambridge Engineering Selector (for CES, Eco Selector & EMIT)
Granta Design Ltd
Rustat House
62 Clifton Road
Cambridge CB1 7EG, UK
T +44 (0)1223 518 895
E info@grantadesign.com
W www.grantadesign.com
CES permits simultaneous selection of material, manufacturing process and

shape from three interlinked comprehensive databases. It has been substantially expanded through the addition of the Eco-Selector component, which brings a range of environmental criteria into the materials selection process. Granta is a member of The Environmental Materials Information Technology (EMIT) Consortium, a collaborative project with the UK's National Physical Laboratory and industrial partners including Eurocopter, Emerson Electric, Rolls-Royce, EADS Astrium, and NASA. EMIT looks at the impact of environmental regulations on the selection of low energy/low carbon footprint eco-materials.

Danish LCA Centre
E info@lca-center.dk
W www.lca-center.dk

Ecodesign PILOT
Vienna University of Technology
Institute for Engineering Design and Logistic Engineering, E307
Getreidemarkt 9
1060 Vienna, Austria
T +43 (0)1 58801 30744
E pilot@ecodesign.at
W www.ecodesign.at/pilot/online/english
A series of online eco-design tools.

PRé Consultants BV
Printerweg 18
3821 AD Amersfoort
the Netherlands
T +31 33 4555022
E support@pre.nl
W www.pre.nl
Suppliers of ECO-it, an entry level LCA software, and SimaPro, a popular professional LCA package evolved from the Eco-indicator 99 methodology. PRé Consultants offer a comprehensive LCA toolbox, as well as training/ consultancy services.

Society of Environmental Toxicology and Chemistry (SETAC)
In Europe:
Avenue de la Toison d'Or, 67
1060 Brussels, Belgium
T +32 2 772 72 81
E setaceu@setac.org
W www.setac.org
In the USA:
1010 North 12th Avenue
Pensacola, FL 32501, USA
T +1 850 469 1500
E setac@setac.org
W www.setac.org

Society of the Promotion of Life-Cycle Assessment Development (SPOLD)
W lca-net.com/spold

RECYCLING

Aluminium Packaging Recycling Organisation (Alupro)
1 Brockhill Court
Brockhill Lane
Redditch
Worcestershire
B97 6RB, UK
T +44 (0)1527 597757
E info@alupro.org.uk
W www.alupro.org.uk
Alupro is a national organization dedicated to the collection and recycling of aluminium drinks cans.

British Metals Recycling Association (BMRA)
16 High Street
Brampton
Huntingdon PE28 4TU, UK
T +44 (0)1480 455249
E admin@recyclemetals.org
W www.recyclemetals.org
The British Metals Recycling Association encourages recycling of ferrous and non-ferrous metals in the UK. It has over 300 members and provides links to other associations and organizations in the metals recycling industry worldwide.

Bureau of International Recycling
Belgium
W www.bir.org
BIR is an international trade association of the recycling industries.

Deutsche Gesellschaft für Kunststoff-Recycling mbH
Frankfurter Strasse 720–726
51145 Cologne, Germany
T +49 (0)22 03937-0
E info@gruener-punkt.de
W www.dkr.de
DKR recycles plastic and other packaging under the Green Dot system (see Duales System Deutschland GmbH, p. 333) and encourages recycling and reuse of this waste. DKR maintains an online database of (mostly German) companies that manufacture materials and products from recycled plastics.

The Environment Directory – Recycling
California, USA
W www.webdirectory.com/Recycling

Industry Council for Electronic Equipment Recycling (ICER)
6 Bath Place
Rivington Street
London EC2A 3JE, UK
T +44 (0)20 7729 4766
E ws03@icer.org.uk
W www.icer.org.uk
The ICER is a cross-industry group examining the best way to improve recycling and reuse of end-of-life electronic equipment.

RECOUP
RECycling Of Used Plastics Ltd
Tower House
Lucy Tower Street
Lincoln LN1 1XW, UK
T +44 (0) 1733 390021
E enquiry@recoup.org
W www.recoup.org
RECOUP is the UK's national plastic-bottle recycling organization with 75 members including plastics manufacturers, beverage companies, retailers and local authorities.

Waste Watch
56–64 Leonard Street
London EC2A 4LT, UK
T +44 (0)20 7549 0300
E info@wastewatch.org.uk
W www.wastewatch.org.uk
Waste Watch is an environmental charity that advises, educates and supports people to make changes that will reduce their environmental impact and improve their quality of life. Waste Watch publishes an online directory of products and materials in the UK made from recycled materials, and manages the independent National Recycling Forum, which promotes recycling.

WRAP
The Old Academy
21 Horse Fair
Banbury OX16 0AH, UK
T +44 (0)1295 819 900
E info@wrap.org.uk
W www.wrap.org.uk
WRAP – the Waste and Resources Action Programme – helps individuals, businesses and local authorities to reduce waste and recycle more, making better use of resources and helping to tackle climate change.

SUSTAINABLE DEVELOPMENT

Centre for Environmental Strategy (CES)
School of Engineering
University of Surrey
Guildford GU2 5XH, UK
T +44 (0)1483 686670
E cesinfo@surrey.ac.uk
W www.surrey.ac.uk/eng/ces/contact.htm

Forum for the Future
Overseas House
19–23 Ironmonger Row
London EC1V 3QN, UK
T +44 (0)20 7324 3630
E info@forumforthefuture.org
W www.forumforthefuture.org

Intergovernmental Panel on Climate Change (IPCC)
T +41 (0)22 730 8208 84
E IPCC-Sec@wmo.int
W www.ipcc.ch

International Institute for Sustainable Development (IISD)
161 Portage Avenue East
6th Floor
Winnipeg, Manitoba
R3B 0Y4 Canada
T +1 204 958-7700
E info@iisd.ca
W www.iisd.org

The Product-Life Institute (Institut de la Durée)
9 chemin des Vignettes
1231 Conches, Switzerland
T +41 (0)22 707 6612
W www.product-life.org
An independent contract research institute developing innovative strategies and policies to encourage reduced resource consumption for a sustainable society. The Institute provides consultancy services to government, industry and universities.

Rocky Mountain Institute
2317 Snowmass Creek Road
Snowmass, CO 81654, USA
T +1 970 927 3851
E info@rmi.org
W www.rmi.org

Sustainability Web Ring
Sustainability Development Communications Network (SDCN), c/o IISD
W sdgateway.net

Tellus Institute
11 Arlington Street
Boston, MA 02116, USA
T +1 617 266 5400
E info@tellus.org
W www.tellus.org

United Nations Department of Economic & Social Affairs: Division for Sustainable Development
Two United Nations Plaza
Room DC2-2220
New York, NY 10017, USA
W www.un.org/esa/dsd

World Resources Institute (WRI)
10 G Street NE, Suite 800
Washington, DC 20002
USA
T +1 202 729 7600
W www.wri.org

Wuppertal Institute for Climate, Environment & Energy
Döppersberg 19
42103 Wuppertal, Germany
T +49 (0)202 2492 0
E info@wupperinst.org
W www.wupperinst.org

TRADE AND BUSINESS ASSOCIATIONS

Alliance for Beverage Cartons and the Environment (ACE)
250 Avenue Louise
Box 106
1050 Brussels, Belgium
T + 32 (0)2 504 07 10
W www.ace.be
ACE is an association of leading producers of beverage cartons and paperboard, which provides information on the impact of these products on the environment.

American Chemistry Council (ACC) Plastics Division
1300 Wilson Boulevard, Suite 800
Arlington, VA 22209, USA
T +1 703 741 5000
W www.americanchemistry.com/plastics

British Plastics Federation
6 Bath Place
Rivington Street
London EC2A 3JE, UK
T +44 (0)20 7457 5000
E reception@www.bpf.co.uk
W www.bpf.co.uk

Composite Panel Association (CPA)
19465 Deerfield Avenue
Suite 306
Leesburg, VA 20176, USA
T +1 703 724 1128
W www.pbmdf.com
The CPA is a US and Canadian organization devoted to promoting the use and acceptance of particleboard, MDF and decorative surface products.

National Association of Paper Merchants (NAPM)
PO Box 2850
Nottingham NG5 2WW, UK
T +44(0)115 841 2129
E info@napm.org.uk
W www.napm.org.uk
NAPM-approved recycled paper and boards are guaranteed to contain 50%, 75% or 100% recycled fibre content from genuine paper and board waste, not mill waste.

PlasticsEurope
Avenue E van Nieuwenhuyse 4
Box 3
1160 Brussels, Belgium
T +32 (2)675 32 97
W www.plasticseurope.org
An extensive resource on plastics including information about plastics and the environment together with detailed eco-profiles of common plastics.

Also refer to Eco-Design Strategies (p. 324)

5Rs is a concept with five cornerstones aimed at reducing the impact of design, manufacturing and products on the environment – to reduce, remanufacture, reuse, recycle and recover (energy by incineration). 'Reduce' implies designing to use fewer raw materials and less energy.

Agenda 21 is a comprehensive blueprint for global action drafted by the 172 governments present at the 1992 Earth Summit organized by the United Nations in Rio de Janiero, Brazil. It is often interpreted and implemented at a local level in 'Local Agenda 21' plans.

Atmosphere refers to the gaseous components at and above the world's surface, including the important gases oxygen, hydrogen, nitrogen, carbon dioxide, methane and ozone.

Bioregion is an identifiable geographical location where integrated human cultural and social systems interact with biotic and abiotic systems to form a distinctive landscape, geology, soils and hydrology, as well as ecosystem relationships.

Biosphere is the term for the living components of the world that meet the seven characteristics of life – movement, feeding, respiration, excretion, growth, reproduction and sensitivity.

Carbon neutral or zero-carbon refers to a net zero emissions of the greenhouse gas carbon dioxide that causes global warming in respect to the lifecycle of the product/artefact/building or service.

Carcinogens are chemicals that are definite or potential agents in causing cancer in humans. They are classified by the World Health Organization according to their perceived risk. Group 1 chemicals carry clear evidence of risk, Group 3 chemicals may have some associated risk.

Carrying capacity is a finite quantity (K) that equates to the ecosystem resources of a defined area such as a locality, habitat, region, country or planet. A given carrying capacity can support a finite population of organisms. Stable populations in harmony with the carrying capacity are sustainable, but excessive population growth can lead to sudden decline and/or permanent reduction in the carrying capacity.

Corporate social responsibility (CSR) is the integration of social and environmental policies into day-to-day corporate business and the involvement of internal and external stakeholders to deliver these policies.

Cradle2Cradle is a protocol developed by McDonough Braungart Design Chemistry (MBDC) USA, for independent certification of recycling, resource reduction, clean production and other strategies that reduce the overall negative environmental impacts of a product or service throughout its lifecycle.

Design for environment (DfE) is the analysis and optimization of the environmental, health and safety issues considered over the entire life of the product. DfE permits resource depletion, waste production and energy usage to be reduced or even eliminated during the manufacture, use and disposal or reuse of the product.

Design for manufacturing (DfM) examines the relationship between resource usage and product design using computer-aided design (CAD) and computer-aided manufacturing (CAM) tools for cost-effectiveness and reduced environmental impacts.

Design for X (DfX) is a generic term where X denotes the specific focus of a design strategy, such as DfD (design for disassembly) or DfE (design for environment).

Downcycling refers to the recycling of a waste stream to create a new material that has properties inferior to those of the original virgin materials. A good example is recycled plastic (HDPE) panels made of multicoloured waste sources.

Eco-design is a design process that considers the environmental impacts associated with a product throughout its entire life from acquisition of raw materials through production, manufacturing and use to end of life. At the same time as reducing environmental impacts, eco-design seeks to improve the aesthetic and functional aspects of the product with due consideration to social and ethical needs. Eco-design is synonymous with the terms design for environment (DfE), often used by the engineering design profession, and lifecycle design (LCD) in North America.

Eco-efficiency embodies the concept of more efficient use of resources with reduced environmental impacts resulting in improved resource productivity, i.e. doing more with less.

Eco-label refers to labels applied to products and materials that conform to standards set by independent organizations to reduce environmental impacts. There are national and international eco-labels – see pages 333–334 for a detailed listing.

Ecological footprint is a measure of the resource use by a population within a defined area of land, including imported resources. Assessment of the ecological footprints of nation states or other defined geographic areas reveals the true environmental impact of those states and their ability to survive on their own resources in the long term. The term ecological footprint can also be applied to products but is more commonly referred to as the environmental 'rucksack' associated with product manufacturing.

Eco-materials are materials that have minimal impact on the environment at the same time as providing maximum performance for the required design task. Eco-materials originating from components from the biosphere are biodegradable and cyclic, whereas eco-materials originating from the technosphere are easily recyclable and can be contained within 'closed-loop' systems.

EcoReDesign (ERD) was first coined by the Royal Melbourne Institute of Technology, Australia, to denote the redesigning of existing products to reduce the environmental impact of one or more components of the product.

Eco-tools A generic name for software or non-software tools that help with the analysis of the environmental impact of products, manufacturing processes, activities and construction. Tools generally fall into several main categories: lifecycle analysis, design, environmental management or eco-audits and energy flow management.

Eco-wheel, or eco-design strategy wheel, is a means of identifying strategies that will assist in making environmental improvements to existing products. It embraces eight strategies: 1) selection of low-impact materials; 2) reduction of materials usage; 3) optimization of production techniques; 4) optimization of distribution system; 5) reduction of impact during use; 6) optimization of initial lifetime; 7) optimization of end-of-life system; and 8) new concept development.

Embodied energy is the total energy stored in a product or material and includes the energy in the raw materials, transport to the place of production, energy in manufacturing and (sometimes) transport energy used in the distribution and retail chain. It is measured in MJ per kilogram or GJ per tonne.

End of life (EoL) describes both the end of the life of the actual product and the cessation of the environmental impacts associated with the product. Disassembly and recycling of components and materials at a product's EoL are preferable to disposal via landfill or incineration.

Environment-conscious manufacturing (ECM) is the application of green engineering techniques to manufacturing to encourage greater efficiency and reduction of emissions and waste.

Environmental management systems (EMS) are aimed at improving the environmental performance of organizations in a systematic way integrated with legislative and compliance requirements. The international benchmark for EMS is the International Standard ISO 14001, which more and more organizations each year are meeting, but national EMS standards also play a significant role, such as the British Standard for Environmental Management, BS 7750. Other independently certified systems exist, such as EMAS, operated in the European Union.

EU Energy label is a classification applied to domestic appliances such as washing machines and refrigerators according to their energy use, expressed as kWh per year. Group A are the most energy-efficient and Group G are the least efficient. This scheme is due to be expanded to other types of appliances.

Geosphere consists of the inorganic, geological components of the world such as minerals, rocks and stone, sea and fresh water.

Green design is a design process in which the focus is on assessing and dealing with individual environmental impacts of a product rather than on the product's entire life.

Greenhouse gases are any man-made gaseous emissions that contribute to a rise in the average temperature of the earth, a phenomenon known as global warming, by trapping the heat of the sun in the earth's atmosphere. The key greenhouse gases include carbon dioxide, mainly from fossil-fuel burning activities; methane from landfill sites, agriculture and coal production; chlorofluorocarbons (CFCs), hydro-chlorofluorocarbons, (HCFCs) and hydrofluoro-carbons (HFCs), used in refrigerants and aerosols; nitrous oxide from nylon and nitric acid production, fossil-fuel burning and agriculture; and sulphur hexafluoride from the chemical industry.

Grey water is the waste water from personal or general domestic washing activities.

Industrial ecology is a holistic approach that considers the interaction between natural, economic and industrial systems. It is also termed industrial metabolism.

Information and Communications Technology (ICT) is the deployment of computers, telecommunications, networks and skills to create systems that deliver more than the sum of their parts.

Intelligent transport system (ITS) is a series of integrated transport networks in which individual networks use specific transport modes but allow easy interconnection to facilitate efficient movement of people.

Lifecycle analysis or Lifecycle assessment (LCA) is the process of analyzing the environmental impact of a product from the cradle to the grave in four major phases: production; transport/distribution/packaging; usage; and disposal or end of life/design for disassembly or recycling.

Lifecycle inventory (LCI) is the practice of analyzing the environmental consequences of inputs required and outputs generated during a product's life.

Lifecycle matrix is a tool or checklist to analyze potential environmental impacts at each phase in the product's lifecycle. Different types of industry create specific lifecycle matrices related to the peculiarities of the manufacturing process of their products.

Lithosphere is the geological strata that makes up the earth's crust.

Mobility path describes a route an individual can take travelling between two points using one or more forms of transport which are, preferably, integrated into a flexible system (see ITS).

Non-renewable resources are those in finite supply that cannot be regenerated or renewed by synthesizing the energy of the sun. Such resources include fossil fuels, metals and plastics. Improving the rate of recycling will extend the longevity of these resources.

Off-gasing is the term for emissions of volatile compounds to the air from synthetic or natural polymers. Emissions usually derive from the additives, elastomers, fillers and residual chemicals from the manufacturing process rather than from the long, molecular-chain polymers.

Open source design is design that is gifted to the Commons and/or Public Domain so that others might collaborate on improving the design for the benefit of society as a whole.

Post-consumer waste is waste that is collected and sorted after the product has been used by the consumer. It includes glass, newspaper and cans from special roadside 'banks' or disposal facilities. It is generally much more variable in composition than pre-consumer waste

Pre-consumer waste is waste generated at the manufacturing plant or production facility.

Producer responsibility (PR) prescribes the legal responsibilities of producers/manufacturers for their products from the cradle to the grave. Recent European legislation for certain product sectors – such as electronic and electrical goods, packaging and vehicles – sets specific requirements regarding 'take-back' of products and targets for recycling components and materials.

Product lifecycle (PLC) is the result of a lifecycle assessment of an individual product, which analyzes its environmental impact.

Product-service-systems (PSS) are designed service systems comprising products, along with supporting infrastructure and networks, that deliver less environmental impact than individually consumed products to meet similar needs.

Product stewardship is the concept of manufacturing responsibility extending beyond the retail or business purchase to include the entire life of the product including its disposal or take-back.

Rebound effect refers to the undesirable environmental impacts that may be generated directly or indirectly by eco-efficient products, such as causing increased demand or new behavioural tendencies for other linked products, services or opportunities.

Renewable resources refer to those resources that originate from storage of energy from the sun by living organisms including plants, animals and humans. Providing that sufficient water, nutrients and sunshine are available, renewable resources can be grown in continuous cycles.

Smart products are those with in-built sensors to control the function of the product automatically or to make the user aware of the condition of the product.

Sustainable is an adjective applied to diverse subjects including populations, cities, development, businesses, communities and habitats; it means that the subject can persist a long time into the future.

Sustainable development: According to the most widely quoted definition, published in the 1987 report 'Our Common Future' by the World Commission on Environment and Development chaired by Gro Harlem Brundtland, the Norwegian Prime Minister, sustainable development is development that meets the needs of the present without compromising the ability of future generations to meet their own needs. The term contains within it two key concepts: the concept of 'needs', in particular the essential needs of the world's poor, to which overriding priority should be given; and the idea of limitations imposed by the state of technology and social organization on the environment's ability to meet present and future needs.

Sustainable products serve human needs without depleting natural and man-made resources, without damage to the carrying capacity of ecosystems and without restricting the options available to present and future generations.

Sustainable product design (SPD) is a design philosophy and practice in which products contribute to social and economic well-being, have negligible impacts on the environment and can be produced from a sustainable resource base. It embodies the practice of eco-design, with due attention to environmental, ethical and social factors, but also includes economic considerations and assessments of resource availability in relation to sustainable production.

Technosphere consists of the synthetic and composite components and materials formed by human intervention in re-ordering and combining components and materials of the biosphere, geosphere and atmosphere. True technosphere materials cannot re-enter the biosphere through the process of biodegradation alone. Synthetic polymers such as plastics are examples of such materials.

Transport energy is the energy expended to transport or distribute a product from the manufacturer to the wholesaler or retailer. Locally manufactured and locally purchased products tend to have much lower transport energies than imported products. The unit of measure is MJ per kilogram.

Use-impact products are consumer products that create (major) environmental impacts, such as cars and electrical appliances.

Volatile organic compounds (VOCs) are natural and synthetic organic chemicals that can easily move between the solid, liquid and gaseous phase.

ACRONYMS AND ABBREVIATIONS

Materials, chemicals

ABS acrylonitrile-butadiene-styrene
CFCs chlorinated fluorocarbons – compounds containing chlorine, fluorine and carbon
CO carbon monoxide
CO_2 carbon dioxide
EPS expanded polystyrene
GRP glass-reinforced plastic (polymer)
HACA high-amylose cornstarch
HC hydrocarbon
HCFCs hydrochlorofluorocarbons – compounds containing hydrogen, chlorine, fluorine and carbon
HDPE high-density polyethylene
HFCs hydrofluorocarbons – compounds containing hydrogen, fluorine and carbon
LDPE low-density polyethylene
MDF medium-density fibreboard
NiCd nickel cadmium
NiMH nickel metal hydride
NO nitrous oxide
NO_x oxides of nitrogen
O_3 ozone
PC polycarbonate
PE polyethylene (polythene)
PET polyethylene terephthalate
PLA polylactide or polylactic acid
PP polypropylene
PS polystyrene
PU polyurethane
PVC polyvinyl chloride
VOC volatile organic compound

Miscellaneous

AC alternating current
CFL compact fluorescent lamp
CNC computer numeric control
DC direct current
EMS environmental management system
EV electric vehicle
LED light emitting diode
MSDS material safety data sheet
PM_{10}s particulate matter (dust, acids and other types) suspended in the air and measuring less than 0.00001 mm diameter
PRN packaging recovery note
PV photovoltaic
UV ultraviolet light

Books

EARLY VISIONARIES

Carson, Rachel, *Silent Spring*, Hamish Hamilton, UK, 1962

Ecologist, The (eds), *A Blueprint for Survival*, Penguin Books, UK and Australia, 1972

Fuller, Richard Buckminster, *Operating Manual for Spaceship Earth*, Feffer & Simons, London and Amsterdam, 1969

Meadows, Donella, Dennis Meadows, Jørgen Randers and William Behrens III, *The Limits to Growth: A Report for the Club of Rome's Project on the Predicament of Mankind*, Earth Island, London, 1972

Meller, James (ed.), *The Buckminster Fuller Reader*, Jonathan Cape, London, 1970

Packard, Vance, *The Hidden Persuaders*, Penguin Books, UK, 1957

Packard, Vance, *The Waste Makers*, Penguin Books, UK and Australia, 1960

Papanek, Victor, *Design for the Real World, Human Ecology and Social Change*, Thames & Hudson, London, 1972

Wright, Frank Lloyd, *The Natural House*, Horizon Press, New York, 1963

ARCHITECTURE

Architecture for Humanity, *Design Like You Give A Damn*, Thames & Hudson, London, 2006

Baggs, Sydney and Joan, *The Healthy House*, Thames & Hudson, London, 1996

Behling, Sophia and Stefan, *Solar Power: The Evolution of Sustainable Architecture*, Prestel Verlag, Munich, 2000

Howard, Nigel and David Sheirs, *The Green Guide to Specification, An Environmental Profiling System for Building Materials and Components*, BRE Report 351, Building Research Establishment, UK, 1998

Jones, David Lloyd, *Architecture and the Environment: Bioclimatic Building Design*, Lawrence King Publishing, London, 1998

Roaf, Susan, *Eco House: A Design Guide*, Architectural Press, Oxford, 2001

Schleifer, Simone, *Small Eco Houses*, Taschen, Cologne, 2007

Strongman, Cathy, *The Sustainable Home: The Essential Guide to Eco Building*, Renovation and Decoration, 2008

Vale, Robert and Brenda, *The New Autonomous House*, Thames & Hudson, London, 2002

Wines, James and Philip Jodidio (eds), *Green Architecture*, Taschen, Cologne, 2000

Woolley, Tom, Sam Kimmins, Paul Harrison and Rob Harrison, *Green Building Handbook, Volume 1: A guide to building products and their impact on the environment*, Taylor & Francis, London, 1997

Woolley, Tom, Sam Kimmins, *Green Building Handbook, Volume 2: A guide to building products and their impact on the environment*, Taylor & Francis, London, 2000

ECO-DESIGN, SUSTAINABLE DESIGN AND DESIGN ACTIVISM

Benson, John F. and Maggie H. Roe (eds), *Landscape and Sustainability*, Spon Press, London, 2000

Beukers, Adriaan and Ed van Hinte, *Lightness: The Inevitable Renaissance of Minimum Energy Structures*, 010 Publishers, Rotterdam, 1999

Bhamra, Tracy and Vicky Lofthouse, *Design for Sustainability: A Practical Approach*, Gower, London, 2007

Birkeland, Janis, *Design for Sustainability: A Sourcebook of Integrated Eco-logical Solutions*, Earthscan Publications, London, 2002

Brezet, Han and Carolein van Hemel, *Ecodesign: A Promising Approach to Sustainable Production and Consumption*, United Nations Environment Programme, Paris, 1997

Chapman, Johnathan and Nick Gant, *Designers, Visionaries & Other Stories*, Earthscan, London, 2007

Commission of the European Communities, *Green Paper on Integrated Product Policy*, COM, Brussels, 2001

Datschefski, Edwin, *The Total Beauty of Sustainable Products*, Rotovision, Brighton, UK, 2001

Fletcher, Kate, *Sustainable Fashion & Textiles*, Earthscan, London, 2008

Fuad-Luke, Alastair, *Design Activism: Beautiful Strangeness for a Sustainable World*, Earthscan, London, 2009

MacKenzie, Dorothy, *Green Design: Design for the Environment*, Rizzoli, New York, 1991

Manzini, Ezio and Francois Jegou, *Sustainable Everyday: Scenarios of Urban Life*, Edizione Ambiente, Milan, Italy, 2003

Papanek, Victor, *The Green Imperative: Ecology & Ethics in Design and Architecture*, Thames & Hudson, London, 1995

Ryan, Chris, *Digital Eco-Sense: Sustainability and ICT: A New Terrain for Innovation*, Lab 3000, Victoria, 2004

Thackara, John, *In the Bubble: Designing in a Complex World*, MIT Press, Cambridge, Massachusetts, 2005

Thorpe, Ann,
The Designer's Atlas of Sustainability, Island Press, Washington, DC, 2007

Van der Ryn, Sim & Stuart Cowan,
Ecological Design, Island Press, Washington, DC, 1996

Van Hinte, Ed and Conny Bakker,
Trespassers: Inspirations for Eco-efficient Design, 010 Publishers, Rotterdam, 1999

Vezzoli, Carlo and Ezio Manzini,
Design for Environmental Sustainability, Springer, London, 2008

Walker, Stuart,
Sustainable by Design, Earthscan, London, 2006

Whiteley, Nigel,
Design For Society, Reaktion Books, London, 1993

SUSTAINABILITY TRANSITION AND CLIMATE CHANGE

Birkeland, Janis,
Positive Development, Earthscan, London, 2008

Goodall, Chris,
How to Live a Low-Carbon Life, Earthscan, London, 2007

Henson, Robert,
The Rough Guide to Climate Change, Rough Guides, London, 2006

Hopkins, Rob,
The Transition Handbook, Green Books, Dartington, 2008

Lynas, Mark,
Six Degrees: Our Future on a Hotter Planet, Harper Perennial, London, 2008

Steffan, Alex,
Worldchanging: A User's Guide for the 21st Century, Abrams, New York, 2008

BUSINESS AND SUSTAINABILITY

Brown, Lester R.,
Eco-economy: Building an Economy for the Earth, W.W. Norton & Co, New York, 2001

Charter, Martin and Ursula Tischner (eds),
Sustainable Solutions: Developing Products and Services for the Future, Greenleaf Publishing, Sheffield, 2001

Fussler, Claude with Peter James,
Driving Eco-innovation, Pitman Publishing, London, 1996

Hawken, P., A. B. Lovins and L. H. Lovins,
Natural Capitalism: Creating the Next Industrial Revolution, Little & Brown, Boston, 1999

Hitchcock, Darcy and Marsha Willard,
The Business Guide to Sustainability, Earthscan, London, 2006

McDonough, William and Michael Braungart,
Cradle to Cradle: Remaking the Way We Make Things, North Point Press, New York, 2002

Porritt, Jonathon,
Capitalism as if the World Matters, Earthscan, London, 2007

Magazines, journals, e-zines, blogs and newsletters

Alastair Fuad-Luke
A website from the author of *The Eco-Design Handbook* that focuses on co-design, eco-design, sustainable design, sustainability transition, slow design and more.
W www.fuad-luke.com

Biothinking
(Biothinking International, UK)
A newsletter that promotes the philosophy of cyclic, solar and safe practices in relation to the design of products and services.
W www.biothinking.com

Design Activism
A blog by author Ann Thorpe that discusses the emerging field of design activism.
W www.designactivism.net

Doors of Perception
John Thackara's blog about design and innovation, and the Doors of Perception conferences, past and present.
W www.doorsofperception.com

Green Futures
(Forum for the Future, UK)
A magazine that focuses on issues of sustainable development using case studies and initiatives in business, industry and local government.
W www.forumforthefuture.org/greenfutures

Inhabitat
Focused on design 'that can save the world', Inhabitat tracks the innovations in technology, practices and materials that are pushing architecture and home design towards a smarter and more sustainable future.
W www.inhabitat.com

Sustainable Business
(USA)
An online monthly magazine, which collates news, features and regular columns from the growing arena of sustainable business.
W www.sustainablebusiness.com

Treehugger
(USA)
Established in 2005 by Graham Hill, Treehugger has become the leading resource for mainstream sustainable thinking and solutions with a contemporary aesthetic. More than 31 new stories are posted daily, and Treehugger has a phenomenal archive of back stories.
W www.treehugger.com

Worldchanging
This organization is a tour de force that addresses our changing world and the ecological, economic and environmental innovation required for the sustainability journey forward. A network of independent journalists, designers and thinkers around the world reveals the ideas, tools and models for building a bright, green future.
W www.worldchanging.com

Sustainable design and design activism networks

Adbusters
W www.adbusters.org

Alternative Technology Association
W www.ata.org.au

Architecture for Humanity
W www.architectureforhumanity.org

Association for Environment Conscious Building
W www.aecb.net

Attainable Utopias
W www.attainable-utopias.org

Centre for Sustainable Design
W www.cfsd.org.uk

Design 21 Social Network
W www.design21sdn.com

O2 Network
W www.o2.org

Open Architecture Network
W openarchitecturenetwork.org

Scottish Ecodesign Association
W www.seda2.org

slowLab
W www.slowlab.net

Social Design Site
W www.socialdesignsite.com

Society for Responsible Design
W www.green.net.au/srd

Sustainable Design Network
W www.sustainable designnet.org.uk

Sustainable Design Research Centre
W www.kingston.ac.uk /design/SDRC/

Sustainable Everyday
W www.sustainable-everyday.net

Illustration Credits

All illustrations are provided courtesy of the designer/designer-maker or manufacturer with specific credits noted below.

The following abbreviations have been used:
t top; *b* bottom; *l* left; *r* right; *c* centre; *c1* column 1; *c2* column 2; *c3* column 3; *c4* column 4.

1 Iøline Crib by Michaele Simmering, Kalon Studios; 2 softseating & softwall by molo design; 9 Aladdin by Stuart Haygarth; 16–17 Ash pendant No.1 by Tom Raffield; 20 *br* Viaduct; 23 *tl* Studio Ilkka Suppanen; 23 *br* Gianni Antoniali, Ikon Udine; 25 *l* Viaduct; 26 Michael Gerlach; 31 *b* Marino Ramazotti; 33 *tl*, Driade, photo by Nacasa & Partners; 34 *bl*, © Pli Design/ © Sprout Design; 37 *t* Angus Mill; 41 *t* Maarten van Houten/ Henk Jan Kamerbeek; 41 *b* Miro Zagnoli; 42 *rt* and *rb* Fran Feijn; 44 *bl* John K. McGregor; 42 *cr* JHenk Jan Kamerbeek; 51 *t* Viaduct; 52 Corné Bastiaansen; 65 *tr* and *br* Dominique Uldry; 71 *r* Dominique Uldry; 73 *tl* Gianni Antoniali, Ikon Udine; 73 *tr* Tom Vack; 80 *b* photo by Robaard/Theuwkens (styling by Majro Karnenborg, CMK) for Droog; 84 *tr* Alastair Fuad-Luke; 84 *br* Sophie Krier & Hans van der Mars; 85 *br* Maarten van Hout, © Moooi B.V.; 86 Gianni Antoniali, Ikon Udine; 87 Design Museum, London; 91 *t* photo by Gerard van Hees for Droog; 97 *t* Viaduct; 98 *r* Viaduct; 103 *t* © Marc Domage; 103 *b* © Véronique Huyghe; 108 *br* Gianni Antoniali; 110 Gabriella Dahlmar; 123 *t* Erik Blinderman; 123 *b* Michaele Simmering; 124 *b* Dominique Uldry; 126 all images Carla Domes Monny, OLPC; 127 Futurefarmers; 134 Franco Vairani/MIT Smart Cities; 143 Christian Rokosch, Waldmeister Bikes; 152 *t* Holger Lonze; 153 *br* Verbüro Olten; 170 *tr* and *tc* Frans Feijn; 171 *tr* Maarten van Houten for Droog; 173 *t* courtesty Aurora Aspen, Haunch of Venison for Stuart Haygarth; 176 *tl* Maarten van Houten; 182 *cl* Alastair Fuad-Luke; 190–191 Marché International Support Office by GLASSX AG; 206 *tr* Alastair Fuad-Luke; 210 *tr* Alastair Fuad-Luke; 211 *tl* Werzalit AG + Co; 211 *tr* Ehlebract AG; 218 *bl* Alastair Fuad-Luke; 220 *t* Renfe and Patier; 220 *b* General Electric ecoimagination; 225 *t* Bernd Kammerer/Kopf AG; 230 *br* Island Wood Center; 231 *tl* Design Museum, London; 232 *b* Peter Grant; 233 Rolf Ditsch SolarArchitektur; 235 *t* Zimmerei Michael Kaufmann; 238 *t* Kevin Lake; 239 *t* Design Museum, London; 239 *b* Zimmerei Michael Kaufmann; 240 Festo Corporate Design; 241 Peabody Trust; 244 Fritz Busam/Thomas Spiegelhalter; 246 t Steffen Jänicke; 247 Peter Bennetts; 253 © Harmut Naegele, Dusseldorf, Germany; 255 *tr* Alastair Fuad-Luke; 256 *bl* Alastair Fuad-Luke; 256 *br* Alastair Fuad-Luke; 261 Rayotec Ltd; 265 BP Solar International; 267 *t* P.J. Hendrikse; 270 *t* Elemental Solutions; 272 *br* Dominique Uldry; 276–277 Alastair Fuad-Luke; 280 *c1* Alastair Fuad-Luke/Kingston University, Rematerialise samples, Jakki Dehn; 283 *c2* Mallinson; 283 *c4* Alastair Fuad-Luke/ Kingston University, Rematerialise samples, Jakki Dehn; 284 *c4* Alastair Fuad-Luke; 285 *c1* Alastair Fuad-Luke; 286 *c2* Alastair Fuad-Luke; 288 *c1* Auro Organic Paint Supplies; 289 *c1* Alastair Fuad-Luke; 290 *c4* Alastair Fuad-Luke; 291 *c2* Richard Learoys; 292 *c1t* Alastair Fuad-Luke; 293 *c1*/*c3*, Lino Codato; 295 *c3t* Alastair Fuad-Luke; 295 *c3b* Alastair Fuad-Luke/Kingston University, Rematerialise samples, Jakki Dehn; 292 *c4b* Alastair Fuad-Luke; 296 *c1* Alastair Fuad-Luke; 296 *c2* Alastair Fuad-Luke; 297 *c2* Alastair Fuad-Luke; 301 *c1* Alastair Fuad-Luke; 301 *c2* Kingston University, Rematerialise samples, Jakki Dehn; 301–302 Franco Vairani/ MIT Smart Cities.